THE SUNDAY TIMES

C R E A T I N G   S U C C E S S

KU-259-248

How to
Write a
Business Plan

Second edition

Brian Finch

KOGAN
PAGE

London and Philadelphia

First published in 2001
Second edition 2006
Reprinted 2006, 2007, 2009

120 Pentonville Road
London N1 9JN
United Kingdom
www.koganpage.com

525 South 4th Street, #241
Philadelphia PA 19147
USA

© Brian Finch, 2001, 2006

The right of Brian Finch to be identified as the author of this work has been asserted by him in accordance with the Copyright, Designs and Patents Act 1988.

ISBN-10   0 7494 4553 X
ISBN-13   978 0 7494 4553 9

The views expressed in this book are those of the author, and are not necessarily the same as those of Times Newspapers Ltd.

British Library Cataloguing-in-Publication Data

A CIP record for this book is available from the British Library.

Library of Congress Cataloging-in-Publication Data

Finch, Brian.
  How to write a business plan / Brian Finch. -- 2nd ed.
    p. cm.
  ISBN 0-7494-4553-X
  1. Business planning. 2. Business writing. I. Title.
HD30.28.F562 2006
658.4'012--dc22
                         2005035380

Typeset by Jean Cussons Typesetting, Diss, Norfolk
Printed and bound in India by Replika Press Pvt Ltd

1 Second-order differential equations in the phase plane

IT IS NOT in general possible to obtain analytic solutions to an arbitrary differential equation. This is not simply because ingenuity fails, but because the repertory of standard functions (polynomials, exp, sin, and so on) in terms of which solutions may be expressed is too limited to accommodate the variety of differential equations encountered in practice. Even if an analytic solution can be found, the 'formula' is often too complicated to display clearly the principal features of the solution; this is particularly true of implicit solutions and of solutions which are in the form of integrals or infinite series.

The qualitative study of differential equations is concerned with how to deduce important characteristics of the solutions of differential equations without actually solving them. Since these considerations apply equally to linear equations and to the much greater variety of nonlinear equations, this book is mainly about nonlinear equations. In this chapter we introduce a geometrical device, the phase plane, which is used extensively for obtaining directly from the differential equation such properties as equilibrium, periodicity, unlimited growth, stability, and so on. The classical pendulum problem shows how the phase plane may be used to reveal all the main features of the solutions of a particular differential equation.

1.1. Phase diagram for the pendulum equation

The simple pendulum consists of a bob of mass m suspended from a fixed point 0 by a light string or rod of length a, which is allowed to swing in a vertical plane. If no friction is present then the equation of motion is

$$ma^2\ddot{\theta} + mga \sin \theta = 0, \tag{1.1}$$

where θ is the inclination of the string to the downward vertical (Fig. 1.1). Since $\ddot{\theta} = \dot{\theta}(\mathrm{d}\dot{\theta}/\mathrm{d}\theta)$, eqn (1.1) becomes

$$ma^2\dot{\theta}\frac{\mathrm{d}\dot{\theta}}{\mathrm{d}\theta} + mga \sin \theta = 0.$$

This equation relates $\dot{\theta}$ and θ instead of θ and t. By integrating with

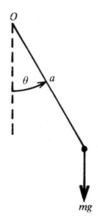

Fig. 1.1. The simple pendulum

respect to θ we obtain

$$ma^2 \int \dot{\theta}\, d\dot{\theta} + mga \int \sin\theta\, d\theta = C$$

where C is a constant. Therefore

$$\tfrac{1}{2}ma^2\dot{\theta}^2 - mga\cos\theta = C. \tag{1.2}$$

Eqn (1.2) expresses conservation of energy, the terms on the left being, in order, the kinetic energy and potential energy of the pendulum. The value of C for a particular motion can be settled by using the initial conditions: the values of θ and $\dot{\theta}$ at $t = 0$. Eqn (1.2) then gives the relation between $\dot{\theta}$ and θ for the motion corresponding to these initial conditions. By choosing different values for C we can obtain this relation for any possible motion.

Eqn (1.2) is also a differential equation for θ in terms of t, but it cannot be solved in terms of elementary functions (see McLachlan, 1956). It is therefore not easy to obtain a useful representation of θ as a function of time. We shall show how it is possible, by working directly with eqn (1.2), to reveal the main characteristics of the solutions.

The relation (1.2) can be represented in a diagram. Set up a cartesian *phase plane* having θ and $\dot{\theta}$ as its axes (Fig. 1.2) and plot the one-parameter family of curves generated by (1.2) for different values of C.

A given pair of values $(\theta, \dot{\theta})$ is called a *state* of the 'system' (in this case a pendulum), and the diagram shows how any state evolves as time progresses. We know that a given state determines all subsequent states,

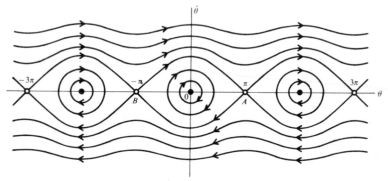

Fig. 1.2. Phase diagram for the simple pendulum

since it serves as initial conditions for the subsequent motion.

The curves depicted in Fig. 1.2 are known as the *phase paths, trajectories,* or *integral curves* corresponding to eqn (1.1), and the complete figure is called the *phase diagram* or *phase portrait* for the system. Each phase path corresponds to a particular possible motion of the system. Associated with each path is a *direction,* indicated by an arrow in Fig. 1.2, showing how the state of the system changes as time *increases*; the direction of the arrows is settled by observing that when $\dot{\theta}$ is positive, θ must be increasing with time and when $\dot{\theta}$ is negative, θ must be decreasing with time.

We shall see that despite the non-appearance of the time variable in the phase plane display, we can deduce several physical features of the pendulum's possible motions from Fig. 1.2. Consider first the possible states of physical equilibrium of the pendulum. The obvious one is when the pendulum hangs without swinging; then $\theta = 0$, $\dot{\theta} = 0$, which corresponds to the origin in Fig. 1.2. The corresponding function $\theta(t) = 0$ is a perfectly legitimate *constant solution* of (1.1): the phase path degenerates to a *single point*. If the suspension consists of a light rod there is a second position of equilibrium, where it is balanced vertically on end. This is the state $\theta = \pi$, $\dot{\theta} = 0$, another constant solution, represented by the point A on the phase diagram. The same physical condition is described by $\theta = -\pi$, $\dot{\theta} = 0$, represented by the point B, and indeed the state $\theta = n\pi$, $\dot{\theta} = 0$, where n is any integer, corresponds to one of these two equilibrium conditions. In fact we have displayed in Fig. 1.2 only part of the phase diagram, whose pattern repeats periodically: there is not in this case a one-to-one relationship between the physical condition of the pendulum and points on its phase diagram.

Since the points O, A, B represent states of physical equilibrium, they are called *equilibrium points* or *critical points* on the phase diagram.

Now consider the family of closed curves immediately surrounding the origin in Fig. 1.2. These indicate periodic motions, in which the pendulum swings to and fro about the vertical. The amplitude of the swing is the maximum value of θ encountered on the curve. For small enough amplitudes, the curves represent the usual 'small amplitude' solutions of the pendulum equation in which eqn (1.1) is simplified by writing $\sin \theta \approx \theta$. The phase paths are nearly ellipses in the small amplitude region.

The wavy lines at the top and bottom of Fig. 1.2, on which $\dot{\theta}$ is of constant sign and θ continuously increases or decreases, correspond to whirling motions of the pendulum. The fluctuations in $\dot{\theta}$ are due to the gravitational influence, and for such phase paths on which $\dot{\theta}$ is very large these fluctuations become imperceptible: the phase paths become nearly straight lines parallel to the θ axis.

We can discuss also the *stability* of the two typical equilibrium points O and A. If the initial state is displaced slightly from O, it goes on to one of the nearby closed curves and the pendulum oscillates with small amplitude about O. We describe the equilibrium point at O as being *stable*. If the initial state is slightly displaced from A (the vertically upward equilibrium position) however, it will normally fall on a phase path which carries the state far from the equilibrium state A into a large oscillation or a whirling condition (see Fig. 1.3). This equilibrium point is therefore described as *unstable*.

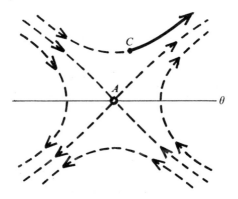

Fig. 1.3. Unstable equilibrium point for the pendulum; displaced state C

1.2. Autonomous equations in the phase plane

The second-order differential equation of general type

$$\ddot{x} = f(x, \dot{x}, t) \tag{1.3}$$

can be interpreted as an equation of motion for a mechanical system, in which x represents displacement of a particle of unit mass, $\dot{x}$ its velocity, $\ddot{x}$ its acceleration, and f the applied force, so that (1.3) expresses Newton's law of motion for the particle:

acceleration = force per unit mass.

A mechanical system is in equilibrium if its state does not change with time. This implies that an equilibrium state corresponds to a constant solution of (1.3), and conversely. A constant solution implies in particular that $\dot{x}$ and $\ddot{x}$ must be simultaneously zero. Note that $\dot{x} = 0$ is not alone sufficient for equilibrium: a swinging pendulum is instantaneously at rest at its maximum amplitude, but this is obviously not a state of equilibrium. Such constant solutions are therefore the constant solutions (if any) of

$$f(x, 0, t) = 0. \tag{1.4}$$

We distinguish between two types of equation:
(i) the *autonomous* type in which $\partial f / \partial t = 0$; that is, f does not depend explicitly on t;
(ii) the *non-autonomous* or *forced* equation where t appears explicitly in the function f.

A typical non-autonomous equation is the linear oscillator with a harmonic forcing term

$$\ddot{x} + k\dot{x} + \omega_0^2 x = F \cos \omega t,$$

in which $f(x, \dot{x}, t) = -k\dot{x} - \omega_0^2 x + F \cos \omega t$. There are no equilibrium states. Equilibrium states are not usually associated with non-autonomous equations although they can occur as, for example, in the equation (Mathieu's equation, Chapter 8)

$$\ddot{x} + (\alpha + \beta \cos t)x = 0,$$

which has an equilibrium state at $x = 0$.

In the present chapter we shall consider only autonomous equations; in this case t is absent and we shall write, instead of (1.3),

$$\ddot{x} = f(x, \dot{x}). \tag{1.5}$$

The constant solutions (which imply equilibrium states) are the solutions (if any) of

$$f(x, 0) = 0. \tag{1.6}$$

For example, the equation $\ddot{x} = (1 - x^2) + x\dot{x}$ has the constant solutions $x = 1$ and $x = -1$.

Now consider the representation of eqn (1.5) on a phase plane. The state of the system at a time $t = t_0$ consists of the pair of numbers $(x(t_0), \dot{x}(t_0))$. This state can be regarded as a pair of initial conditions, and therefore determines subsequent (and earlier) states. The succession of states can be depicted on a phase plane, as in Section 1.1 for the pendulum, in which the axes are for x and $\dot{x}$. We will relabel the $\dot{x}$ axis as y, thus defining y by

$$\dot{x} = y; \tag{1.7}$$

then $\ddot{x} = \dot{y}$ and (1.5) becomes

$$\dot{y} = f(x, y). \tag{1.8}$$

Equations (1.7) and (1.8) can be looked on as two simultaneous first-order equations for the functions $x(t)$, $y(t)$ (or $\dot{x}(t)$). To obtain the relation between y and x which will give the phase paths, eliminate t by dividing (1.8) by (1.7): $\dot{y}/\dot{x} = f(x, y)/y$, or

$$\frac{dy}{dx} = \frac{f(x, y)}{y}. \tag{1.9}$$

The solutions of this equation are the phase paths.

Example 1.1   *Find the phase paths for the equation $\ddot{x} + \alpha \sin x = 0$.*

This is essentially eqn (1.1). We define

$$\dot{x} = y$$

and the equation gives

$$\dot{y} = -\alpha \sin x.$$

The equation for the phase paths is

$$\frac{dy}{dx} = -\frac{\alpha \sin x}{y},$$

which separates to give

$$\int y \, dy = -\alpha \int \sin x \, dx + C,$$

or
$$\tfrac{1}{2}y^2 - \alpha \cos x = C,$$

where C is the parameter of the phase paths. The diagram in the plane is as Fig. 1.2.

The process of obtaining constant solutions of eqn (1.5) amounts to putting $\dot{x} = 0$, $\ddot{x} = 0$, or to putting $\dot{x} = 0$, $\dot{y} = 0$ in (1.7) and (1.8). An equilibrium point (x_0, y_0) on the phase diagram is therefore any solution of the pair

$$y = 0, f(x, y) = 0. \qquad (1.10)$$

Note that
(a) *equilibrium points are always situated on the x axis;*
(b) *except at equilibrium points the phase paths cut the x axis at right angles* (from (1.7), (1.8) and (1.10));
(c) *closed paths represent periodic solutions, since on completing a circuit the original state is returned to, and the motion simply repeats itself indefinitely.*

Now consider the time taken between two points on a phase path. Figure 1.4 shows a segment $\mathscr{C}$ of the phase path joining two points A and B. P represents any intermediate state: we call it a *representative point* of the path. P moves along $\mathscr{C}$ with velocity $(\dot{x}, \dot{y})$, or $(\dot{x}, \ddot{x})$ by (1.7). The time

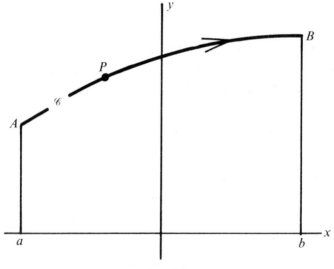

FIG. 1.4.

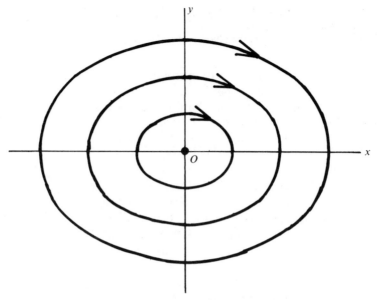

FIG. 1.5. Centre for the simple harmonic oscillator

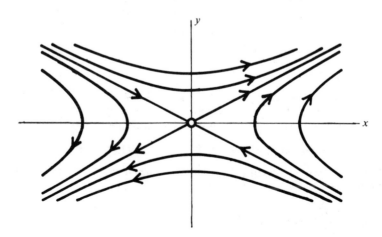

FIG. 1.6. Saddle point

taken, T_{AB}, is therefore given by

$$T_{AB} = \int_{\mathscr{C}} dt \equiv \int_{\mathscr{C}} \left(\frac{dx}{dt}\right)^{-1} \left(\frac{dx}{dt}\right) dt = \int_{\mathscr{C}} \frac{dx}{y}, \qquad (1.11)$$

which is calculable when $\mathscr{C}$ is given. The phase diagram therefore merely suppresses and does not completely lose track of the time variable since (1.11) depends only on the geometry of the phase path.

Example 1.2 *Construct the phase diagram for the simple harmonic oscillator equation* $\ddot{x} + \omega^2 x = 0.$

This approximates to the pendulum equation for small-amplitude swings. Corresponding to eqns (1.7) and (1.8) we have

$$\dot{x} = y, \quad \dot{y} = -\omega^2 x.$$

There is one equilibrium point at

$$x = 0, \quad y = 0.$$

The phase paths are the solutions of

$$\frac{dy}{dx} = -\omega^2 \frac{x}{y},$$

which is separable, leading to

$$y^2 + \omega^2 x^2 = C.$$

The phase portrait therefore consists of a family of ellipses concentric with the origin (Fig. 1.5). All solutions are therefore periodic. The equilibrium point is stable.

An equilibrium point surrounded in its immediate neighbourhood (not necessarily over the whole plane) by closed paths is called a *centre*. A centre is a *stable* equilibrium point.

Example 1.3 *Construct the phase diagram for the equation* $\ddot{x} - \omega^2 x = 0.$

The equivalent first-order pair ((1.7) and (1.8)) are

$$\dot{x} = y, \quad \dot{y} = \omega^2 x.$$

There is a single equilibrium point at

$$x = 0, \quad y = 0.$$

The phase paths are the solutions of

$$\frac{dy}{dx} = \omega^2 \frac{x}{y}$$

therefore their equations are

$$y^2 - \omega^2 x^2 = C.$$

These paths are hyperbolas with asymptotes $y = \pm \omega x$ as shown in Fig. 1.6.

An equilibrium point with paths of this type in its neighbourhood is called a *saddle point*. Such a point is *unstable* since a small displacement from the equilibrium state will generally involve a solution which goes far from this state. (The question of stability is discussed precisely in Chapter 9.) In the figures, stable equilibrium points are usually indicated by a full dot ●, and unstable ones by an 'open' dot ○.

1.3 Conservative systems

In mechanical terms, consider a system with one degree of freedom and let x be a generalized coordinate (position, angle, etc.). Let $\mathcal{T}$ and $\mathcal{V}$ be the kinetic and potential energy functions, and assume that they take the form

$$\mathcal{T} = \tfrac{1}{2} m(x) \dot{x}^2, \quad \mathcal{V} = \mathcal{V}(x) \tag{1.12}$$

where m is another function, $m(x) > 0$. If the system is *conservative*, then the total energy, $\mathcal{E}$, is constant during motion:

$$\tfrac{1}{2} m(x) \dot{x}^2 + \mathcal{V}(x) = \mathcal{E}, \quad \text{constant}, \tag{1.13}$$

which gives the phase paths. The type of equation of motion which leads to (1.13) can be obtained by taking the time derivative of (1.13):

$$m(x) \ddot{x} + \tfrac{1}{2} m'(x) \dot{x}^2 + \mathcal{V}'(x) = 0. \tag{1.14}$$

Equation (1.14) can be simplified by introducing a new variable, u, in place of x (in effect another generalized coordinate) by

$$u = \int \sqrt{\{m(x)\}} \, dx.$$

Equation (1.14) becomes

$$\ddot{u} + \mathcal{Q}'(u) = 0, \tag{1.15}$$

where $\mathcal{Q}'(u) = \mathcal{V}'(x) m^{-1/2}(x)$, and the corresponding energy-type equation for the phase paths is

$$\tfrac{1}{2} \dot{u}^2 + \mathcal{Q}(u) = C. \tag{1.16}$$

We shall need only the form (1.15), but it should be remembered that

equations of the form (1.14) also represent conservative systems.

Translating (1.15) and (1.16) back into the usual notation, we take the *standard equation for a conservative system* to be

$$\ddot{x} = f(x),\tag{1.17}$$

where the 'force per unit mass' f, is independent of $\dot{x}$. Suppose that f is continuous. Define for the new problem a function $\mathscr{V}$ by

$$\mathscr{V}(x) = - \int f(x)\,\mathrm{d}x, \quad \text{or} \quad \mathscr{V}'(x) = -f(x).\tag{1.18}$$

Then $\mathscr{V}$ may be spoken of the potential energy function for (1.17). The equilibrium points of (1.17) are then given by

$$-f(x) = \mathscr{V}'(x) = 0\tag{1.19}$$

(that is, the stationary points of the potential energy, as expected). By writing $\dot{x} = y$ as in (1.7), the phase paths are given by the 'energy equation'

$$\tfrac{1}{2}y^2 + \mathscr{V}(x) = C\tag{1.20}$$

for various values of C, or by

$$y = \pm\sqrt{(2C - 2\mathscr{V}(x))}.\tag{1.21}$$

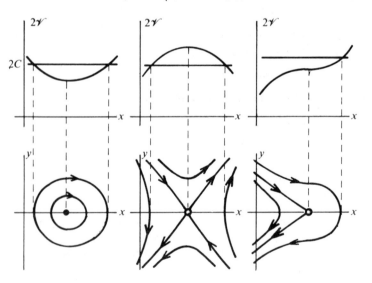

FIG. 1.7. ● stable equilibrium point, ○ unstable equilibrium point

Figure 1.7 shows the types of equilibrium point arising from the three types of turning point of $\mathscr{V}$: a local minimum always leads to a centre, a maximum to a saddle point, and a point of inflexion to a cusp.

The construction can be thought of in the following way. For a fixed C, $2C - 2\mathscr{V}(x)$ can be read off from the top frame for the range of x for which this is non-negative, y calculated from (1.21) for each x, and the symmetrically-placed pair of points inserted in the lower frame.

Alternatively we can view the matter in terms of $f(x)$, regarded as the 'restoring force' for a particle on a nonlinear spring. If $f(x)$ changes sign from positive to negative on passing through the equilibrium point ($\mathscr{V}$ having a minimum), the system behaves like an ordinary linear spring and oscillates, implying a centre. If it changes from negative to positive, the particle is repelled from the equilibrium point and there is a saddle.

Example 1.4   *Sketch the phase plane for the equation* $\ddot{x} = x^3 - x$.

This represents a conservative system (in fact the pendulum equation (1.1) for moderate amplitudes, writing $\sin \theta \approx \theta - \frac{1}{6}\theta^3$, leads to an equation reducible to this one). We have $\mathscr{V}'(x) = x - x^3$ and $\mathscr{V}(x) = -\frac{1}{4}x^4 + \frac{1}{2}x^2$ (to an additive constant). Figure 1.8 shows the construction of the phase diagram.

There are three equilibrium points: at $(0, 0)$, a centre; at $(1, 0)$, a saddle; and at $(-1, 0)$, a saddle. The reconciliation between the types of phase path originating round these points is achieved by special paths called the *separatrices* (shown as broken lines). These correspond to values of C of 0 and $\frac{1}{4}$, the ordinates of the maxima and minimum of $\mathscr{V}$. They start or end at equilibrium points—they must not be mistaken for closed paths.

1.4   The damped linear oscillator

Generally speaking (apart from (1.14)) equations of the form

$$\ddot{x} = f(x, \dot{x}) \tag{1.22}$$

do not arise from conservative systems, and can be expected to show new phenomena. The simplest such system is a linear oscillator with linear damping, having the equation

$$\ddot{x} + k\dot{x} + cx = 0 \tag{1.23}$$

where $c > 0$, $k > 0$. An equation of this form describes, for example, a spring-mass system with a dashpot or a circuit containing inductance, capacitance, and resistance, and serves as a model for many other oscillating systems. We shall show how the familiar features of damped oscillations show up on the phase plane.

Equation (1.23) is a standard type of linear equation. The nature of its

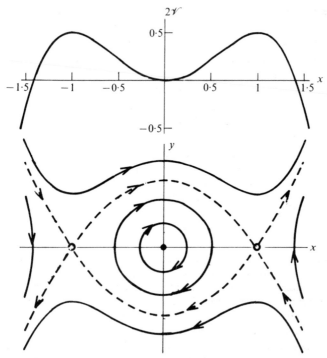

FIG. 1.8. Centre and saddle points. _ _ _ _ _ separatrices

solutions depend on whether the roots of the auxiliary equation

$$m^2 + km + c = 0$$

are real and different, complex, or coincident and real. The roots are given by

$$\begin{matrix} m_1 \\ m_2 \end{matrix} = \tfrac{1}{2}\{-k \pm \sqrt{(k^2 - 4c)}\},$$

and the discriminant, Δ,

$$\Delta = k^2 - 4c \tag{1.24}$$

is therefore the parameter which determines the general type of motion.

Strong damping ($\Delta > 0$)

   The solutions are given by

$$x(t) = A\,e^{m_1 t} + B\,e^{m_2 t}, \tag{1.25}$$

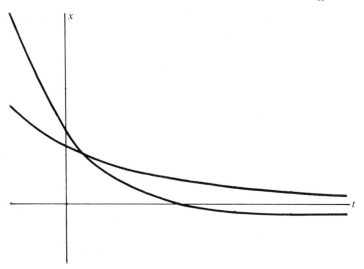

FIG. 1.9. Solution curves for a heavily-damped simple harmonic oscillator

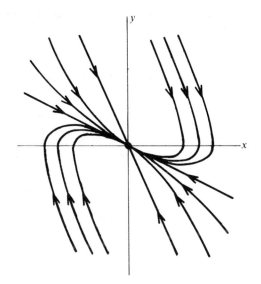

FIG. 1.10. Stable node

where m_1 and m_2 are real and negative, A and B being any constants. Figure 1.9 shows two typical solutions. There is no oscillation and the t axis is cut at most once.

To construct the phase paths we could write as usual

$$\dot{x} = y, \quad \dot{y} = -cx - ky. \tag{1.26}$$

There is a single equilibrium point at $x = 0, y = 0$. The equation for the phase paths is

$$\frac{dy}{dx} = -c\frac{x}{y} - k. \tag{1.27}$$

A general approach to linear systems such as (1.26) will be described in Chapter 2. For the present we remark that the solutions of (1.27) are too complicated for simple interpretation. We therefore proceed in the following way. From (1.25)

$$x = A\,e^{m_1 t} + B\,e^{m_2 t}, \quad \text{so} \quad y = \dot{x} = Am_1\,e^{m_1 t} + Bm_2\,e^{m_2 t}, \tag{1.28}$$

and for fixed A and B, (1.28) constitutes a parametric representation of a phase path. The phase paths in Fig. 1.10 are plotted in this way for certain values of k and c.

This shows a new type of equilibrium point, called a *node*. For all slight displacements from $x = 0$, $\dot{x} = 0$, the state returns to the equilibrium point. It is therefore a *stable node*. That all the phase paths terminate at the origin can be seen by letting $t \to \infty$ in (1.28).

Weak damping $(\Delta < 0)$

The exponents are complex with negative real part, and the solutions are

$$x(t) = A\exp(-\tfrac{1}{2}kt)\cos\{\tfrac{1}{2}\sqrt{(-\Delta)}t + \alpha\} \tag{1.29}$$

where A, α are arbitrary constants. A typical solution is shown in Fig. 1.11(a): it represents an oscillation with exponentially decreasing amplitude, decaying more rapidly for larger k. Its image on the phase plane, plotted parametrically as before, is shown in Fig. 1.11(b).

The equilibrium point at the origin is called a *stable spiral* or a *stable focus*.

Critical damping $(\Delta = 0)$

In this case $m_1 = m_2 = -\tfrac{1}{2}k$ and the solutions are

$$x(t) = (A + Bt)\exp(-\tfrac{1}{2}kt).$$

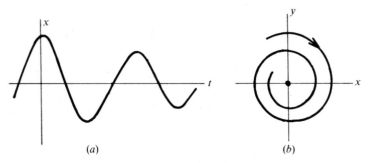

FIG. 1.11. Stable spiral

The solutions resemble those for strong damping and the phase diagram shows a stable node.

Permutations of signs of the parameters k and c are possible.

(i) $c < 0, k \neq 0$. m_1 and m_2 are real but have different signs. The phase diagram shows a saddle point.

(ii) $c > 0, k < 0$. This case is called *negative damping*. Instead of energy being lost by the equivalent of a resistance or friction, energy is generated constantly in the system. The node or spiral is now *unstable* as shown in Fig. 1.12, since a slight disturbance from equilibrium leads to the system being carried far from the equilibrium state.

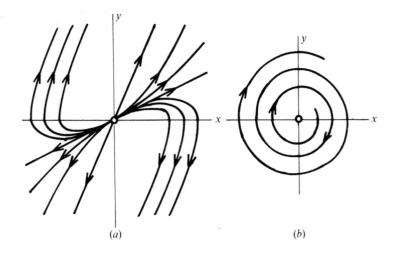

FIG. 1.12. (a) Unstable node; (b) Unstable spiral

1.5 Nonlinear damping

Returning to the system

$$\ddot{x} = f(x, \dot{x}),$$

assume that f takes the form

$$f(x, \dot{x}) = -h(x, \dot{x}) - g(x), \tag{1.30}$$

where h does not contain an additive function of x. Then

$$\ddot{x} + h(x, \dot{x}) + g(x) = 0. \tag{1.31}$$

In mechanical terms the equation describes the displacement x of a particle of unit mass under a force system containing a conservative element, $g(x)$, and a dissipative or energy-generating component, $h(x, \dot{x})$. In the case of a spring, $-g(x)$ is the restoring force. If g is of the appropriate type we should expect a tendency to oscillate, modified by the presence of the term $h(x, \dot{x})$.

The kinetic energy $\mathscr{T}$ is equal to $\frac{1}{2}\dot{x}^2$. We define a potential energy function for (1.31) by

$$\mathscr{V}(x) = \int g(x)\,dx, \quad \text{so that} \quad g(x) = \mathscr{V}'(x). \tag{1.32}$$

The total energy $\mathscr{E}$ defined by

$$\mathscr{E} = \mathscr{T} + \mathscr{V} = \frac{1}{2}\dot{x}^2 + \int g(x)\,dx \tag{1.33}$$

is not in general constant. Consider how $\mathscr{E}$ changes as a particular motion progresses: that is, along a phase path.

$$\frac{d\mathscr{E}}{dt} = \frac{d\mathscr{T}}{dt} + \frac{d\mathscr{V}}{dt} = \dot{x}\ddot{x} + \mathscr{V}'(x)\dot{x}$$

$$= \dot{x}\{-f(x, \dot{x}) + g(x)\} \quad \text{(on the path)}$$

$$= -\dot{x}h(x, \dot{x}) \tag{1.34}$$

by (1.31). Integrate (1.34) with respect to t from $t = \tau_0$ to $t = \tau$; then

$$\mathscr{E}(\tau) - \mathscr{E}(\tau_0) = -\int_{\tau_0}^{\tau} \dot{x}h(x, \dot{x})\,dt \tag{1.35}$$

(where $x(t)$ is the solution of (1.30) on the chosen phase path).

It may be possible to say, for a phase path lying in some region of the phase plane, either that

(i) $\dot{x}h(x, \dot{x}) = yh(x, y) > 0$, in which case $\mathscr{E}(\tau) < \mathscr{E}(\tau_0)$: that is, the energy

decreases and h has a damping effect, contributing to a general decrease in amplitude; or that

(ii) $\dot{x}h(x, \dot{x}) = yh(x, y) < 0$, in which case $\mathscr{E}(\tau) > \mathscr{E}(\tau_0)$: the effect is of an internal source injecting energy into the system.

A system may contain both characteristics. For example, a pendulum clock has energy stored in a weight which is supplied to the pendulum, and a balance is achieved between the rate of energy supplied and the loss through friction to maintain a steady oscillation.

Example 1.5   *Examine the equation $\ddot{x} + |\dot{x}|\dot{x} + x = 0$ for damping effects.*

$$h(x, \dot{x}) = |\dot{x}|\dot{x},$$

and

$$yh(x, y) = |y|y^2 > 0, \quad y \neq 0.$$

Except for the equilibrium point $(0, 0)$ there is a loss of energy along every phase path no matter where it goes in the phase plane. We should therefore not be surprised if, from any initial state, the corresponding phase path eventually entered the origin, and motion ceased.

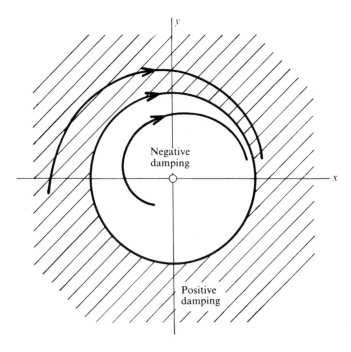

FIG. 1.13. Approach of two phase paths to a stable limit-cycle

Example 1.6 *Examine the equation* $\ddot{x} + (x^2 + \dot{x}^2 - 1)\dot{x} + x = 0$ *for damping or energy input effects.*

Putting $\dot{x} = y$,

$$yh(x, y) = (x^2 + y^2 - 1)y^2.$$

When $x^2 + y^2 < 1$, $yh(x, y) < 0$ and when $x^2 + y^2 > 1$, $yh(x, y) > 0$.

The regions of energy loss and energy input on the phase plane are shown in Fig. 1.13. It can be verified that $x = \cos t$ is a solution of the differential equation, and this is represented by the circle $x^2 + y^2 = 1$ on the phase plane. The phase diagram consists of this circle, together with paths spiralling on to it from the outside, paths spiralling on to it from the inside, and the equilibrium point at the origin. The *isolated* closed path $x^2 + y^2 = 1$; isolated in the sense that there is no other closed path in its neighbourhood; is called a *limit cycle*. All paths approach the circle as $t \to \infty$. A limit cycle is therefore one of the most important features of a physical system.

A useful way to approach this and similar equations is to use polar coordinates (r, θ) where $x = r\cos\theta$, $y = r\sin\theta$. It follows that, since $r^2 = x^2 + y^2$ and $\tan\theta = y/x$,

$$\dot{r} = (x\dot{x} + y\dot{y})/r, \quad \dot{\theta} = (x\dot{y} - \dot{x}y)/r^2.$$

Thus

$$\dot{r} = -r(r^2 - 1)\sin^2\theta, \tag{1.36}$$

$$\dot{\theta} = -1 - (r^2 - 1)\sin\theta\cos\theta. \tag{1.37}$$

A particular solution of (1.36) is given by $r = 1$, the unit circle previously obtained. Also from (1.36) $\dot{r} > 0$ for $r < 1$ (sin $\theta \neq 0$), and $\dot{r} < 0$ for $r > 1$ (sin $\theta \neq 0$). We infer that outside the limit cycle r decreases and inside r increases with time, implying that the limit cycle is stable, the paths approaching it from both inside and outside.

In the system of Example 1.6 the regions of positive and negative damping are separated by the circle $x^2 + y^2 = 1$. In the important equation (van der Pol's equation)

$$\ddot{x} + e(x^2 - 1)\dot{x} + x = 0, \quad e > 0, \tag{1.38}$$

negative damping occurs in the strip $|x| < 1$ and positive damping in the half-planes $|x| > 1$. We shall see later that this type of damping also leads to a limit cycle. Van der Pol's equation originally arose as an idealization of a spontaneously oscillating, or self-excited, valve circuit (Minorsky, 1962). It will provide for us the simplest form of equation having this pattern of positive and negative damping which gives rise to a limit cycle.

A closely-connected equation is Rayleigh's equation

$$\ddot{x} + e(\tfrac{1}{3}\dot{x}^2 - 1)\dot{x} + x = 0, \tag{1.39}$$

which originated in connection with a theory of the oscillation of a violin string. It has a similar distribution of positive and negative damping regions in the phase plane, and in fact reduces to van der Pol's equation by differentiating and putting $\dot{x} = z$.

1.6 Some applications

(i) *Dry friction*

Dry (or Coulomb) friction occurs when the surfaces of two solids are in contact and in relative motion without lubrication. The system shown in Fig. 1.14 illustrates dry friction. A continuous belt is driven by rollers

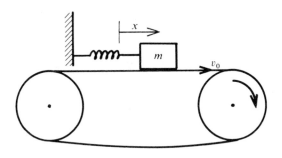

FIG. 1.14.

at a constant speed v_0. A block of mass m connected to a fixed support by a spring of stiffness c rests on the belt. If F is the frictional force between the block and the belt and x is the extension of the spring, then the equation of motion is

$$m\ddot{x} + cx = F.$$

Assume that F depends on the slip velocity, $v_0 - \dot{x}$; a typical relation is shown in Fig. 1.15(*a*). We will replace this function by a simpler one having a discontinuity at the origin (Fig. 1.15(*b*)):

$$F = F_0 \operatorname{sgn}(v_0 - \dot{x})$$

where F_0 is a positive constant (see Fig. 1.15(*b*)) and the sgn (signum) function is defined by

$$\operatorname{sgn}(u) = \begin{cases} 1, & u > 0, \\ 0, & u = 0, \\ -1, & u < 0. \end{cases}$$

Contents

oscillations: amplitude limitation, harmonic and subharmonic response, and jump phenomena and entrainment. Stability and instability of the solutions is determined by analysis of the van der Pol plane for these cases. Chapters 8 to 10 are devoted to formal stability questions for systems. Chapter 8 presents the necessary linear theory, including Floquet theory, for the chapters on stability and Liapunov functions. In Chapter 11 the existence of periodic solutions for some relatively simple representative cases is discussed.

There are over 400 examples, some worked in full in the text and others left as exercises, often with solutions and hints. The exercises provide a way of indicating developments for which there is no room in the text, and sometimes to present more specialized illustrative material. On the whole we have tried to keep the text itself free from scientific technicality and to present equations in a simple reduced form, believing that students have enough to do to follow the underlying arguments.

Our thanks are due to Olwen Brindley and Doreen Large for cheerfully typing the whole book twice over. We are also grateful for the computing facilities provided by the Computer Centre of the University of Keele, and for advice given by colleagues in the Department of Mathematics.

Keele, 1976   D.W.J.
                                                                      P.S.

Preface

SOME IDEA of the size of the literature on nonlinear systems can be obtained from the bibliography given by Cesari (1971). The wide applicability of the subject to the physical sciences, and more recently to the biological and social sciences, generates a continuous supply of new problems of practical and theoretical interest.

The present book grew from courses in nonlinear differential equations given over several years at the University of Keele. It is an introduction to dynamical systems in the context of systems of differential equations, intended for students of mathematics and science who are mainly interested in the more direct applications of the subject. The level is about that of a second- or third-year undergraduate course in Britain. It has been found that much of the material in Chapters 1 to 4 and 9 to 11 can be covered in a one-term course by students having a background of elementary differential equations and some linear algebra. The book is designed to accommodate courses of varying emphasis, the chapters forming fairly self-contained groups from which a coherent selection can be made without large parts of the argument being lost.

We have given the subject a qualitative slant throughout, emphasising how the presence of a nonlinear element may introduce totally novel phenomena. In order that these new features should not seem too shadowy, many of the diagrams are drawn from computed data and are not merely sketches. In Chapter 3 we advise the reader on how to do his own computing using the simplest means, a practice which adds a dimension of conviction to the whole subject.

Chapters 1 to 4 treat mainly two-dimensional systems and second-order equations in the phase plane. The treatment is kept at an intuitive and elementary level, but we try to encourage the reader to feel that almost immediately he has available useful new investigative techniques. The main features of the phase plane, equilibrium points, periodic solutions, limit-cycles, domains of attraction and the associated stability ideas are treated informally. Quantitative estimates for periodic solutions are obtained by energy balance and harmonic balance methods.

Chapters 5 to 7 deal with expansions in terms of a small parameter, singular perturbations, and the phenomena associated with forced

D. W. JORDAN
AND
P. SMITH
University of Keele

Nonlinear ordinary differential equations

CLARENDON PRESS · OXFORD

Oxford University Press, Walton Street, Oxford OX2 6DP

OXFORD   LONDON   GLASGOW   NEW YORK
TORONTO   MELBOURNE   WELLINGTON   CAPE TOWN
IBADAN   NAIROBI   DAR ES SALAAM   LUSAKA
KUALA LUMPUR   SINGAPORE   JAKARTA   HONG KONG   TOKYO
DELHI   BOMBAY   CALCUTTA   MADRAS   KARACHI

First published 1977
Reprinted with corrections 1979, 1983

British Library Cataloguing in Publication Data

Jordan, Dominic William
    Nonlinear ordinary differential equations.
    —(Oxford applied mathematics and computing
    science series).
    1. Differential equations, Nonlinear
    I. Title   II. Smith, Peter
    III. Series
    515'.352   QA372   77-30065

    ISBN 0-19-859620-0
    ISBN 0-19-859621-9   Pbk

Reproduced from copy supplied
printed and bound in Great Britain
by Billing and Sons Limited
Guildford, London, Oxford, Worcester

Oxford Applied Mathematics
and Computing Science Series

General Editors
J. Crank, H. G. Martin, D. M. Melluish

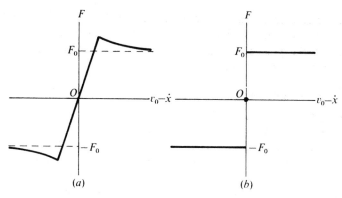

FIG. 1.15.

The equation of motion becomes

$$m\ddot{x} + cx = F_0 \, \text{sgn} \, (v_0 - \dot{x}).$$

The term on the right is equal to F_0 when $v_0 > \dot{x}$, and $- F_0$ when $v_0 < \dot{x}$, and we obtain the following solutions for the phase paths in these regions:

$$y = \dot{x} > v_0: \quad my^2 + (x + F_0/c)^2 c = \text{constant},$$
$$y = \dot{x} < v_0: \quad my^2 + (x - F_0/c)^2 c = \text{constant}.$$

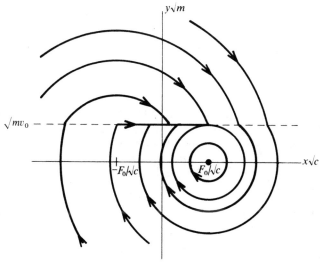

FIG. 1.16.

These are families of ellipses, the first having its centre at $(-F_0/c, 0)$ and the second at $(F_0/c, 0)$. Figure 1.16 shows the corresponding phase diagram, plotted as $\sqrt{m}y$ against $\sqrt{c}x$ to give circular paths.

There is a single equilibrium point, at $(F_0/c, 0)$, which is a centre. Points on y (or $\dot{x}$) $= v_0$ are not covered by the differential equation since this is where F is discontinuous, so the behaviour must be deduced from other physical arguments. On encountering the state $\dot{x} = v_0$ for $|x| < F_0/c$ the block will move with the belt until the maximum available friction, F_0, is insufficient to resist the increasing spring tension. This is when $x = F_0/c$; the block then goes into an oscillation represented by the closed path through $(F_0/c, v_0)$. In fact, for any initial conditions lying outside this ellipse, the system ultimately settles into this oscillation. A computed phase diagram corresponding to a frictional force as in Fig. 1.15(*a*) is displayed in Exercise 50, Chapter 3.

(ii) *The brake*

Consider a simple brake shoe applied to the hub of a wheel as shown in Fig. 1.17. The friction force will depend on the pressure and the angular velocity of the wheel, $\dot{\theta}$. We assume again a simplified dry-friction relation corresponding to constant pressure

$$F = -F_0 \operatorname{sgn}(\dot{\theta})$$

so if the wheel is otherwise freely spinning its equation of motion is

$$I\ddot{\theta} = -F_0 a \operatorname{sgn}(\dot{\theta}),$$

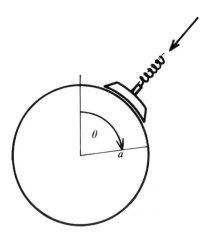

FIG. 1.17. A brake

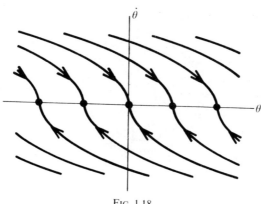

FIG. 1.18.

where I is the moment of inertia of the wheel and a the radius of the brake drum. The phase paths are found by rewriting the differential equation:

$$I\dot{\theta}\frac{d\dot{\theta}}{d\theta} = -F_0 a \operatorname{sgn}(\dot{\theta}),$$

whence for $\dot{\theta} > 0$

$$\tfrac{1}{2}I\dot{\theta}^2 = -F_0 a\theta + C$$

and for $\dot{\theta} < 0$

$$\tfrac{1}{2}I\dot{\theta}^2 = F_0 a\theta + C.$$

These represent two families of parabolas as shown in Fig. 1.18. $(\theta, 0)$ is an equilibrium point for every θ.

(iii) *The pendulum clock: a limit cycle*

Figure 1.19 shows the main features of the pendulum clock. The 'escape wheel' is a toothed wheel, which drives the hands of the clock through a succession of gears. It has a spindle around which is wound a wire with a weight at its free end. The escape wheel is arrested by the 'anchor' which has two teeth. The anchor is attached to the shaft of the pendulum and rocks with it, controlling the rotation of the escape wheel. Its teeth are so designed that as the pendulum reaches its maximum amplitude on one side one tooth of the escape wheel is released, and the escape wheel is then stopped again by the other tooth on the anchor. Every time this happens the anchor receives a small impulse or pressure. It is this impulse which maintains the oscillation of the pendulum, which

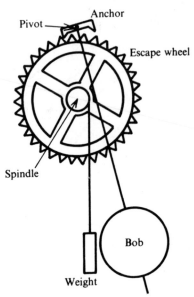

Fig. 1.19. A weight-driven clock mechanism

would otherwise die away. The loss of potential energy due to the weight's descent is therefore fed periodically into the pendulum via the anchor mechanism.

It can be shown that the system will settle into steady oscillations of fixed amplitude independently of sporadic disturbance and of initial conditions. If the pendulum is swinging with too great an amplitude its loss of energy per cycle due to friction is large, and the impulse supplied by the escapement is insufficient to offset this. The amplitude consequently decreases. If the amplitude is too small, the frictional loss is small; the impulses will over-compensate and the amplitude will build up. A balanced state is therefore approached, which appears in the $\theta, \dot{\theta}$ plane (Fig. 1.20) as an isolated closed curve $\mathscr{C}$. Such an isolated periodic oscillation, or *limit cycle* (see Example 1.6) can occur only in systems described by nonlinear equations, and the following simple model shows where the nonlinearity is located. The motion can be approximated by the equation

$$I\ddot{\theta} + k\dot{\theta} + c\theta = f(\theta), \tag{1.40}$$

where I is the moment of inertia of the pendulum, k is a small damping constant, c is another constant determined by gravity, θ is the angular

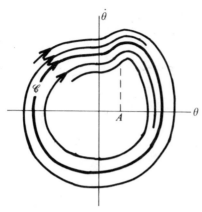

Fig. 1.20. Phase diagram for a clock, impulse applied near A

displacement, and $f(\theta)$ is the moment, supplied once per cycle by the escapement mechanism. Since f is periodic in θ, it must be a nonlinear function of θ.

Such an oscillation, generated by an energy source whose input is not regulated externally, but which *automatically* synchronizes with the existing oscillation, is called *self-excited*. Here the build-up is limited by the friction.

1.7 Parameter-dependent conservative systems

Suppose $x(t)$ satisfies

$$\ddot{x} = f(x, \lambda)$$

where λ is a parameter. The equilibrium points of the system are given by $f(x, \lambda) = 0$, and in general their location will depend on the parameter λ. In mechanical terms, for a particle of unit mass with displacement x, $f(x, \lambda)$ represents the force experienced by the particle. Suppose there exists a function $\mathscr{V}(x, \lambda)$ such that $f(x, \lambda) = -\partial\mathscr{V}/\partial x$ for each value of λ; then $\mathscr{V}(x, \lambda)$ is the potential energy of the system and equilibrium points correspond to stationary values of the potential energy. As indicated in Section 1.3, we expect a minimum of potential energy to correspond to a stable equilibrium point, and other stationary values (the maximum and point of inflexion) to be unstable. In fact, $\mathscr{V}$ is a minimum at $x = x_1$ if $\partial\mathscr{V}/\partial x$ changes from negative to positive on passing through x_1; this implies that $f(x, \lambda)$ changes sign from positive to negative as x increases through $x = x_1$.

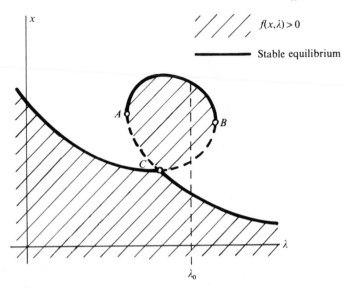

FIG. 1.21. Stability curves for the equilibrium points of $\ddot{x} = f(x, \lambda)$

There exists a simple method of displaying the stability of equilibrium points for parameter-dependent systems in which both the number and stability of equilibrium points may vary with λ. We assume $f(x, \lambda)$ to be continuous in both x and λ. Plot the curve $f(x, \lambda) = 0$ in the λ, x plane; this curve represents the equilibrium points. Shade the domains in which $f(x, \lambda) > 0$ as shown in Fig. 1.21. If a segment of the curve has shading below it, the corresponding equilibrium points are stable, since for fixed λ, f changes from positive to negative as x increases.

For example, the solid line between A and B corresponds to stable equilibrium points. A and B are unstable: C is also unstable since f is positive on both sides of C. The nature of equilibrium points can easily be read from the figure; when $\lambda = \lambda_0$ as shown, the system has three equilibrium points, two of which are stable. A, B and C are known as *bifurcation points*. As λ varies through such points the equilibrium point may split into two or more, or several equilibrium points may appear or merge into a single one.

Example 1.7 *A bead slides on a smooth circular wire of radius a which is constrained to rotate about a vertical diameter with constant angular velocity ω. Analyse the stability of the bead.*

The bead has a velocity component $a\dot{\theta}$ tangential to the wire and a component

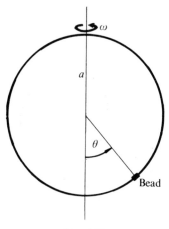

FIG. 1.22.

$a\omega \sin \theta$ perpendicular to the wire, where θ is the inclination of the radius to the bead to the downward vertical as shown in Fig. 1.22. The kinetic energy $\mathcal{T}$ and potential energy $\mathcal{V}$ are given by

$$\mathcal{T} = \tfrac{1}{2}ma^2(\dot{\theta}^2 + \omega^2 \sin^2 \theta), \quad \mathcal{V} = -mga \cos \theta.$$

Since the system is subject to a moving constraint (that is, the angular velocity of the wire is imposed) the usual energy equation does not hold. Lagrange's equation for the system is

$$\frac{d}{dt}\left(\frac{\partial \mathcal{T}}{\partial \dot{\theta}}\right) - \frac{\partial \mathcal{T}}{\partial \theta} = -\frac{\partial \mathcal{V}}{\partial \theta},$$

which gives

$$a\ddot{\theta} = a\omega^2 \sin \theta \cos \theta - g \sin \theta.$$

Set $a\omega^2/g = \lambda$. Then

$$a\ddot{\theta} = a\dot{\theta}\frac{d\dot{\theta}}{d\theta} = g \sin \theta(\lambda \cos \theta - 1)$$

which, after integration, becomes

$$\tfrac{1}{2}a\dot{\theta}^2 = g(1 - \tfrac{1}{2}\lambda \cos \theta)\cos \theta + C,$$

the equation of the phase paths.

In the previous theory

$$f(\theta, \lambda) = g \sin \theta(\lambda \cos \theta - 1)/a$$

and the equilibrium points are given by $f(\theta, \lambda) = 0$, which is satisfied when $\sin \theta$

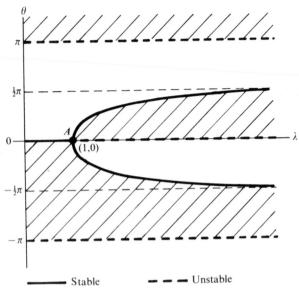

FIG. 1.23.

$= 0$ or $\cos\theta = \lambda^{-1}$. From the periodicity of the problem, $\theta = \pi$ and $\theta = -\pi$ correspond to the same state of the system.

The regions where $f < 0$ and $f > 0$ are separated by curves where $f = 0$, and can be located, therefore, by checking the sign at particular points; for example, $f(\tfrac{1}{2}\pi, 1) = -g/a < 0$. Figure 1.23 shows the stable and unstable equilibrium positions of the bead. A is a stable bifurcation point.

Phase diagrams for the system may be constructed as in Section 1.3 for fixed values of λ. Two possibilities are shown in Fig. 1.24. Note that they confirm the stability predictions of Fig. 1.23.

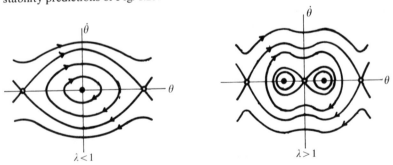

FIG. 1.24.

Exercises

1. Locate the equilibrium points and sketch the phase diagrams in their neighbourhood for the following equations:

   (i) $\ddot{x} - k\dot{x} = 0$.

   (ii) $\ddot{x} - 8x\dot{x} = 0$.

   (iii) $\ddot{x} = k\,(|x| > 1), \quad \ddot{x} = 0\,(|x| < 1)$.

   (iv) $\ddot{x} + 3\dot{x} + 2x = 0$.

   (v) $\ddot{x} - 4\dot{x} + 40x = 0$.

   (vi) $\ddot{x} + 3|\dot{x}| + 2x = 0$.

   (vii) $\ddot{x} + k\,\mathrm{sgn}\,(\dot{x}) + c\,\mathrm{sgn}\,(x) = 0, \quad c > k$. Show that the path starting at $(x_0, 0)$ reaches $((c-k)^2 x_0/(c+k)^2, 0)$ after one circuit of the origin. Deduce that the origin is a spiral point.

   (viii) $\ddot{x} + x\,\mathrm{sgn}\,(x) = 0$.

2. Sketch the phase diagram for the equation $\ddot{x} = -x - \alpha x^3$, considering all values of α. Check the stability of the equilibrium points by the method of Section 1.7.

3. A certain dynamical system is governed by the equation $\ddot{x} + \dot{x}^2 + x = 0$. Show that the origin is a centre in the phase plane, and that the open and closed paths are separated by the path $2y^2 = 1 - 2x$.

4. Sketch the phase diagrams for the equation $\ddot{x} + e^x = a$, for $a < 0$, $a = 0$, and $a > 0$.

5. Sketch the phase diagram for the equation $\ddot{x} - e^x = a$, for $a < 0$, $a = 0$, and $a > 0$.

6. The potential energy $\mathscr{V}(x)$ of a conservative system is continuous, and is strictly increasing for $x < -1$, zero for $|x| \leqslant 1$, and strictly decreasing for $x > 1$. Locate the equilibrium points and sketch the phase diagram for the system.

7. Figure 1.25 shows a pendulum striking an inclined wall. Sketch the phase diagram, for α positive and α negative, when (i) there is no loss of energy at impact, (ii) the magnitude of the velocity is halved on impact.

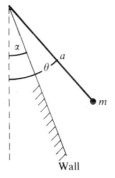

FIG. 1.25.

Wall

8. Show that the time elapsed, T, along a phase path $\mathscr{C}$ of the system $\dot{x} = y$, $\dot{y} = f(x, y)$ is given, in a form alternative to (1.11), by

$$T = \int_{\mathscr{C}} (y^2 + f^2)^{-1/2} \, ds,$$

where ds is an element of distance along $\mathscr{C}$.

By writing $\delta s \simeq (y^2 + f^2)^{1/2} \delta t$, indicate, very roughly, equal time intervals along the phase paths of the system $\dot{x} = y$, $\dot{y} = 2x$.

9. On the phase diagram for the equation $\ddot{x} + x = 0$, the phase paths are circles. Use (1.11) in the form $\delta t \simeq \delta x / y$ to indicate, roughly, equal time steps along several phase paths.

10. Repeat Exercise 9 for the equation $\ddot{x} + 9x = 0$, in which the phase paths are ellipses.

11. The pendulum equation, $\ddot{x} + \omega^2 \sin x = 0$, can be approximated for moderate amplitudes by the equation $\ddot{x} + \omega^2 (x - \frac{1}{6} x^3) = 0$. Sketch the phase diagram for the latter equation, and explain the differences between it and Fig. 1.2.

12. The displacement, x, of a spring-mounted mass under the action of Coulomb dry friction is assumed to satisfy

$$m\ddot{x} + cx = -F_0 \operatorname{sgn}(\dot{x}),$$

where m, c and F_0 are positive constants (Section 1.6). The motion starts at $t = 0$, with $x = x_0 > 3F_0/c$ and $\dot{x} = 0$. Subsequently, whenever $x = -\alpha$, where $2F_0/c - x_0 < -\alpha < 0$ and $\dot{x} > 0$, a trigger operates, to increase suddenly the forward velocity so that the kinetic energy increases by a constant amount E. Show that if $E > 8F_0^2/c$, a periodic motion is approached, and show that the largest value of x in the periodic motion is equal to $F_0/c + E/4F_0$.

13. In Exercise 12, suppose that the energy is increased by E at $x = -\alpha$ for both $\dot{x} < 0$ and $\dot{x} > 0$; that is, there are two injections of energy per cycle. Show that periodic motion is possible if $E > 6F_0^2/c$, and find the amplitude of the oscillation.

14. The 'friction pendulum' consists of a pendulum attached to a sleeve, which embraces a close-fitting cylinder as shown. The cylinder is turned at a constant rate Ω. The sleeve is subject to Coulomb dry friction through the couple $G = -F_0 \operatorname{sgn}(\dot{\theta} - \Omega)$. Write down the equation of motion, find the equilibrium states, and sketch the phase diagram.

15. By plotting the 'potential energy' of the nonlinear conservative system $\ddot{x} = x^4 - x^2$, construct the phase diagram of the system. A particular path has the initial conditions $x = \frac{1}{2}$, $\dot{x} = 0$ at $t = 0$. Is the subsequent motion periodic?

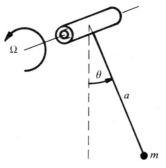

Fig. 1.26.

16. The system $\ddot{x}+x = -F_0 \operatorname{sgn}(\dot{x})$, $F_0 > 0$, has the initial conditions $x = x_0 > 0$, $\dot{x} = 0$. Show that the phase path will spiral exactly n times before entering equilibrium (Section 1.6) if $(4n-1)F_0 < x_0 < (4n+1)F_0$.

17. A pendulum of length a has a bob of mass m which is subject to a horizontal force $m\omega^2 a \sin\theta$, where θ is the inclination to the downward vertical. Show that the equation of motion is $\ddot{\theta} = \omega^2(\cos\theta - \lambda)\sin\theta$, where $\lambda = g/\omega^2 a$. Investigate the stability of the equilibrium states by the method of Section 1.7 for parameter-dependent systems. Sketch the phase diagrams for various λ.

18. Investigate the stability of the equilibrium points of the parameter-dependent system $\ddot{x} = (x - \lambda)(x^2 - \lambda)$.

19. If a bead slides on a smooth parabolic wire rotating with constant angular velocity ω about a vertical axis, then the distance x of the particle from the axis of rotation satisfies $(1 + x^2)\ddot{x} + (g - \omega^2 + \dot{x}^2)x = 0$. Analyse the motion of the bead in the phase plane.

20. A particle is attached to a fixed point 0 on a smooth horizontal plane by an elastic string. When unstretched, the length of the string is a. The equation of motion of the particle, which is constrained to move on a straight line through 0, is

$$\ddot{x} = -x + a\operatorname{sgn}(x), \quad |x| > a \quad \text{(when the string is stretched)}$$

$$\ddot{x} = 0, \quad |x| < a \quad \text{(when the string is slack)},$$

x being the displacement from 0. Find the equilibrium points and the equations of the phase paths, and sketch the phase diagram.

21. The equation of motion of a conservative system is $\ddot{x} + g(x) = 0$, where $g(0) = 0$; $g(x) < 0$ for $x < 0$ and $g(x) > 0$ for $x > 0$; and

$$\int_0^x g(u)\,du \to \infty \quad \text{as} \quad x \to \pm\infty. \tag{a}$$

Show that the motion is always periodic.

By considering $g(x) = xe^{-x^2}$, show that if (a) does not hold, the motions are not all necessarily periodic.

22. The wave function $u(x, t)$ satisfies the partial differential equation

$$\frac{\partial^2 u}{\partial x^2} + \alpha \frac{\partial u}{\partial x} + \beta u^3 + \gamma \frac{\partial u}{\partial t} = 0,$$

where α, β and γ are positive constants. Show that there exist permanent wave solutions of the form $u(x, t) = U(x - ct)$ for any c, where $U(\zeta)$ satisfies

$$\frac{d^2 U}{d\zeta^2} + (\alpha - \gamma c)\frac{dU}{d\zeta} + \beta U^3 = 0.$$

Using Exercise 21, show that when $c = \alpha/\gamma$, all such waves are periodic.

23. The linear oscillator $\ddot{x} + \dot{x} + x = 0$ is set in motion with initial conditions $x = 0, \dot{x} = v$, at $t = 0$. After the first and each subsequent cycle the kinetic energy is instantaneously increased by a constant, E, in such a manner as to increase $\dot{x}$. Show that if $E = \frac{1}{2}v^2(1 - e^{4\pi/\sqrt{3}})$, a periodic motion occurs. Find the maximum value of x in a cycle.

24. Show how solutions of Exercise 23 having arbitrary initial conditions spiral on to the periodic solution (which is a limit cycle). Sketch the phase diagram.

25. The kinetic energy, $\mathcal{T}$, and the potential energy, $\mathcal{V}$, of a system with one degree of freedom are given by

$$\mathcal{T} = T_0(x) + \dot{x}T_1(x) + \dot{x}^2 T_2(x), \quad \mathcal{V} = \mathcal{V}(x).$$

Use Lagrange's equation

$$\frac{d}{dt}\left(\frac{\partial \mathcal{T}}{\partial \dot{x}}\right) - \frac{\partial \mathcal{T}}{\partial x} = -\frac{\partial \mathcal{V}}{\partial x}$$

to obtain the equation of motion of the system. Show that the equilibrium points are stationary points of $T_0(x) - \mathcal{V}(x)$, and that the phase paths are given by the energy equation

$$T_2(x)\dot{x}^2 - T_0(x) + \mathcal{V}(x) = \text{constant}.$$

26. Sketch the phase diagram for the equation $\ddot{x} = -f(x + \dot{x})$, where

$$f(u) = \begin{cases} f_0, & u \geqslant c, \\ f_0 u/c, & |u| \leqslant c, \\ -f_0, & u \leqslant -c, \end{cases}$$

where f_0, c are constants, $f_0 > 0$, and $c > 0$. How does the system behave as $c \to 0$?

27. Sketch the phase diagram for the equation $\ddot{x} = u$, where

$$u = -\mathrm{sgn}\,(\sqrt{2}|x|^{1/2}\,\mathrm{sgn}\,(x) + \dot{x}).$$

(u is an elementary control variable which can switch between $+1$ and -1. The curve $\sqrt{2}|x|^{1/2}\,\mathrm{sgn}\,(x) + y = 0$ is called the switching curve.)

28. The relativistic equation for an oscillator is

$$\frac{d}{dt}\left\{\frac{m_0\dot{x}}{\sqrt{[1 - (\dot{x}/c)^2]}}\right\} + kx = 0$$

where m_0, c, and k are positive constants. Show that the phase paths are given by

$$\frac{m_0 c^2}{\sqrt{[1 - (y/c)^2]}} + \tfrac{1}{2}kx^2 = \text{constant}.$$

If $y = 0$ when $x = a$, show that the period, T, of an oscillation is given by

$$T = \frac{4}{c\sqrt{\varepsilon}} \int_0^a \frac{[1 + \varepsilon(a^2 - x^2)]\,dx}{\sqrt{(a^2 - x^2)}\sqrt{[2 + \varepsilon(a^2 - x^2)]}}, \qquad \varepsilon = k/2m_0 c^2.$$

The constant ε is small; by expanding the integrand in powers of ε show that

$$T \approx 2\pi\sqrt{(m_0/k)}\,(1 + \tfrac{3}{8}\varepsilon a^2).$$

29. A mass m is attached to the mid-point of an elastic string of length $2a$ and stiffness λ. There is no gravity acting, and the tension is zero in the equilibrium

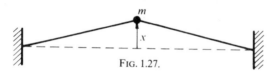

Fig. 1.27.

position. Obtain the equation of motion for transverse oscillations and sketch the phase paths.

30. Analyse, from the point of view of Section 1.5, the equation of motion for the dry-friction problem of Section 1.6, and reconcile this with the phase diagram Fig. 1.16.

31. Show that, in the phase diagram for a pendulum clock (Section 1.6), there is a net loss of energy along paths far enough from the origin, and a net gain along paths close enough to the origin. (Put eqn (1.40) into the form (1.30) and use (1.35).)

32. A pendulum with a magnetic bob oscillates in a vertical plane over a magnet, which repels the bob according to the inverse square law, so that the equation of motion is

$$ma^2\ddot{\theta} = -mga\sin\theta + Fh\sin\phi,$$

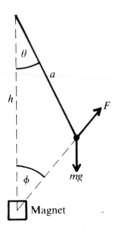

FIG. 1.28.

where $h > a$ and $F = c/(a^2 + h^2 - 2ah\cos\theta)$ and c is a constant. Find the equilibrium positions of the bob, and classify them as centres and saddle-points according to the parameters of the problem. Describe the motion of the pendulum.

33. A pendulum with equation $\ddot{x} + \sin x = 0$ oscillates with amplitude a. Show that its period, T, is equal to $4K(\beta)$, where $\beta = \sin^2\frac{1}{2}a$ and

$$K(\beta) = \int_0^{\pi/2} \frac{d\phi}{\sqrt{(1 - \beta\sin^2\phi)}}.$$

$K(\beta)$ has the power series representation

$$K(\beta) = \frac{1}{2}\pi[1 + (\tfrac{1}{2})^2\beta + (\tfrac{1\cdot3}{2\cdot4})^2\beta^2 + \ldots], \quad |\beta| < 1.$$

Deduce that, for small amplitudes,

$$T = 2\pi(1 + \tfrac{1}{16}a^2 + \tfrac{11}{3072}a^4) + O(a^6).$$

34. Repeat Exercise 33 with the equation $\ddot{x} + x - \varepsilon x^3 = 0$, and show that

$$T = \frac{4\sqrt{2}}{\sqrt{(2 - \varepsilon a^2)}} K(\beta), \quad \beta = \frac{\varepsilon a^2}{2 - \varepsilon a^2},$$

and that

$$T = 2\pi(1 + \tfrac{3}{8}\varepsilon a^2 + \tfrac{57}{256}\varepsilon^2 a^4) + O(a^6), \quad \varepsilon a^2 < 2.$$

35. Show that equations of the form $\ddot{x} + g(x)\dot{x}^2 + h(x) = 0$ are effectively conservative. (Find a transformation of x which puts the equation into the usual conservative form. Cf. eqn (1.14).)

36. Compare the phase diagrams of the following systems and the second order equations in x to which they give rise:
(i) $\dot{x} = 1, \dot{y} = 0$,   (ii) $\dot{x} = y, \dot{y} = 0$,   (iii) $\dot{x} = xy, \dot{y} = 0$.

37. Show that the phase plane for the equation

$$\ddot{x} - \varepsilon x \dot{x} + x = 0$$

has a centre at the origin, by first calculating the phase paths.

38. Show that the equation $\ddot{x} + x + \varepsilon x^3 = 0$ with $x(0) = a, \dot{x}(0) = 0$ has phase paths given by

$$\dot{x}^2 + x^2 + \tfrac{1}{2}\varepsilon x^4 = (1 + \tfrac{1}{2}\varepsilon a^2)a^2.$$

Show that the origin is a centre. Are all phase paths closed, and hence all solutions periodic?

39. Locate the equilibrium points of the equation

$$\ddot{x} + \lambda + x^3 - x = 0,$$

in the x, λ plane. Show that the phase paths are given by

$$\tfrac{1}{2}\dot{x}^2 + \lambda x + \tfrac{1}{4}x^4 - \tfrac{1}{2}x^2 = \text{constant}.$$

Sketch, in the $x, \dot{x}, \lambda$ space, the surface which represents the set of phase diagrams for which the constant (the energy) is zero. Suppose the system is oscillating with zero energy and, by means of two controllers, λ is slowly reduced and the energy is kept zero. What happens as λ passes through the value $-\tfrac{1}{3}\sqrt{(\tfrac{2}{3})}$?

40. Burgers' equation

$$\frac{\partial \phi}{\partial t} + \phi \frac{\partial \phi}{\partial x} = c \frac{\partial^2 \phi}{\partial x^2}$$

shows diffusion and nonlinear effects. Find the equation for permanent waves by putting $\phi = U(x - ct)$, where c is the constant wave speed. Find the equilibrium points and the phase paths for the resulting equation and interpret the phase diagram.

41. Show that the oscillations of the system

$$\ddot{x} + 1 - (1 + 2x)^{-1/2} = 0$$

in the neighbourhood of the origin have period independent of the amplitude.

2 First-order systems in two variables and linearization

CHAPTER 1 describes the application of phase-plane methods to the equation $\ddot{x} = f(x, \dot{x})$ through the equivalent first-order system $\dot{x} = y$, $\dot{y} = f(x, y)$. This approach permits a useful line of argument based on a mechanical interpretation of the original equation. Frequently, however, the appropriate formulation of mechanical, biological, and geometrical problems is not through a second-order equation at all, but directly as a more general type of first-order system of the form $\dot{x} = X(x, y)$, $\dot{y} = Y(x, y)$. The appearance of these equations is an invitation to construct a phase plane with x, y coordinates in which solutions are represented by curves $(x(t), y(t))$ where $x(t), y(t)$ are the solutions. The constant solutions are represented by equilibrium points obtained by solving the equations $X(x, y) = 0$, $Y(x, y) = 0$, and these may now occur anywhere in the plane. Near the equilibrium points we may make a linear approximation to $X(x, y), Y(x, y)$, solve the simpler equations obtained and so determine the local character of the paths. This enables the stability of the equilibrium states to be settled and is a starting-point for global investigations of the solutions. This chapter is mainly concerned with presenting examples and with classifying equilibrium points according to their linear approximations.

2.1 The general phase plane

Consider the general autonomous first-order system

$$\dot{x} = X(x, y), \quad \dot{y} = Y(x, y) \tag{2.1}$$

of which the type considered in Chapter 1,

$$\dot{x} = y, \quad \dot{y} = f(x, y), \tag{2.2}$$

is a special case. As in Section 1.2, a system is called *autonomous* when the time variable does not appear in the right-hand side of (2.1). We shall give examples later of how such systems arise.

The solutions $x(t), y(t)$ of (2.1) may be represented on a plane with cartesian axes x, y. Then as t increases $(x(t), y(t))$ traces out a directed curve in the plane called a *phase path*.

The appropriate form for the initial conditions of (2.1) is

$$x = x_0, \quad y = y_0 \quad \text{at} \quad t = t_0$$

where x_0 and y_0 are the initial values at time t_0; by the existence and uniqueness theorem (Appendix A) there is one and only one solution satisfying this condition when (x_0, y_0) is an 'ordinary point'. This does not at once mean that there is one and only one phase path through the point (x_0, y_0) on the phase diagram, because this same point could serve as the initial conditions for other starting times. Therefore it might seem that other phase paths through the same point could result: the phase diagram would then be a tangle of criss-crossed curves. We shall see that this will not be so by forming the differential equation for the phase paths. Since $\dot{y}/\dot{x} = dy/dx$ on a path the required equation is

$$\frac{dy}{dx} = \frac{Y(x, y)}{X(x, y)}. \tag{2.3}$$

There are two classes of exceptional points where solution curves of (2.3) (the phase paths) may meet. First, there may be points where Y/X has some manifest singularity so that the uniqueness theorem fails: these are *singular points*. Of more immediate interest, there are usually points where

$$X(x, y) = 0, \quad Y(x, y) = 0, \tag{2.4}$$

which are the *equilibrium points*. If a solution of (2.4) is x_1, y_1, then $x(t) = x_1, y(t) = y_1$ are *constant* solutions of (2.1), and are degenerate phase paths. There is one and only one solution curve passing through any ordinary point of (2.3), which may have singular points where (2.1) does not. Therefore there is one and only one phase path containing the ordinary point (x_0, y_0), no matter at what time the initial condition (x_0, y_0) is prescribed. Thus, infinitely many solutions of (2.1), differing only by time displacements, map onto a single phase path.

Equation (2.3) does not give any indication of the direction to be associated with a phase path, and this must be settled by reference to (2.1): the signs of X and Y at a point determine the direction through the point, and generally the directions at all other points can be settled by continuity.

The diagram depicting the phase paths is called the phase diagram. The point (x, y) is called a *state* of the system, and the phase diagram depicts the evolution of the states of the system from arbitrary initial states.

Example 2.1   *Compare the phase diagrams of the systems*
(i) $\dot{x} = y, \dot{y} = -x$;   (ii) $\dot{x} = xy, \dot{y} = -x^2$.
    The equation for the paths is the same for both, namely

$$\frac{dy}{dx} = -\frac{x}{y}$$

(strictly, for $y \neq 0$, and for $x \neq 0$ in the second case), giving a family of circles in both cases. However, in case (i) there is an equilibrium point only at the origin, but in case (ii) every point on the y axis is an equilibrium point. The directions, too, are different. By considering the signs of $\dot{x}, \dot{y}$ in the various quadrants the phase diagram of Figure 2.1 is produced.

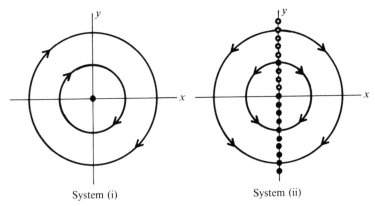

System (i)   System (ii)

FIG. 2.1.

    In the construction of phase paths note that the equilibrium points are the points of intersection of $X(x, y) = 0$ and $Y(x, y) = 0$. It is often useful to plot these curves since, at ordinary points on them, the phase paths cut $X(x, y) = 0$ parallel to the y-axis and cut $Y(x, y) = 0$ parallel to the x-axis. The *vector* $(X(x, y), Y(x, y))$ indicates the direction of a phase path at a point.
    A second-order differential equation can be reduced to the general form (2.2) in an arbitrary number of ways and this occasionally has advantages akin to changing the variable to simplify a differential equation. For example the reduction (2.1) applied to $x\ddot{x} - \dot{x}^2 - x^3 = 0$ leads to the system

$$\dot{x} = y, \quad \dot{y} = y^2/x + x^2. \tag{2.5}$$

Suppose we define, instead of y, another variable y_1 by $y_1(t) = \dot{x}(t)/x(t)$,

variables and linearization

$$\dot{x} = xy_1. \tag{2.6}$$

Then from (2.6), $\ddot{x} = x\dot{y}_1 + \dot{x}y_1 = x\dot{y}_1 + xy_1^2$ (using (2.6) again). But from the differential equation, $\ddot{x} = \dot{x}^2/x + x^2 = xy_1^2 + x^2$. Therefore

$$\dot{y}_1 = x. \tag{2.7}$$

The pair of equations (2.6) and (2.7) afford a representation alternative to (2.5). The phase diagram x, y_1 will, of course, be different in appearance from the x, y diagram.

The time T elapsing along a segment $\mathscr{C}$ of a phase path connecting two states (see Fig. 1.4) is given by

$$T = \int_{\mathscr{C}} dt = \int_{\mathscr{C}} \left(\frac{dx}{dt}\right)^{-1}\left(\frac{dx}{dt}\right)dt = \int_{\mathscr{C}} \frac{dx}{X(x, y)}. \tag{2.8}$$

Alternatively, let ds be a length element of $\mathscr{C}$. Then $ds^2 = dx^2 + dy^2$ on the path, and

$$T = \int_{\mathscr{C}} \left(\frac{ds}{dt}\right)^{-1}\left(\frac{ds}{dt}\right)dt = \int_{\mathscr{C}} \frac{ds}{(X^2 + Y^2)^{1/2}}. \tag{2.9}$$

The integrals above depend only on X and Y and the geometry of the phase path; therefore the time scale is implicit in the phase diagram.

2.2 Some population models

In the following examples systems of the type (2.1) arise naturally. Further examples from biology can be found in Pielou (1969) and Rosen (1973).

Example 2.2 *A predator–prey problem* (Volterra's model)

In a lake there are two species of fish: A, which lives on plants of which there is a plentiful supply, and B (the predator) which subsists by eating A (the prey). We shall construct a crude model for the interaction of A and B.

Let $x(t)$ be the population of A and $y(t)$ that of B. We assume that A is relatively long-lived and rapidly breeding if left alone. Then in time δt there is a population increase given by

$$ax\delta t, \quad a > 0$$

due to births and 'natural' deaths, and a 'negative increase'

$$-cxy\delta t, \quad c > 0$$

owing to A's being eaten by B (the number being eaten in this time being assumed proportional to the number of encounters between types A and B). The net population increase of A, δx, is given by

$$\delta \dot{x} = ax\delta t - cxy\delta t,$$

so that in the limit $\delta t \to 0$

$$\dot{x} = ax - cxy. \tag{2.10}$$

We now assume that, in the absence of prey, the starvation rate of B predominates over the birth rate, but that the compensating growth of B is again proportional to the number of encounters with A. This gives

$$\dot{y} = -by + dxy \tag{2.11}$$

with $b > 0$, $d > 0$. Equations (2.10) and (2.11) are a pair of simultaneous nonlinear equations of the form (2.2).

We now plot the phase diagram in the x, y plane. Only the quadrant

$$x \geqslant 0, \quad y \geqslant 0$$

is of interest. The equilibrium points are where

$$X(x, y) \equiv ax - cxy = 0, \quad Y(x, y) \equiv -by + dxy = 0;$$

that is at $(0, 0)$ and $(b/d, a/c)$. The phase paths are given by $dy/dx = Y/X$, or

$$\frac{dy}{dx} = \frac{(-b + dx)y}{(a - cy)x},$$

which is a separable equation leading to

$$\int \frac{(a - cy)}{y} dy = \int \frac{(-b + dx)}{x} dx,$$

or

$$a \log_e y + b \log_e x - cy - dx = C \tag{2.12}$$

where C is an arbitrary constant, the parameter of the family. Writing (2.12) in the form $(a \log_e y - cy) + (b \log_e x - dx) = C$, the result of Exercise 27 shows that this is a system of closed curves centred on the equilibrium point $(b/d, a/c)$.

Figure 2.2 shows a detailed plot of the phase paths for a particular case. The direction on the paths, indicating the change of state as t increases, can be obtained by, say, finding the sign of $\dot{x}$ on $x = b/d$. In fact, the direction at a single point, even on $x = 0$ or $y = 0$, determines the directions at all points by continuity.

Since the paths are closed, the fluctuations of $x(t)$ and $y(t)$, starting from any initial population, are predicted as periodic, the maximum population of A being about a quarter of a period behind the maximum population of B. As A gets eaten, causing B to thrive, the population x of A is reduced, causing eventually a drop in that of B. The shortage of predators then leads to a resurgence of A and

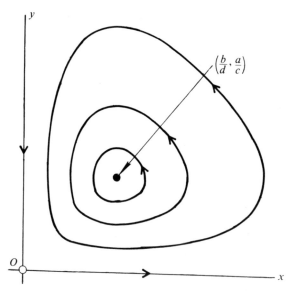

FIG. 2.2. Typical phase diagram for the predator–prey problem

the cycle starts again. A sudden change in state due to external causes, such as a bad season for the plants, puts the state on to another closed curve, but no tendency to an equilibrium population, nor for the population to disappear, is predicted. If we expect such a tendency, then we must construct a different model (see Exercises 12 and 13).

In Section 2.3 we shall show how the nature of equilibrium points, and hence the general character of the solutions (periodic, tending to equilibrium, unstable, etc.), can be determined without solving the equations explicitly.

Example 2.3 *A general epidemic model*

Consider the spread of a non-fatal disease in a population which is assumed to have constant size over the period of the epidemic. At time t suppose the population consists of

$x(t)$ susceptibles: those so far uninfected and therefore liable to infection;

$y(t)$ infectives: those who have the disease and are still at large;

$z(t)$ who are isolated, or who have recovered and are therefore immune.

Assume there is a steady contact rate between susceptibles and infectives and that a constant proportion of these contacts result in transmission. Then in time δt, δx of the susceptibles become infective, where

$$\delta x = -\beta x y \delta t,$$

and β is constant.

If γ is the rate at which current infectives become isolated, then

$$\delta y = \beta xy \delta t - \gamma y \delta t.$$

The number of new isolates δz is given by

$$\delta z = \gamma y \delta t.$$

Now let $\delta t \to 0$. Then the system

$$\dot{x} = -\beta xy, \quad \dot{y} = \beta xy - \gamma y, \quad \dot{z} = \gamma y, \qquad (2.13)$$

with suitable initial conditions, determines the progress of the disease. Note that the result of adding the equations is

$$\frac{d}{dt}(x+y+z) = 0;$$

that is to say, the assumption of a constant population is built in to the model. The analysis of this problem in the phase plane is left as an exercise (Exercise 32). We shall instead look in detail at a more complicated situation:

Example 2.4 *Recurrent epidemic*

Suppose that the problem is as before, except that the stock of susceptibles $x(t)$ is being added to at a constant rate μ per unit time. This condition could be the result of fresh births in the presence of a childhood disease such as measles in the absence of vaccination. In order to balance the population in the simplest way we shall assume that deaths occur naturally and only among the immune, that is, among the $z(t)$ older people most of whom have had the disease. For a constant population the equations become

$$\dot{x} = -\beta xy + \mu, \qquad (2.14)$$

$$\dot{y} = \beta xy - \gamma y, \qquad (2.15)$$

$$\dot{z} = \gamma y - \mu; \qquad (2.16)$$

(note that $d/dt(x+y+z) = 0$: the population is steady).

Consider the variation of x and y, the active participants, represented on the x, y phase plane. We need only (2.14) and (2.15), which show an equilibrium point $(\gamma/\beta, \mu/\gamma)$.

Instead of trying to solve the equation for the phase paths we shall try to get an idea of what the phase diagram is like by forming linear approximations to the right-hand sides of (2.14), (2.15) in the neighbourhood of the equilibrium point. Near the equilibrium point we write

$$x = \gamma/\beta + \xi, \quad y = \mu/\gamma + \eta \qquad (2.17)$$

(ξ, η small) so that $\dot{x} = \dot{\xi}$ and $\dot{y} = \dot{\eta}$. Retaining only the linear terms in the expansion of the right sides of (2.14), (2.15) we obtain

$$\dot{\xi} = -\frac{\beta\mu}{\gamma}\xi - \gamma\eta, \tag{2.18}$$

$$\dot{\eta} = \frac{\beta\mu}{\gamma}\xi. \tag{2.19}$$

We are said to have *linearized* (2.14) and (2.15) near the equilibrium point. Elimination of ξ gives

$$\gamma\ddot{\eta} + (\beta\mu)\dot{\eta} + (\beta\mu\gamma)\eta = 0. \tag{2.20}$$

This is the equation for the damped linear oscillator (Section 1.4), and we may compare (2.20) with eqn (1.23) (or the system (1.26)) of Chapter 1, but it is necessary to remember that it only holds as an approximation close to the equilibrium point of (2.14) and (2.15). When the 'damping' is light ($\beta\mu/\gamma^2 < 4$) the phase path is a spiral. Figure 2.3 shows some phase paths for a particular case. All starting conditions lead to the stable equilibrium point E: this point is called the 'endemic state' for the disease.

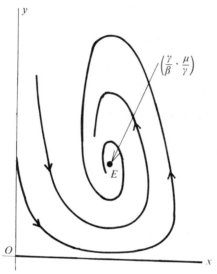

FIG. 2.3. Typical phase diagram for the recurrent epidemic

2.3 Linear approximation at equilibrium points

Approximation to a nonlinear system by linearizing it at an equilibrium point, as in the last example, is a most important and generally used technique. If the geometrical nature of the equilibrium points can be settled in this way the general character of the phase diagram is often clear. Consider the system

$$\dot{x} = X(x, y), \quad \dot{y} = Y(x, y). \tag{2.21}$$

Suppose that the equilibrium point to be studied has been moved to the origin by a translation of axes, if necessary, so that $X(0, 0) = Y(0, 0) = 0$. We can therefore write, by a Taylor expansion,

$$X(x, y) = ax + by + P(x, y), \quad Y(x, y) = cx + dy + Q(x, y),$$

where $P(x, y) = O(r^2)$ and $Q(x, y) = O(r^2)$ as $r = \sqrt{(x^2 + y^2)} \to 0$, and

$$a = \frac{\partial X}{\partial x}(0, 0), \quad b = \frac{\partial X}{\partial y}(0, 0), \quad c = \frac{\partial Y}{\partial x}(0, 0), \quad d = \frac{\partial Y}{\partial y}(0, 0). \tag{2.22}$$

The *linear approximation* to (2.21) in the neighbourhood of the origin is defined as the system

$$\dot{x} = ax + by, \quad \dot{y} = cx + dy. \tag{2.23}$$

We expect that the solutions of (2.23) will be geometrically similar to those of (2.21) near the origin, an expectation fulfilled in most cases (but see Exercise 3).

It is known that there are non-trivial solutions of (2.23) of the form

$$x = re^{\lambda t}, \quad y = se^{\lambda t} \tag{2.24}$$

where r and s are related constants, and λ is another constant. All these constants may be complex. To find values for λ, put (2.24) into (2.23); we obtain

$$(a - \lambda)r + bs = 0,$$
$$cr + (d - \lambda)s = 0. \tag{2.25}$$

Non-trivial solutions exist if and only if

$$\begin{vmatrix} a - \lambda & b \\ c & d - \lambda \end{vmatrix} = 0,$$

or

$$\lambda^2 - (a + d)\lambda + (ad - bc) = 0, \tag{2.26}$$

which is called the characteristic equation. When this equation has two different roots, λ_1, λ_2, two linearly independent families of solutions are generated by (2.24), corresponding to $\lambda = \lambda_1$ and $\lambda = \lambda_2$ respectively. We shall consider only this case here. Now write

$$p = a + d, \quad q = ad - bc. \tag{2.27}$$

The characteristic equation becomes

$$\lambda^2 - p\lambda + q = 0. \tag{2.28}$$

Let Δ be the discriminant:

$$\Delta = p^2 - 4q, \tag{2.29}$$

then the roots of (2.28) are λ_1, λ_2 given by

$$\begin{matrix} \lambda_1 \\ \lambda_2 \end{matrix} = \tfrac{1}{2}(p \pm \Delta^{1/2}). \tag{2.30}$$

We are considering only the case

$$\Delta \neq 0, \tag{2.31}$$

so that $\lambda_1 \neq \lambda_2$.

By substituting first λ_1 and then λ_2 into the equations (2.25) we find the permissible corresponding non-zero pairs r, s. Since the equations are homogeneous, if r, s is one solution, all solutions are given by Cr, Cs, where C is any constant. Let $r = r_1$, $s = s_1$ be any one fixed solution corresponding to $\lambda = \lambda_1$, and similarly for λ_2. Then, because of the linearity of the system (2.23) the general solution is

$$\begin{aligned} x(t) &= C_1 r_1 e^{\lambda_1 t} + C_2 r_2 e^{\lambda_2 t}, \\ y(t) &= C_1 s_1 e^{\lambda_1 t} + C_2 s_2 e^{\lambda_2 t}, \end{aligned} \tag{2.32}$$

where C_1 and C_2 are arbitrary constants. Note that when λ_1, λ_2 are complex, r_1, s_1, r_2, s_2 are complex; therefore, if the solutions x, y are to be real, we must allow C_1, C_2 to be complex in general.

Example 2.5 *Find the general solution of the system*

$$\dot{x} = x + y, \quad \dot{y} = -5x - 3y.$$

The characteristic equation (2.28) is

$$\lambda^2 + 2\lambda + 2 = 0$$

so that

$$\lambda_1 = -1 + i, \quad \lambda_2 = -1 - i.$$

Equations (2.25) are, for λ_1

$$(2 - i)r_1 + s_1 = 0, \quad -5r_1 - (2 + i)s_1 = 0.$$

(Note that these equations are equivalent.) A particular solution is

$$r_1 = 1, \quad s_1 = -2 + i.$$

Since $\lambda_2 = \bar\lambda_1$ a solution of the equations corresponding to λ_2 is $r_2 = \bar r_1, s_2 = \bar s_1$ or

$$r_2 = 1, \quad s_2 = -2 - i.$$

In the form (2.32) the general solution is therefore

$$x(t) = C_1 e^{(-1+i)t} + C_2 e^{(-1-i)t},$$
$$y(t) = C_1(-2+i)e^{(-1+i)t} + C_2(-2-i)e^{(-1-i)t}.$$

If we choose $C_2 = \bar C_1$ with C_1 arbitrary, all the real solutions are obtained. Put $C_1 = \frac{1}{2}c_1 + \frac{1}{2}ic_2$ where c_1, c_2 are arbitrary. Then this simplifies to

$$x(t) = e^{-t}(c_1 \cos t - c_2 \sin t),$$
$$y(t) = -e^{-t}\{(2c_1 + c_2)\cos t + (c_1 - 2c_2)\sin t\}.$$

We are interested not so much in the solutions (2.32) as in the nature of the paths on the phase plane x, y. These are given parametrically by (2.32), the parameter being t. On the phase plane we have also

$$\frac{dy}{dx} = \frac{\dot y}{\dot x} = \frac{\lambda_1 C_1 s_1 e^{\lambda_1 t} + \lambda_2 C_2 s_2 e^{\lambda_2 t}}{\lambda_1 C_1 r_1 e^{\lambda_1 t} + \lambda_2 C_2 r_2 e^{\lambda_2 t}}. \tag{2.33}$$

We consider several cases.

(i) λ_1, λ_2 real, same sign, $\lambda_1 \neq \lambda_2$ (*a node*)

Suppose that $\lambda_2 < \lambda_1 < 0$. From (2.33),

$$\frac{dy}{dx} = \frac{\lambda_1 C_1 s_1 + \lambda_2 C_2 s_2 e^{-(\lambda_1 - \lambda_2)t}}{\lambda_1 C_1 r_1 + \lambda_2 C_2 r_2 e^{-(\lambda_1 - \lambda_2)t}}. \tag{2.34}$$

Solutions for which $C_1 = 0$ approach the origin for $t \to +\infty$, (from (2.32)), along the line

$$\frac{y}{x} = \frac{s_2}{r_2}, \quad \text{(constant).} \tag{2.35}$$

Notice that the origin is approached along this line from two opposed directions: there is not merely one path going *through* the origin.

Solutions for which $C_2 = 0$ give another pair of straight-line paths into the origin, in opposite directions along

$$\frac{y}{x} = \frac{s_1}{r_1}, \quad \text{(constant).} \tag{2.36}$$

Now consider the family for which neither C_1 nor C_2 are zero. These paths have slopes such that, from (2.34),

$$\frac{dy}{dx} \to \frac{s_2}{r_2} \quad \text{as} \quad t \to -\infty,$$

and as $t \to \infty$, x and $y \to 0$ in directions such that

$$\frac{dy}{dx} \to \frac{s_1}{r_1}, \quad t \to \infty. \tag{2.37}$$

All solutions except the first, (2.35), are therefore tangential to the straight line (2.36) at the origin as $t \to \infty$, and parallel to the straight line (2.35) as $t \to -\infty$. The equilibrium point at $x = y = 0$ is called a *node*, and is illustrated in Fig. 2.4 for the case where λ_1, λ_2 are both negative. In this case the node is *stable*. If λ_1, λ_2 are real and *positive*, with $\lambda_1 \neq \lambda_2$, the figure is of the same type with the arrows reversed: the equilibrium point is *unstable*.

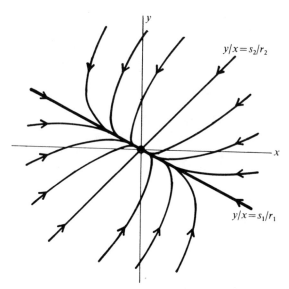

FIG. 2.4. Stable node

These cases occur when

$$\Delta > 0, \quad q > 0,$$

and

$$p < 0 \text{ (stable node)}, \quad p > 0 \text{ (unstable node).} \tag{2.38}$$

In the special case where one of the roots is zero, say λ_2, we obtain

$dy/dx =$ (constant) for all solutions, which describes a family of straight lines.

(ii) λ_1, λ_2 real, with different signs (a saddle point)

Suppose $\lambda_1 > 0, \lambda_2 < 0$. Referring to (2.32) it can be seen that only solutions having $C_1 = 0$ approach the origin as $t \to \infty$. These solutions are represented by straight-line paths with

$$\frac{y}{x} = \frac{s_2}{r_2}, \quad \text{(constant).} \tag{2.39}$$

Solutions with $C_2 = 0$ similarly go into

$$\frac{y}{x} = \frac{s_1}{r_1}, \quad \text{(constant),} \tag{2.40}$$

and the representative point departs from the origin to ∞ as $t \to \infty$.

Reading from eqn (2.34), in other cases

$$\frac{dy}{dx} \to \frac{s_1}{r_1} \quad \text{as} \quad t \to \infty,$$

approaching the straight line (2.40); and

$$\frac{dy}{dx} \to \frac{s_2}{r_2}, \quad \text{as} \quad t \to -\infty,$$

departing from the straight line (2.39).

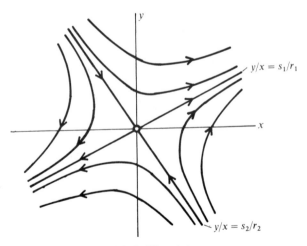

FIG. 2.5. Saddle point

The straight-line paths $y/x = s_1/r_1, y/x = s_2/r_2$ are asymptotes for the other solutions and are called the *separatrices*. The equilibrium point is a *saddle point* (see Fig. 2.5), which is always *unstable*.

The condition for this case is

$$\Delta > 0, \quad q < 0. \tag{2.41}$$

(iii) λ_1, λ_2 *complex with non-zero real part* (*a spiral or focus*)

We have $\lambda_1 = \bar{\lambda}_2, r_1 = \bar{r}_2, s_1 = \bar{s}_2$ (from (2.25)); and we let $C_1 = \bar{C}_2$ in order that (2.32) should represent real solutions. Put also

$$\lambda_1 = \bar{\lambda}_2 = \alpha + i\beta,$$

where α and β are real. Then (2.32) takes the form

$$\begin{aligned}
x &= C e^{\alpha t} \cos(\beta t + \gamma), \\
y &= CK e^{\alpha t} \cos(\beta t + \gamma + \kappa),
\end{aligned} \tag{2.42}$$

where C, γ are arbitrary and K, κ are constants depending only on the coefficients of the system. Since the effect of varying γ is simply to alter the time origin the family of curves on the phase plane is essentially one-parameter. The phase diagram consists of contracting spirals surrounding the origin and approaching it if $\alpha < 0$ (as shown in Fig. 2.6), and spirals expanding from the origin if $\alpha > 0$.

The equilibrium point is therefore a *stable spiral* if $\text{Re}(\lambda_1) < 0$ and an *unstable spiral* if $\text{Re}(\lambda_1) > 0$.

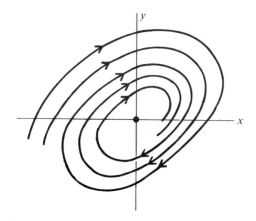

Fig. 2.6. Stable spiral

The conditions on p, q, Δ for a spiral are

$$\Delta < 0,$$
$$p < 0 \ (stable), \quad p > 0 \ (unstable). \tag{2.43}$$

(iv) $\lambda_1 = \lambda_2$ *(real)*

This is the case where there is only one family of solutions of the form (2.24). The nature of this degenerate case can be ascertained as follows. When λ_1 is very close to λ_2 the solutions r_1, s_1 and r_2, s_2 respectively, obtained from (2.25), are nearly the same. Now, in the analysis of the node (Fig. 2.4), the straight line $y/x = s_1/r_1$ is a path. All paths are tangential to this line at the origin, and parallel to the line $y/x = s_2/r_2$ at infinity. In the present case the lines $y/x = s_1/r_1, y/x = s_2/r_2$ become coincident.

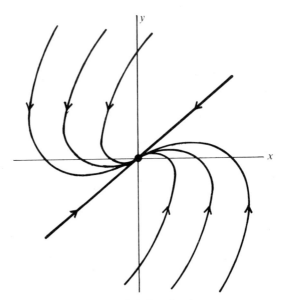

FIG. 2.7. Stable inflected node

The phase diagram is shown in Fig. 2.7 for the case $\lambda_1 = \lambda_2 < 0$. This is called a *stable inflected node*. The conditions are

$$\Delta = 0,$$
$$p < 0 \ (stable\ node), \quad p > 0 \ (unstable\ node). \tag{2.44}$$

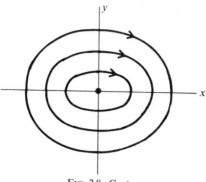

FIG. 2.8. Centre

(v) λ_1, λ_2 *pure imaginary*

Write $\lambda_1 = i\beta, \lambda_2 = -i\beta, \beta$ real. Then as in the case of the spiral, we find

$$x = C \cos(\beta t + \gamma), \quad y = CK \cos(\beta t + \gamma + \kappa),$$

where C and γ are arbitrary. These represent closed curves surrounding the origin. The equilibrium point is a *centre* (Fig. 2.8), and the paths are ellipses. The conditions for this are

$$p = 0, \quad q > 0. \tag{2.45}$$

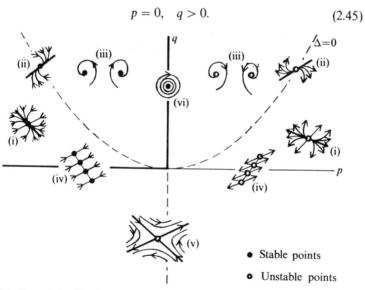

● Stable points

○ Unstable points

FIG. 2.9. General classification for the linear system $\dot{x} = ax + by, \dot{y} = cx + dy$ ($p = a + d$, $q = ad - bc, \Delta = p^2 - 4q$)

The following table lists the cases described above.

(i) λ_1, λ_2 real, unequal, same sign $\Delta > 0$, $q > 0$ Node
(ii) $\lambda_1 = \lambda_2$ (real) $b \neq 0$, $c \neq 0$ $\Delta = 0$, $p \neq 0$ Inflected node

(For the case $b = c = \Delta = 0$ see Exercise 11.)

(iii) λ_1, λ_2 complex, non-
                               zero real part $\Delta < 0$, $p \neq 0$ Spiral
(iv) $\lambda_1 \neq 0$, $\lambda_2 = 0$ $q = 0$ Parallel lines
(iv) λ_1, λ_2 real, different sign $q < 0$ Saddle point
(vi) λ_1, λ_2 pure imaginary $q > 0$, $p = 0$ Centre

Figure 2.9 shows the nature of the equilibrium points on the p, q plane. The *stable* equilibrium points lie in the quadrant $p \leqslant 0, q \geqslant 0$ except $(p, q) = (0, 0)$.

Example 2.6   *Classify the equilibrium point at* $(0,0)$ *for the system*
$$\dot{x} = e^{-x-3y} - 1, \quad \dot{y} = -x(1 - y^2).$$

The linear approximation about $(0,0)$ is
$$\dot{x} = -x - 3y, \quad \dot{y} = -x,$$

by using a Taylor expansion for the exponential function. Comparison with (2.23) gives
$$a = -1, \quad b = -3, \quad c = -1, \quad d = 0;$$
therefore
$$p = -1, \quad q = -3, \quad \Delta = 13.$$

From Fig. 2.9, the equilibrium point is a saddle point.

Example 2.7   *Classify the equilibrium points of*
$$\dot{x} = x - y, \quad \dot{y} = x^2 - 1.$$

The equilibrium points are at $(1, 1)$ and $(-1, -1)$. Linearize the equation in the neighbourhood of $(1, 1)$ by writing
$$x = 1 + \xi, \quad y = 1 + \eta.$$

Retaining only first-order terms, the equations become
$$\dot{\xi} = \xi - \eta, \quad \dot{\eta} = 2\xi.$$

This is a linear system: comparison with (2.23) gives
$$a = 1, \quad b = -1, \quad c = 2, \quad d = 0.$$

Therefore

$$p = 1, \quad q = 2, \quad \Delta = -7.$$

From Fig. 2.9, the phase diagram near $(1, 1)$ is an unstable spiral.

Near the point $(-1, -1)$ put

$$x = -1 + \xi, \quad y = -1 + \eta,$$

leading to

$$\dot{\xi} = \xi - \eta, \quad \dot{\eta} = -2\eta$$

Here,

$$a = 1, \text{'} \quad b = -1, \quad c = -2, \quad d = 0,$$

so that

$$p = 1, \quad q = -2, \quad \Delta = 9.$$

From Fig. 2.9, this equilibrium point is a saddle.

In classifying the equilibrium points as in the preceding examples we have taken for granted an unproved assumption, that the phase paths of the original equation and those of the linearized equation near the equilibrium point are of the same character. This is true in general for spirals, nodes, and saddle points, but not for a centre: Exercise 3 shows, for example, that the linear approximation may predict a centre where the original equation had a spiral. Conversely, the system $\dot{x} = y, \dot{y} = -x^3$ has a centre at the origin, but the linear approximation $\dot{x} = y, \dot{y} = 0$ has not. The theory is fully presented by Andronov *et al.* (1973).

2.4 Linear systems in matrix form

We briefly describe an alternative approach to the analysis of equilibrium points in the last section. Write the system (2.23) in the form

$$\dot{\mathbf{x}} = A\mathbf{x} \tag{2.46}$$

where

$$A = \begin{pmatrix} a & b \\ c & d \end{pmatrix}, \quad \mathbf{x} = \begin{pmatrix} x \\ y \end{pmatrix}, \quad \dot{\mathbf{x}} = \begin{pmatrix} \dot{x} \\ \dot{y} \end{pmatrix}. \tag{2.47}$$

Look for two linearly independent solutions of the form

$$\mathbf{x} = \mathbf{u}\,e^{\lambda t}, \tag{2.48}$$

where

$$\mathbf{u} = \begin{pmatrix} r \\ s \end{pmatrix} \neq \mathbf{0}. \tag{2.49}$$

Then $\dot{\mathbf{x}} = \lambda \mathbf{u}\,e^{\lambda t}$, and (2.46) gives

$$(A - \lambda I)\mathbf{u} = \mathbf{0} \tag{2.50}$$

where I is the identity matrix. Non-zero $\mathbf{u}$ satisfies (2.50) if and only if

$$\det(A - \lambda I) = 0,$$

or

$$\begin{vmatrix} a - \lambda & b \\ c & d - \lambda \end{vmatrix} = 0. \qquad (2.51)$$

(2.51) is (2.26) again, and (2.50) is equation (2.25) again.

Therefore λ_1, λ_2, the solutions of (2.51) are the *eigenvalues* of A; and $\mathbf{u} = \mathbf{u}_1$ and $\mathbf{u} = \mathbf{u}_2$, the solutions of (2.50), are a pair of *eigenvectors* corresponding respectively to λ_1 and λ_2. Provided $\lambda_1 \neq \lambda_2$, the general solution of (2.46) is

$$\mathbf{x} = C_1 \mathbf{u}_1 \, e^{\lambda_1 t} + C_2 \mathbf{u}_2 \, e^{\lambda_2 t}, \qquad (2.52)$$

which is eqn (2.32).

Now reconsider the problem of finding the phase paths. In general their equation is

$$\frac{\mathrm{d}y}{\mathrm{d}x} = \frac{ax + by}{cx + dy}. \qquad (2.53)$$

This can always be solved but the reader can soon convince himself that, except in special cases, the results are too complicated to give a simple discussion of the geometrical nature of the solutions.

We can obtain linear transformations which reduce the equation (2.54) to manageable form. A nonsingular linear transformation from x, y to x_1, y_1:

$$\mathbf{x}_1 = S\mathbf{x} \qquad (2.54)$$

does not change the nature of the singular points; for example spirals in the x, y plane remain spirals in the x_1, y_1 plane, and so on. Writing

$$\mathbf{x} = S^{-1}\mathbf{x}_1, \quad \dot{\mathbf{x}} = S^{-1}\dot{\mathbf{x}}_1,$$

(2.46) becomes $S^{-1}\dot{\mathbf{x}}_1 = AS^{-1}\mathbf{x}_1$, or

$$\dot{\mathbf{x}}_1 = SAS^{-1}\mathbf{x}_1 = B\mathbf{x}_1, \qquad (2.55)$$

say. It is known from algebraic theory (Ayres, 1962) that S can be chosen so that B takes one of several 'canonical forms', which in general are simpler than A, the particular form depending on the nature of the eigenvalues of A. In the new coordinates x_1, y_1, the equations are therefore simpler although the topological character of the transformed equilibrium point at the origin is not affected. We do not need to calculate the S for this purpose: we only need the fact that the relevant S exists.

Neglecting certain degenerate cases, the calculations are as follows:

(i) λ_1, λ_2 *real, different, and non-zero*
We can choose S so that

$$\dot{x}_1 = \lambda_1 x_1, \quad \dot{y}_1 = \lambda_2 y_1.$$

Then the equation for the phase paths is

$$\frac{dy_1}{dx_1} = \frac{\lambda_2}{\lambda_1} \frac{y_1}{x_1}$$

whose solutions are given by

$$y_1 = C|x_1|^{\lambda_2/\lambda_1}$$

where C is arbitrary. $x_1 = 0$ is also a phase path. This is a node when $\lambda_2/\lambda_1 > 0$, and consideration of the signs of $\dot{x}_1, \dot{y}_1$ show that it is stable when $\lambda_1, \lambda_2 < 0$, and unstable when $\lambda_1, \lambda_2 > 0$. If $\lambda_2/\lambda_1 < 0$ (λ_1, λ_2 of opposite sign), the origin is a saddle-point.

(ii) $\lambda_1 = \lambda_2, = \lambda$, *say* (*b and c not both zero*)
We can choose S so that

$$\dot{x}_1 = \lambda x_1 + y_1, \quad \dot{y}_1 = \lambda y_1$$

(with λ necessarily real). The equation for the phase paths is

$$\frac{dy_1}{dx_1} = \frac{\lambda y_1}{\lambda x_1 + y_1},$$

and the solutions are $y_1 = 0$, and

$$x_1 = \frac{1}{\lambda} y_1 \log_e |y_1| + C y_1$$

where C is arbitrary. $x_1 = 0$ is not a phase path. All solutions are tangential to $y_1 = 0$ at $(0,0)$, and ultimately parallel to $y_1 = 0$ in the opposite direction; hence it is an inflected node, stable if $\lambda < 0$ and unstable if $\lambda > 0$.

(iii) $\lambda_1 = \bar{\lambda}_2, = \alpha + i\beta, \beta \neq 0$
S can be chosen so that the equations become

$$\dot{x}_1 = \alpha x_1 - \beta y_1, \quad \dot{y}_1 = \beta x_1 + \alpha y_1.$$

Put $z = x_1 + iy_1$; then $\dot{z} = (\alpha + i\beta)z$, and by writing $z(t) = r(t)e^{i\theta(t)}$, where $r(t) = |z(t)|$, we obtain the equations in polar coordinates

$$\dot{r} = \alpha r, \quad \dot{\theta} = \beta.$$

The origin is therefore a stable spiral if $\alpha < 0, \beta \neq 0$; and an unstable spiral if $\alpha > 0, \beta \neq 0$.

If $\alpha = 0, \beta \neq 0$, the origin is a centre.

These results agree with those obtained in the preceding section.

Exercises

1. Locate the equilibrium points and find the equations of the phase paths for the following systems:

(i) $\dot{x} = x + y, \quad \dot{y} = x - y + 1$;

(ii) $\dot{x} = x + y, \quad \dot{y} = x + y + 2$;

(iii) $\dot{x} = y^3, \quad \dot{y} = x^3$;

(iv) $\dot{x} = \sin y, \quad \dot{y} = \cos x$;

(v) $\dot{x} = y^3, \quad \dot{y} = y(x + y^3)$.

2. Locate the equilibrium points of the following nonlinear systems and classify them according to their linear approximations:

(i) $\dot{x} = -6y + 2xy - 8, \quad \dot{y} = y^2 - x^2$;

(ii) $\dot{x} = -2x - y + 2, \quad \dot{y} = xy$;

(iii) $\dot{x} = 4 - 4x^2 - y^2, \quad \dot{y} = 3xy$;

(iv) $\dot{x} = \sin y, \quad \dot{y} = x + x^3$;

(v) $\dot{x} = -x + e^{-y} - 1, \quad \dot{y} = 1 - e^{x-y}$.

3. Show that the origin is a spiral point of the system $\dot{x} = -y - x\sqrt{(x^2 + y^2)}$, $\dot{y} = x - y\sqrt{(x^2 + y^2)}$, but a centre for its linear approximation.

4. Obtain all solutions of the system $\dot{x} = ax + by, \dot{y} = cx + dy$ when the eigenvalues are equal.

5. Show that the systems $\dot{x} = y, \dot{y} = -x - y^2$, and $\dot{x} = x + y_1, \dot{y}_1 = -2x - y_1 - (x + y_1)^2$, both represent the equation $\ddot{x} + \dot{x}^2 + x = 0$ in different phase planes.

6. Use eqn (2.9) in the form $\delta s \simeq \delta t \sqrt{(X^2 + Y^2)}$ to mark off approximately equal time steps on some of the phase paths of $\dot{x} = xy, \dot{y} = xy - y^2$.

7. Obtain approximations to the phase paths described by (2.12) in the neighbourhood of the equilibrium point $x = b/d, y = a/c$. (Write $x = b/d + \xi$, $y = a/c + \eta$, and expand the logarithms to second order terms in ξ and η.)

8. Show that when the system $\dot{x} = y, \dot{y} = cx + dy$ has a saddle-point, the separatrices have slopes λ_1 and λ_2, the roots of (2.26).

9. Show that if the system $\dot{x} = y, \dot{y} = cx + dy$ has a stable node, then the straight line paths (2.35) and (2.36) have slopes λ_1 and λ_2 (the roots of (2.26)), both slopes being positive.

10. For the system $\dot{x} = ax + by, \dot{y} = cx + dy$, where $ad - bc = 0$, show that all points on the line $cx + dy = 0$ are equilibrium points.

11. Show that the inflected node for the system $\dot{x} = ax + by$, $\dot{y} = cx + dy$ becomes 'star-shaped' when $a = d$ and $b = c = 0$.

12. The interaction between two species is governed by the deterministic model $\dot{H} = (a_1 - b_1 H - c_1 P)H$, $\dot{P} = (-a_2 + c_2 H)P$, where H is the population of the host (or prey), and P is that of the parasite (or predator), all constants being positive. (Compare Example 2.2: the term $-b_1 H^2$ represents interference with the host population when it gets too large.) Find the equilibrium states for the populations, and find how they vary with time from various initial populations.

13. With the same terminology as in Exercise 12, analyse the system $\dot{H} = (a_1 - b_1 H - c_1 P)H$, $\dot{P} = (a_2 - b_2 P + c_2 H)P$, all the constants being positive. (In this model the parasite can survive on alternative food supplies, although the prevalence of the host encourages growth in the population.) Find the equilibrium states. Confirm that the parasite population can persist even if the host dies out.

14. Consider the host-parasite population model $\dot{H} = (a_1 - c_1 P)H$, $\dot{P} = (a_2 - c_2 P/H)P$, where the constants are positive. Analyse the system in the H, P plane.

15. In the population model $\dot{F} = -\alpha F + \beta \mu(M)F$, $\dot{M} = -\alpha M + \gamma \mu(M)F$, where $\alpha > 0, \beta > 0, \gamma > 0$, F and M are the female and male populations. In both cases the death-rates are α. The birth-rate is governed by the coefficient $\mu(M) = 1 - e^{-kM}$, $k > 0$, so that for large M the birth-rate of females is βF and that for males is γF, the rates being unequal in general. Show that if $\beta > \alpha$ then there are two equilibrium states, at $(0, 0)$ and at $([-\beta/\gamma k] \log [(\beta - \alpha)/\beta], [-1/k] \log [(\beta - \alpha)/\beta])$.

Show that the origin is a stable inflected node and the other equilibrium point is a saddle-point, according to their linear approximations. Verify that $M = \gamma F/\beta$ is a particular solution. Sketch the phase diagram and discuss the stability of the populations.

16. A rumour spreads through a closed population of constant size $N + 1$. At time t the total population can be classified into three categories:

x persons who are ignorant of the rumour;
y persons who are actively spreading the rumour;
z persons who have heard the rumour but have stopped spreading it: if two persons who are spreading the rumour meet then they stop spreading it.
The contact rate between any two categories is constant, μ.
Show that the equations

$$\dot{x} = -\mu xy, \quad \dot{y} = \mu[xy - y(y-1) - yz]$$

give a deterministic model of the problem. Find the equations of the phase paths and sketch the phase diagram.

Show that, when initially $y = 1$ and $x = N$, the number of people who ultimately never hear the rumour is x_1, where

$$2N + 1 - 2x_1 + N \log (x_1/N) = 0.$$

17. The one-dimensional steady flow of a gas with viscosity and heat conduction satisfies the equations

$$\frac{\mu_0}{\rho c_1} \frac{dv}{dx} = \sqrt{(2v)} [2v - \sqrt{(2v)} + \theta]$$

$$\frac{k}{gR\rho c_1} \frac{d\theta}{dx} = \sqrt{(2v)} \left[\frac{\theta}{\gamma - 1} - v + \sqrt{(2v)} - c \right]$$

where $v = u^2/2c_1^2$, $c = c_2^2/c_1^2$ and $\theta = gRT/c_1^2 = p/\rho c_1^2$. In this notation, x is measured in the direction of flow, u is the velocity, T is the temperature, ρ is the density, p the pressure, R the gas constant, k the coefficient of thermal conductivity, μ_0 the coefficient of viscosity, γ the ratio of the specific heats, and c_1, c_2 are arbitrary constants. Find the equilibrium states of the system.

18. A particle moves under a central attractive force γ/r^α per unit mass, where r, θ are the polar coordinates of the particle in its plane of motion. Show that

$$\frac{d^2u}{d\theta^2} + u = \frac{\gamma}{h^2} u^{\alpha - 2},$$

where $u = r^{-1}$, h is the angular momentum about the origin per unit mass of the particle, and γ is a constant. Find the non-trivial equilibrium point in the u, $du/d\theta$ plane and classify it according to its linear approximation. What can you say about the stability of the circular orbit under this central force?

19. The relativistic equation for the central orbit of a planet is

$$\frac{d^2u}{d\theta^2} + u = \alpha + \varepsilon u^2$$

where $u = 1/r$, and r, θ are the polar coordinates of the planet in the plane of its motion. The term εu is the 'Einstein correction', and α and ε are positive constants, with ε very small. Find the equilibrium point which corresponds to a perturbation of the Newtonian orbit. Show that the equilibrium point is a centre in the u, $du/d\theta$ plane according to the linear approximation. Confirm this by using the method of Section 1.3.

20. A top is set spinning at an axial rate n about its pivotal point, which is fixed in space. The equations for its motion, in terms of the angles θ and μ are (see Fig. 2.10)

$$A\ddot{\theta} - A(\Omega + \dot{\mu})^2 \sin \theta \cos \theta + Cn(\Omega + \dot{\mu}) \sin \theta - Mgh \sin \theta = 0,$$

$$A\dot{\theta}^2 + A(\Omega + \dot{\mu})^2 \sin^2 \theta + 2Mgh \cos \theta = E;$$

where (A, A, C) are the principal moments of inertia about O, M is the mass of the

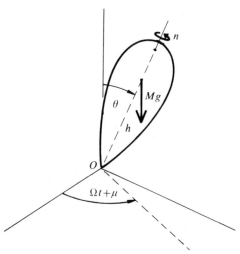

FIG. 2.10.

top, h is the distance between the mass-centre and the pivot, and E is a constant. Show that one equilibrium state is given by $\theta = \alpha$, provided E and μ satisfy

$$A\Omega^2 \cos\alpha - Cn\Omega + Mgh = 0, \quad A\Omega^2 \sin^2\alpha + 2Mgh\cos\alpha = E.$$

Discuss the motion of the top in this state.

Suppose that $E = 2Mgh$, so that $\theta = 0$ is an equilibrium state. Show that, close to this state, θ satisfies

$$A\ddot{\theta} + (Cn\Omega - A\Omega^2 - Mgh)\theta = 0.$$

For what condition on Ω is the motion stable?

21. Sketch the phase diagram for the system $\dot{x} = -y\sqrt{(1-x^2)}$, $\dot{y} = x\sqrt{(1-x^2)}$.

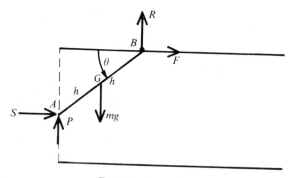

FIG. 2.11. Garage door

22. Figure 2.11 shows an 'up-and-over' garage door AB, of height $2h$, mass m, and moment of inertia I about G. The force P is applied by springs and weights, and it can be designed to have suitable characteristics. The unknown reactions S and R, and the frictional force F are as shown.

(i) If the runner B is smooth, show that it is possible to design the door so that it is in equilibrium at every angle θ, when P is a suitable constant. Sketch the phase diagram.

(ii) Suppose now that a frictional force F, with $|F/R| = \mu$, is introduced at B. What values should $P(\theta)$ have at $\theta = 0$ and $\theta = \frac{1}{2}\pi$ in order that the open and closed positions should be in equilibrium?

(iii) A larger design problem can be attempted. Construct a function $P(\theta)$ which will ensure that the open and closed positions are stable nodes (why nodes?) according to the linear approximations. Do you expect the door to have a further equilibrium position? (This problem does not, of course, have a unique solution.)

23. A disc of radius a is freely pivoted at its centre A so that it can turn in a vertical plane. An elastic string, of natural length $2a$ and stiffness λ connects a point on the circumference of the disc to a fixed point O, distance $2a$ from A. Show that θ satisfies

$$I\ddot{\theta} = -Ta\sin\phi, \quad T = \lambda a[(5 - 4\cos\theta)^{1/2} - 2],$$

where I is the moment of inertia of the disc about A. Find the equilibrium states of the disc.

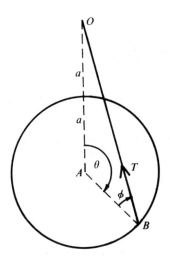

FIG. 2.12.

24. A man rows a boat across a river of width a occupying the strip $0 \leqslant x \leqslant a$ in the x, y plane, always rowing towards a fixed point on one bank, say $(0, 0)$. He rows at a constant speed u relative to the water, and the river flows at a constant speed v. Show that

$$\dot{x} = -ux/\sqrt{(x^2 + y^2)}, \quad \dot{y} = v - uy/\sqrt{(x^2 + y^2)},$$

where (x, y) are the coordinates of the boat. Show that the phase paths are given by $y + \sqrt{(x^2 + y^2)} = Ax^{1-\alpha}$, where $\alpha = v/u$. Sketch the phase diagram for $\alpha < 1$ and interpret it. What kind of point is the origin? What happens to the boat if $\alpha > 1$?

25. In a simple model of a national economy, $\dot{I} = I - \alpha C$, $\dot{C} = \beta(I - C - G)$, where I is the national income, C is the rate of consumer spending, and G the rate of government expenditure; the constants α and β satisfy $1 < \alpha < \infty$, $1 \leqslant \beta < \infty$. Show that if the rate of government expenditure G_0 is constant there is an equilibrium state. Classify the equilibrium state and show that the economy oscillates when $\beta = 1$.

Consider the situation when government expenditure is related to the national income by the rule $G = G_0 + kI$, where $k > 0$. Show that there is no equilibrium state if $k \geqslant (\alpha - 1)/\alpha$. How does the economy then behave?

Discuss an economy in which $G = G_0 + kI^2$, and show that there are two equilibrium states.

26. Let $f(x)$ and $g(y)$ have local minima at $x = a$ and $y = b$ respectively. Show that $f(x) + g(y)$ has a minimum at (a, b). Deduce that there exists a neighbourhood of (a, b) in which all solutions of the family of equations

$$f(x) + g(y) = \text{constant}$$

represent closed curves surrounding (a, b).

Show that $(0, 0)$ is a centre for the system $\dot{x} = y^5$, $\dot{y} = -x^3$, and that all paths are closed curves.

27. For the predator-prey problem in Section 2.2, show by using Exercise 26 that all solutions in $H > 0$, $P > 0$ are periodic.

28. Show that the equilibrium points of the system $\dot{x} = \sin y$, $\dot{y} = -\sin x$, are centres or saddle-points, and that, except for the separatrices of the saddle-points, all the paths are closed.

29. Show that the three-dimensional system

$$\dot{x} = y(z-2), \quad \dot{y} = x(2-z)+1, \quad x^2 + y^2 + z^2 = 1$$

has exactly two equilibrium points and that both are centres according to their linear approximations. Sketch the phase diagram as it appears on the unit sphere.

30. Show that the phase paths of the Hamiltonian system $\dot{x} = -\partial H/\partial y$, $\dot{y} = \partial H/\partial x$ are given by $H(x, y) = $ constant. (This is a generalization of the conservative system of Section 1.3.) Equilibrium points occur at the stationary points of $H(x, y)$. If (x_0, y_0) is an equilibrium point, show that (x_0, y_0) is stable according to the linear approximation if $H(x, y)$ has a maximum or a minimum at the point. (Assume that all the second derivatives of H are non-zero at x_0, y_0.)

31. The equilibrium points of the nonlinear parameter-dependent system $\dot{x} = y$, $\dot{y} = -f(x, y, \lambda)$ lie on the curve $f(x, 0, \lambda) = 0$ in the x, λ plane. Show that an equilibrium point x_1, λ_1 is stable and that all neighbouring solutions tend to this point (according to the linear approximation) if $f_x(x_1, 0, \lambda_1) < 0$ and $f_y(x_1, 0, \lambda_1) < 0$.
   Interpret stability in terms of the surface $f(x, y, \lambda) = 0$ in the x, y, λ space and the regions in which $f(x, y, \lambda) > 0$.

32. Find the equations for the phase paths for the general epidemic described (Section 2.2) by the system

$$\dot{x} = -\beta xy, \quad \dot{y} = \beta xy - \gamma y, \quad \dot{z} = \gamma y.$$

Sketch the phase diagram in the x, y plane. Confirm that the number of infectives reaches its maximum when $x = \gamma/\beta$.

33. Two species x and y are competing for a common limited food supply. Their growth equations are given by

$$\dot{x} = xP(x, y), \quad \dot{y} = yQ(x, y).$$

Express in mathematical terms the following constraints on the populations:
   (i) if either species increases, the growth rate of the other goes down;
   (ii) if one species is absent, the other shows limited growth characteristics;
   (iii) if either population is very large, then neither species can multiply.
   Assuming that P and Q have the necessary properties, sketch a first-quadrant phase diagram in which also $P = 0$ and $Q = 0$ do not intersect, and $P = 0$ is 'below' $Q = 0$. Indicate the signs of $\dot{x}$ and $\dot{y}$ in the regions into which the quadrant is now divided. Deduce the character of the phase paths by qualitative arguments.

34. Sketch the phase diagram for the competing species x and y for which

$$\dot{x} = (1 - x^2 - y^2)x, \quad \dot{y} = (1 \cdot 1 - x - y)y.$$

35. A space satellite is in free flight on the line joining, and between, a planet (mass m_1) and its moon (mass m_2), which are at a fixed distance a apart. Show that

$$-\frac{\gamma m_1}{x^2} + \frac{\gamma m_2}{(a-x)^2} = \ddot{x}$$

where x is the distance of the satellite from the planet and γ is the gravitational constant. Show that the equilibrium point is unstable according to the linear approximation.

36. The system

$$\dot{V}_1 = -\sigma V_1 + f(E - V_2), \quad \dot{V}_2 = -\sigma V_2 + f(E - V_1), \quad \sigma > 0, \quad E > 0$$

represents (Andronov and Chaikin, 1949) a model of a triggered sweeping circuit for an oscilloscope. The conditions on $f(u)$ are: $f(u)$ continuous, $-\infty < u < \infty, f(-u) = -f(u), f(u)$ tends to a limit as $u \to \infty$, and $f'(u)$ is monotonic decreasing (see Figure 3.19).

Show by a geometrical argument that there is always at least one equilibrium point, (V_0, V_0) say, and that when $f'(E - V_0) < \sigma$ it is the only one; and deduce by taking the linear approximation that it is a stable node. (Note that $f'(E - v) = -df(E - v)/dv$.)

Show that when $f'(E - V_0) > \sigma$ there are two others, at $(V', (1/\sigma)f(E - V'))$ and $((1/\sigma)f(E - V'), V')$ respectively for some V'. Show that these are stable nodes, and that the one at (V_0, V_0) is a saddle-point.

37. Show that the equation $\ddot{x} + \omega^2(t)x = 0$ is equivalent to the first-order system

$$\begin{pmatrix} \dot{x}_1 \\ \dot{x}_2 \end{pmatrix} = \begin{pmatrix} -\dot{\omega}/\omega & -\omega \\ \omega & 0 \end{pmatrix} \begin{pmatrix} x_1 \\ x_2 \end{pmatrix},$$

where $x_1 = \dot{x}/\omega$ and $x_2 = x$.

38. The response of a certain biological oscillator, (x, y), $x \geq 0$, $y \geq 0$, to a stimulus measured by a constant b satisfies the system

$$\dot{x} = x - ay + b, \quad \dot{y} = x - cy \text{ for } x \geq 0, \quad y \geq 0;$$

$$\dot{y} = -cy \text{ for } x = 0.$$

Show that when $c < 1$ and $4a > (1 + c)^2$ then there exists a limit cycle, part of which lies on the y axis, whose period is independent of b. Sketch the corresponding solution function. (Varying the amount of stimulus translates the response in time without affecting the period.)

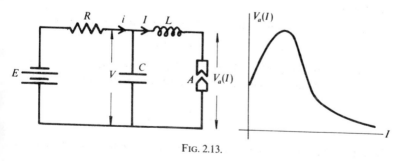

Fig. 2.13.

39. Figure 2.13 represents a circuit for activating an electric arc A which has the voltage-current characteristic shown. Show that $L\dot{I} = V - V_a(I)$, $RC\dot{V} = -RI - V + E$. By forming the linear approximating equations near the equilibrium points find the conditions on E, L, C, R and V_a' for stable working.

40. The equation for the current in the circuit of Fig. 2.14(a) is

$$LC\ddot{x} + RC\dot{x} + x = I.$$

Neglect the grid current, and assume that I depends only on the relative grid potential e_g: $I = I_s$ (saturation current) for $e_g > 0$ and $I = 0$ for $e_g < 0$. Assume also that $M > 0$, so that $e_g \gtrless 0$ according as $\dot{x} \gtrless 0$. Find the nature of the phase paths. By considering their successive intersections with the x axis show that a limit-cycle is approached from all initial conditions.

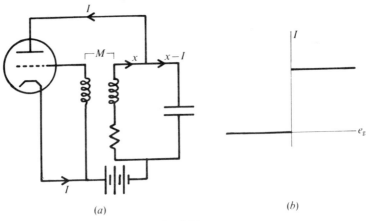

(a)   (b)

FIG. 2.14.

41. For the circuit in Figure 2.14(a) assume that the relation between I and e_g is as in Fig. 2.15: $I = f(e_g + ke_p)$, where e_g and e_p are the relative grid and plate potentials, $k > 0$ is a constant, and in the neighbourhood of the point of inflection, $f(u) = I_0 + au - bu^3$, where $a > 0$, $b > 0$. Deduce the equation for x when the operating point is the point of inflection. Find when the origin is a stable or an unstable point of equilibrium. (A form of Rayleigh's equation (1.39) is obtained, implying an unstable or a stable limit-cycle respectively.)

42. Figure 2.16(a) represents two identical D.C. generators connected in parallel, with inductance and resistance L, r. R is the resistance of the load. Show that the equations for the currents are

$$L\frac{di_1}{dt} = -(r+R)i_1 - Ri_2 + E(i_1), \quad L\frac{di_2}{dt} = -Ri_1 - (r+R)i_2 + E(i_2).$$

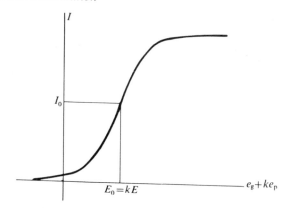

FIG. 2.15.

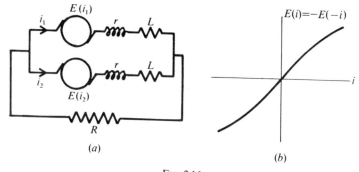

(a)

(b)

FIG. 2.16.

Assuming that $E(i)$ has the characteristics indicated by Fig. 2.16(b) show that
   (i) when $E'(0) < r$ the state $i_1 = i_2 = 0$ is stable and is otherwise unstable;
   (ii) when $E'(0) > r$ there is a stable state $i_1 = -i_2$ (no current flows to R);
   (iii) when $E'(0) > r+2R$ there is a state with $i_1 = i_2$, which may be unstable.

43. Show that the Emden–Fowler equation of astrophysics

$$(\xi^2\eta')' + \xi^\lambda\eta^n = 0$$

is equivalent to the predator-prey model

$$\dot{x} = -x(1+x+y), \quad \dot{y} = y(\lambda+1+nx+y)$$

after the change of variable

$$x = \xi\eta'/\eta, \quad y = \xi^{\lambda-1}\eta^n/\eta', \quad t = \log|\xi|.$$

44. Show that Blasius' equation

$$\eta''' + \eta\eta'' = 0$$

is transformed by

$$x = \eta\eta'/\eta'', \quad y = \eta'^2/\eta\eta'', \quad t = \log|\eta'|$$

into

$$\dot{x} = x(1+x+y), \quad \dot{y} = y(2+x-y).$$

45. Consider the family of linear systems

$$\dot{x} = X\cos\alpha - Y\sin\alpha, \quad \dot{y} = X\sin\alpha + Y\cos\alpha$$

where

$$X = ax+by, \quad Y = cx+dy,$$

and a, b, c, d are constants and α is a parameter. Show that the equilibrium point at the origin passes through the sequence stable node, stable spiral, centre, unstable spiral, unstable node, as α varies over range π.

46. Show that, given $X(x, y)$, the system equivalent to the equation $\ddot{x} + h(x, \dot{x}) = 0$ is

$$\dot{x} = X(x, y), \quad \dot{y} = -\left\{ h(x, X) + X\frac{\partial X}{\partial x} \right\} \bigg/ \frac{\partial X}{\partial y}.$$

47. The following system models two species with populations N_1 and N_2 competing for a common food supply:

$$\dot{N}_1 = \{a_1 - d_1(bN_1 + cN_2)\}N_1,$$

$$\dot{N}_2 = \{a_2 - d_2(bN_1 + cN_2)\}N_2.$$

Classify the equilibrium points of the system. Show that if $a_1 d_2 > a_2 d_1$ then the species N_2 dies out and the species N_1 approaches a limiting size (Volterra's Exclusion Principle).

3 Geometrical and computational aspects of the phase diagram

IN THIS CHAPTER we discuss several topics which are useful in sketching phase diagrams. The 'index' of an equilibrium point provides supporting information on its nature and complexity which is particularly useful in strongly nonlinear cases where the linear approximation is zero. Secondly, the phase diagram does not give a complete picture of the solutions; it is not sufficiently specific about the behaviour of paths 'at infinity' beyond the boundaries of the diagram; but we show various projections which include 'points at infinity' and give the required overall view. Thirdly, a difficult question is to determine whether there are any limit cycles and roughly where they are; this question is treated again in Chapter 11, but here we give some elementary conditions for their existence or non-existence. Finally, having obtained all the information our methods allow about the geometrical layout of the phase diagram we may want to compute a number of typical paths, and some suggestions for carrying this out are made in Section 3.4.

3.1 The index of a point

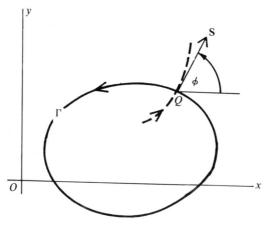

FIG. 3.1.

Given the system

$$\dot{x} = X(x, y), \quad \dot{y} = Y(x, y), \tag{3.1}$$

let Γ be a smooth closed curve consisting of ordinary points of (3.1). Let Q be a point on Γ (Fig. 3.1); then there is one and only one phase path passing through Q. The paths (without implication as to direction) belong to the family described by the equation

$$\frac{dy}{dx} = \frac{Y(x, y)}{X(x, y)}. \tag{3.2}$$

In time $\delta t > 0$ the coordinates of the point Q, (x_Q, y_Q), will increase by $\delta x, \delta y$ respectively, where

$$\delta x \approx X(x_Q, y_Q)\delta t, \quad \delta y \approx Y(x_Q, y_Q)\delta t.$$

Therefore the vector $\mathbf{S} = (X, Y)$ is tangential to the path through the point, and points in the direction of increasing t. Its inclination can be measured by the angle ϕ measured anti-clockwise from the positive direction of the x axis to the direction of $\mathbf{S}$, so that

$$\tan \phi = Y/X. \tag{3.3}$$

When the value of ϕ at one point has been decided, the value for other points is settled by requiring ϕ to be a continuous function of position, except that ϕ will not in general have its original value on returning to its starting point after a full cycle: the value may differ by $2\pi n$, where n is an integer.

Example 3.1   *Trace the variation of the vector $\mathbf{S}$ and the angle ϕ when $X(x, y)$ $= (2x^2 - 1)$, $Y(x, y) = 2xy$, and Γ is the unit circle centred at the origin.*

Let θ be the polar angle of the representative point, and put $x = \cos \theta$, $y = \sin \theta$ on Γ. Then $(X, Y) = (\cos 2\theta, \sin 2\theta), 0 \leqslant \theta \leqslant 2\pi$. This vector is displayed in Fig. 3.2. The angle ϕ takes, for example, the values $0, \frac{1}{2}\pi, 2\pi, 4\pi$ at $A, B, C,$ and A' as we track anticlockwise round the circle.

In any complete revolution, ϕ increases by 4π.

In every case the change in ϕ must be a multiple of 2π;

$$[\phi]_\Gamma = 2\pi I_\Gamma \tag{3.4}$$

say, where I_Γ is an integer, positive, negative, or zero. I_Γ *is called the index of Γ with respect to the vector field (X, Y), Γ being described anticlockwise.*

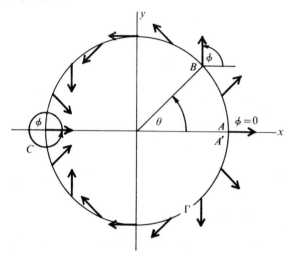

Fig. 3.2.

An algebraical representation of I_Γ is obtained as follows. Suppose that the curve Γ is described anticlockwise once by the position vector $\mathbf{r}$, where

$$\mathbf{r}(s) = (x(s), y(s)), \quad s_0 \leqslant s \leqslant s_1, \tag{3.5}$$

and s is a parameter. From (3.3),

$$\frac{\mathrm{d}}{\mathrm{d}s}(\tan \phi) = \frac{\mathrm{d}}{\mathrm{d}s}\left(\frac{Y}{X}\right),$$

or, after some reduction,

$$\frac{\mathrm{d}\phi}{\mathrm{d}s} = \frac{X\dot{Y} - Y\dot{X}}{X^2 + Y^2}. \tag{3.6}$$

(The dot denotes differentiation with respect to s.) Then from (3.4)

$$I_\Gamma = \frac{1}{2\pi} \int_{s_0}^{s_1} \frac{\mathrm{d}\phi}{\mathrm{d}s}\,\mathrm{d}s = \frac{1}{2\pi} \int_{s_0}^{s_1} \frac{X\dot{Y} - Y\dot{X}}{X^2 + Y^2}\,\mathrm{d}s. \tag{3.7}$$

Alternatively, we can write (see Fig. 3.3)

$$I_\Gamma = \frac{1}{2\pi} \oint_{\Gamma_\mathbf{R}} \frac{X\,\mathrm{d}Y - Y\,\mathrm{d}X}{X^2 + Y^2}. \tag{3.8}$$

As $\mathbf{r}(s) = (x(s), y(s))$ describes Γ, $\mathbf{R}(s) = (X, Y)$, regarded as a position vector on a plane with axes X, Y, describes another curve, Γ_R say. Γ_R is closed, since **R** returns to its original value after a complete cycle. From eqn (3.4) Γ_R encircles the origin I_Γ times, anti-clockwise when I_Γ is positive and clockwise when it is negative. This is illustrated in Fig. 3.3 for a particular case.

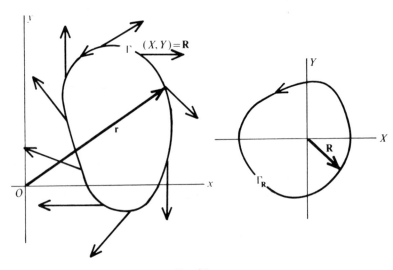

FIG. 3.3.

THEOREM 3.1   *Suppose that, on and inside* Γ, X, Y *and their first derivatives are continuous and* X *and* Y *are not simultaneously zero. (In other words there is no equilibrium point there.) Then* I_Γ *is zero.*

Proof.   Green's theorem for plane curves applied to (3.8) gives

$$I_\Gamma = \frac{1}{2\pi} \int\int_{S_R} \left[\frac{\partial}{\partial X} \frac{X}{X^2 + Y^2} + \frac{\partial}{\partial Y} \frac{Y}{X^2 + Y^2} \right] dX \, dY,$$

where S_R is the region enclosed by Γ_R. The application of the theorem is justified by the given conditions. It is easily verified that the integrand is identically zero.

COROLLARY TO THEOREM 3.1   *Let* Γ *be a closed curve, and* Γ' *a closed curve inside* Γ, *then if on* Γ, Γ' *and the region between them there is no equilibrium point, and if* X, Y *and their first derivatives are continuous there, then* $I_\Gamma = I_{\Gamma'}$.

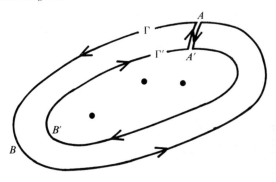

FIG. 3.4.

Proof Let AA' be a line connecting Γ and Γ', and consider the composite contour, $\mathscr{C}$, described by $ABAA'B'A'A$ in Fig. 3.4. Since $\mathscr{C}$ contains no equilibrium points, $I_{\mathscr{C}} = 0$ by the Theorem. But by (3.4)

$$0 = I_{\mathscr{C}} = \frac{1}{2\pi} \oint_{\mathscr{C}} \mathrm{d}\phi = \frac{1}{2\pi} \left(\oint_{\Gamma} \mathrm{d}\phi + \oint_{AA'} \mathrm{d}\phi - \oint_{\Gamma'} + \oint_{A'A} \mathrm{d}\phi \right)$$

where

$$\oint_{\Gamma'} \mathrm{d}\phi \quad \text{and} \quad \oint_{\Gamma} \mathrm{d}\phi$$

represent integrals taken in the anti-clockwise direction. Since

$$\oint_{A'A} \mathrm{d}\phi = - \oint_{AA'} \mathrm{d}\phi,$$

we obtain

$$0 = \frac{1}{2\pi} \left(\oint_{\Gamma} \mathrm{d}\phi - \oint_{\Gamma'} \mathrm{d}\phi \right)$$

so that

$$I_{\Gamma} = I_{\Gamma'}.$$

This theorem shows that the index I_{Γ} of the vector field (X, Y) with respect to Γ is to a large extent independent of Γ, and enables the index to be associated with special points rather than with contours. *If the smoothness conditions on the field (X, Y) are satisfied in a region containing a single equilibrium point, then any simple closed curve Γ surrounding the point generates the same number I_{Γ}. We therefore drop the suffix Γ, and say that I is the index of the equilibrium point.*

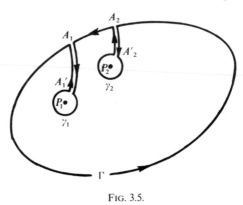

FIG. 3.5.

THEOREM 3.2   *If* Γ *surrounds n equilibrium points* $P_1, P_2, \ldots, P_n$ *then*

$$I_\Gamma = \sum_{i=1}^{n} I_i$$

where I_i *is the index of the point* P_i, $i = 1, 2, \ldots, n$.

Proof   We illustrate the proof for the case of two equilibrium points, at P_1, P_2. Construct a contour $\mathscr{C}$ consisting of Γ, two 'bridges' $A_1 A_1'$ and $A_2 A_2'$, and curves γ_1 and γ_2 surrounding P_1 and P_2 respectively, as shown in Fig. 3.5. The interior of $\mathscr{C}$ does not contain any equilibrium points, so $I_{\mathscr{C}} = 0$. But also

$$I_{\mathscr{C}} = \frac{1}{2\pi} \oint_\Gamma d\phi - \frac{1}{2\pi} \oint_{\gamma_1} d\phi - \frac{1}{2\pi} \oint_{\gamma_2} d\phi,$$

since the integrals over the 'bridges' cancel to zero. Therefore

$$I_\Gamma = I_1 + I_2.$$

The easiest way to calculate the index in a particular case is to use the following result.

THEOREM 3.3   *Let p be the number of times* $Y(x, y)/X(x, y)$ *changes from* $+\infty$ *to* $-\infty$, *and q the number of times it changes from* $-\infty$ *to* $+\infty$, *on* Γ. *Then* $I_\Gamma = \frac{1}{2}(p - q)$.

Proof   Since $\tan \phi = Y/X$, we are simply counting the number of times the direction (X, Y) is vertical (up or down), and associating a direction of rotation. We could instead examine the points where $\tan \phi$ is zero, that is, where Y is zero. Then if P and Q are numbers of changes in $\tan \phi$

from (negative/positive) to (positive/negative) respectively across the zero of Y, $I_\Gamma = \frac{1}{2}(P-Q)$.

Example 3.2 *Find the index of the equilibrium point* $(0,0)$ *of* $\dot{x} = y^3$, $\dot{y} = x^3$.

By stretching the theory a little we can let $\overset{1}{\Gamma}$ be the square with sides $x = \pm 1$, $y = \pm 1$. Start from $(1,1)$ and go round anti-clockwise. On $y = 1$, $\tan \phi = x^3$, with no infinities; similarly on $y = -1$. On $x = -1$, $\tan \phi = -y^{-3}$ with a change from $-\infty$ to $+\infty$, and on $x = 1$, $\tan \phi = y^{-3}$, with a change from $-\infty$ to $+\infty$. Thus $p = 0$, $q = 2$ and the index is -1.

Example 3.3 *Find the index of the equilibrium point at* $(0,0)$ *of the system* $\dot{x} = y - \frac{1}{2}xy - 3x^2$, $\dot{y} = -xy - \frac{3}{2}y^2$.

There is another equilibrium point at $(-\frac{1}{4}, \frac{1}{6})$. The contour must not enclose this, so choose a square bounded by $x = \pm \frac{1}{8}$, $y = \pm \frac{1}{8}$. We have

$$\tan \phi = -y(x + \tfrac{3}{2}y)/(y - \tfrac{1}{2}xy - 3x^2).$$

and the numerator is simple, so it is easy to consider the zeros of $\tan \phi$.

On AB and CD in Fig. 3.6, $\tan \phi$ is never zero. On BC, $\tan \phi$ is zero at $y = 1/12$, passing from negative to positive, and at $y = 0$, passing from negative to positive. On DA, $\tan \phi$ is zero at $y = -1/12$, passing from positive to negative, and at $y = 0$, passing from negative to positive. Therefore $P = 3$ and $Q = 1$, so $I = 1$. The equilibrium point is a type for which linearization gives no information.

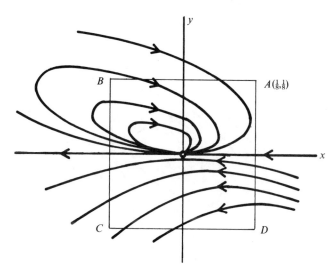

Fig. 3.6.

If we already know the nature of the equilibrium point, the index is readily found by simply drawing a figure and following the angle round. The following shows the indices of the elementary types met in Chapter 2.

(i) *A saddle point*   The change in ϕ in a single circuit of the curve Γ surrounding the saddle point is -2π, and the index is therefore -1.

(ii) *A centre*   The index is $+1$. (This is irrespective of the direction of the paths.)

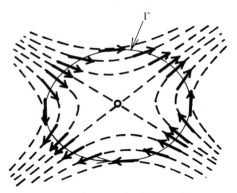

FIG. 3.7. Saddle point, index (-1)

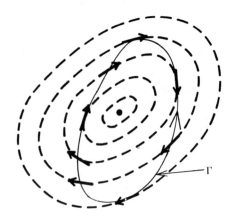

FIG. 3.8. Centre, index 1

(iii) *A spiral* (*stable or unstable*)   The index is $+1$.

(iv) *A node* (*stable or unstable*)   The index is $+1$.

A simple practical way of finding the index of a curve Γ uses a watch. Set the watch to 12.00 (say) and put the watch at a point on Γ with the minute hand pointing in the direction of the phase path there. Move the

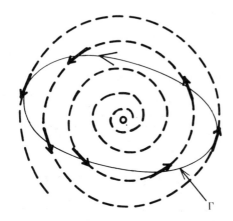

FIG. 3.9. Spiral, index 1

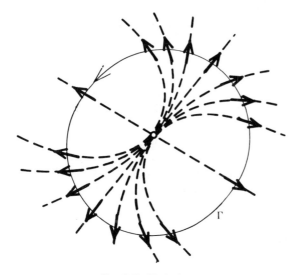

FIG. 3.10. Node, index 1

watch without rotation around Γ always turning the direction of the
minute hand along the phase path. After one counterclockwise circuit
Γ, the index is given by the gain (-1 for each hour gained) or loss, of
the hour hand. For example the 'time' after one circuit of the centre is
11.00, an index of $+1$; for the saddle-point the 'time' is 1.00, an index of
-1.

3.2   The index at infinity

In the next two sections we show some techniques for getting a more
global view of the pattern of trajectories.

DEFINITION   *The index, I_∞, of the point at infinity. Introduce new
coordinates x_1, y_1 by the transformation (inversion in the origin)*

$$x_1 = \frac{x}{x^2 + y^2}, \quad y_1 = -\frac{y}{x^2 + y^2}$$

*(or, in polar coordinates, let $\theta_1 = -\theta$ and $r_1 = 1/r$). The index of the
origin, $x_1 = y_1 = 0$, for the transformed equation is called the index at
infinity for the original equation.*

In order to prove a result concerning the index of the point at infinity,
it is convenient to write

$$z = x + iy, \quad Z = X + iY,$$

and the differential equation system becomes

$$\frac{dz}{dt} = Z. \tag{3.9}$$

(Z is not, of course, a regular function of z in general.) Note that ϕ (eqn
(3.3)) is given by

$$\phi = \arg Z. \tag{3.10}$$

The transformation in the definition is equivalent to the transformation

$$z_1 = z^{-1}$$

and (3.9) then becomes

$$-\frac{1}{z_1^2} \frac{dz_1}{dt} = Z,$$

or

$$\frac{dz_1}{dt} = -z_1^2 Z = Z_1 \qquad (3.11)$$

say. We require the index of (3.11) at $z_1 = 0$.

THEOREM 3.4 *The index I_∞ for the system $\overset{\circ}{x} = X(x, y)$, $\overset{\circ}{y} = Y(x, y)$, having a finite number n of equilibrium points with indices I_i, $i = 1, 2, \ldots, n$, is given by*

$$I_\infty = 2 - \sum_{i=1}^{n} I_i. \qquad (3.12)$$

Proof To obtain the index at the origin, surround $z_1 = 0$ by a simple closed contour $\mathscr{C}_1$ containing no other equilibrium points of the system. Under the transformation $z = z_1^{-1}$, $\mathscr{C}_1$ described anti-clockwise maps into $\mathscr{C}$ described clockwise, and the exterior of $\mathscr{C}_1$ maps into the interior of $\mathscr{C}$ (Fig. 3.11). Thus $\mathscr{C}$ embraces all the equilibrium points of the original system.

Now let

$$z_1 = r_1 e^{i\theta_1} \qquad (3.13)$$

on $\mathscr{C}_1$; and on $\mathscr{C}_1$ and $\mathscr{C}$ respectively let

$$Z_1 = \rho_1 e^{i\phi_1}, \quad Z = \rho e^{i\phi},$$

where $\rho_1 = |Z_1|$, $\rho = |Z|$. Then

$$I_\infty = \frac{1}{2\pi} [\phi_1]_{\mathscr{C}_1}.$$

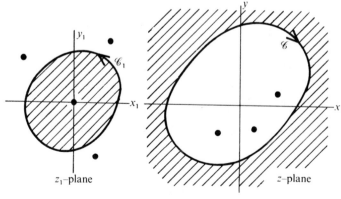

FIG. 3.11.

But from (3.11)

$$Z_1 = \rho_1 e^{i\phi_1} = -r_1^2 e^{2i\theta_1} \rho e^{i\phi} = r_1^2 \rho \, e^{i(2\theta_1 + \phi + \pi)}.$$

Therefore

$$I_\infty = \frac{1}{2\pi} [2\theta_1 + \phi + \pi]_{\mathscr{C}_1}$$

$$= \frac{1}{2\pi} \{2[\theta_1]_{\mathscr{C}_1} + [\phi]_{\mathscr{C}_1}\} = \frac{1}{2\pi} \{2[\theta_1]_{\mathscr{C}_1} - [\phi]_{\mathscr{C}}\}$$

$$= \frac{1}{2\pi} \left\{ 2.2\pi - \sum_{i=1}^{n} I_i \right\}$$

by Theorem 3.2, since also $\mathscr{C}$ is described clockwise. The result follows.

COROLLARY TO THEOREM 3.4   *Under the conditions stated in the theorem, the sum of all the indices, including that of the point at infinity, is 2.*

Example 3.4   *Find the index at infinity of the system*

$$\dot{x} = x - y, \quad \dot{y} = x - y^2.$$

We show a different approach from that of Examples 3.2 and 3.3. Putting $x = r \cos \theta, y = r \sin \theta$ we have

$$Z = (r \cos \theta - r \sin \theta) + i(r \cos \theta - r^2 \sin^2 \theta),$$

and the transformation $z = z_1^{-1}$ gives $r = r_1^{-1}, \theta = -\theta_1$, so from (3.11)

$$Z_1 = -r_1^2 (\cos 2\theta_1 + i \sin 2\theta_1) \{ r_1^{-1} (\cos \theta_1 + \sin \theta_1)$$
$$+ i r_1^{-1} (\cos \theta_1 - r_1^{-1} \sin^2 \theta_1) \}. \quad (3.14)$$

We require the index of eqn (3.11) at the point $r_1 = 0$. To evaluate it choose Γ_1 to be a circle centered at the origin, having arbitrarily small radius equal to r_1. Then we need to consider, in (3.14), only the terms of $O(1)$, the terms of $O(r_1)$ making no contribution in the limit $r_1 \to 0$. Then on Γ

$$Z_1 = (\cos 2\theta_1 + i \sin 2\theta_1) \sin^2 \theta_1 + O(r_1) = \sin^2 \theta_1 \, e^{2i\theta_1} + O(r_1).$$

As shown in Fig. 3.12 the direction, ϕ_1, of Z_1 is equal to $2\theta_1$, and the index at infinity is therefore equal to 2.

It can be confirmed that the other equilibrium points are a spiral or centre at $x = 0, y = 0$ with index 1, and a saddle at $x = 1, y = 1$ with index -1. The sum of the indices is therefore 2, as required by Theorem 3.4.

3.3   The phase diagram at infinity

Phase diagrams such as we have shown are incomplete since there is always an area outside the picture which we cannot see. When the phase

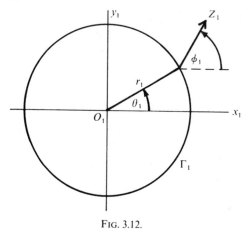

FIG. 3.12.

diagram has a simple structure this may not be a serious loss, but in complicated cases the overall structure may be obscure. We therefore show how the whole phase diagram, including the infinitely remote parts, can be displayed.

Projection on to a sphere

In Fig. 3.13, $\mathscr{S}$ is a sphere of unit radius touching the phase plane $\mathscr{P}$ at O. Its centre is O^*. A representative point P on $\mathscr{P}$ projects from O^* on to P' on the sphere. Paths on $\mathscr{P}$ become paths on $\mathscr{S}$; such features as

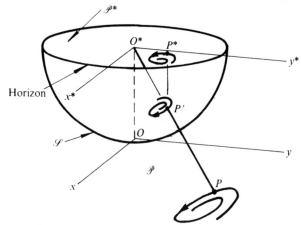

FIG. 3.13. Projection of the phase plane $\mathscr{P}$ on to the hemisphere $\mathscr{S}$, and from $\mathscr{S}$ on to the diametrical plane $\mathscr{P}^*$

spirals, nodes and saddle points on $\mathscr{P}$ appear as such on $\mathscr{S}$. The orientation of curves is also unchanged provided the sphere is viewed from inside. Local distortion occurs increasingly for features far from the origin, radial distance on $\mathscr{P}$ becoming greatly scaled down on $\mathscr{S}$. The horizontal great circle is called the *horizon*, and the happenings 'at infinity' on $\mathscr{P}$ occur near the horizon of $\mathscr{S}$. Radial straight lines on $\mathscr{P}$ project on to great circles of $\mathscr{S}$ in a vertical plane, cutting the horizon at right angles. Non-radial straight lines project on to great circles which cut the horizon at other angles. To represent on another plane the resulting pattern of lines on the lower hemisphere, we may simply take its orthogonal projection on to the diametrical plane $\mathscr{P}^*$ containing the horizon. A topological analysis (Andronov *et al.*, 1973) requires points on $\mathscr{P}$ to be mapped *twice* onto $\mathscr{S}$: to a point P on $\mathscr{P}$ there are two intersections of O^*P with $\mathscr{S}$, and the upper hemisphere is the reflection of the lower in the centre. Our interest here, however, is chiefly with obtaining pictures.

Example 3.5 $\dot{x} = y, \dot{y} = -4x - 5y$

This represents a heavily damped linear oscillator. On the phase plane $\mathscr{P}$, $(0, 0)$ is a stable node. $y = -x$, $y = -4x$ are two phase paths, and the results of Chapter 2 show that all other paths are tangent to the first at the origin and parallel to the second at infinity. The appearance of the plane $\mathscr{P}^*$ with axes x^*, y^* is shown in Fig. 3.14.

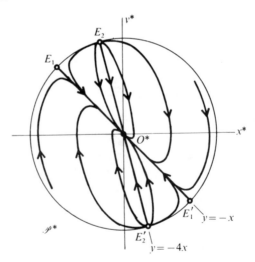

Fig. 3.14.

The paths $y = -x$, $y = -4x$ on the phase plane $\mathscr{P}$ are radial straight lines and project into lines of longitude through O on the sphere. Their projections on to the diametrical plane are therefore again straight lines $E_1O, E_1'O, E_2O, E_2'O$ of Fig. 3.14.

Since all other paths become parallel to $y = -4x$ at large distance from the origin, their projections enter the points E_2, E_2' on the sphere as shown. Other mappings of the hemisphere on to a plane can be devised.

Detail on the horizon

In order to study the paths on the hemisphere near the horizon we will project the paths found on $\mathscr{S}$ from O^* on to the plane $\mathscr{U}: x = 1$. We thus have, simultaneously, a projection of the paths in the phase plane $\mathscr{P}$ on to $\mathscr{U}$. u, z are axes with origin at N, where $\mathscr{S}$ touches $\mathscr{U}: u$ is parallel to y and z is positive downwards, so u, z is right-handed viewed from O^*. Points in $\mathscr{P}$ for $x < 0$ go into the left half of the lower hemisphere, then through O^* on to $\mathscr{U}$ for $z < 0$; and $\mathscr{P}$ for $x > 0$ goes into the right half of $\mathscr{S}$, then on to $\mathscr{U}$ for $z > 0$. Points at infinity on $\mathscr{P}$ go to the horizon of $\mathscr{S}$, then to $z = 0$ on $\mathscr{U}$. The points at infinity on $\mathscr{P}$ in direction $\pm y$ go to points at infinity on $\mathscr{U}$, as does the point O. The topological features of $\mathscr{P}$ are preserved on $\mathscr{U}$, except that the orientation of closed paths on the part of $\mathscr{S}$ corresponding to $x < 0$ is reversed (see Fig. 3.15).

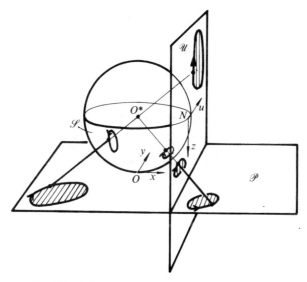

Fig. 3.15. Projection of $\mathscr{P}$ and $\mathscr{S}$ on to the vertical plane $\mathscr{U}$

It can be shown easily that the transformation from $\mathscr{P}$ to $\mathscr{U}$ is given by

$$u = y/x, \quad z = 1/x, \tag{3.15}$$

with inverse

$$x = 1/z, \quad y = u/z.$$

Example 3.6 *Examine the system $\dot{x} = y$, $\dot{y} = -4x - 5y$ at infinity (excluding the directions $\pm y$).*

(This is the system of Example 3.5.) The transformation (3.15) leads to

$$\dot{u} = (x\dot{y} - y\dot{x})/x^2 = -(4x^2 + 5xy + y^2)/x^2 = -4 - 5u - u^2$$

and, similarly,

$$\dot{z} = -\dot{x}/x^2 = -uz.$$

There are equilibrium points at $E_1 : u = -1, z = 0$, and at $E_2 : u = -4, z = 0$ and it can be confirmed in the usual way that E_1 is a saddle and E_2 a node. The pattern of the paths is shown in Fig. 3.16, and should be compared with Fig. 3.14.

To examine the phase plane at infinity in the directions $\pm y$, $\mathscr{S}$ is projected on another plane $\mathscr{V} : y = 1$. The corresponding axes v, z, right-handed viewed from O^*, have their origin at the point of tangency; v points in the direction of $-x$ and z downward. The required transformation is

$$v = x/y, \quad z = 1/y, \tag{3.16}$$

or

$$x = v/z, \quad y = 1/z.$$

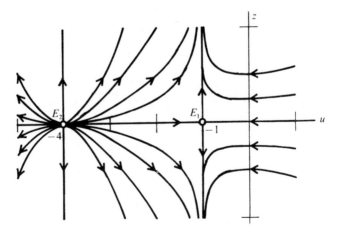

Fig. 3.16. Phase paths near the horizon

3.4 Limit cycles and other closed paths

As we have seen (Section 1.5), a *limit cycle* is an isolated periodic solution of an autonomous system, represented in the phase plane by an isolated closed path. The neighbouring paths are, by definition, not closed, but spiral into or away from the limit cycle $\mathscr{C}$ as shown in Fig. 3.17. In the case illustrated, which is a *stable limit cycle*, the device represented by the system (which might be, for example, an electrical circuit), will spontaneously drift into the corresponding periodic oscillation from a wide range of initial states. The existence of limit cycles is therefore a feature of great practical importance.

Linear systems with constant coefficients cannot exhibit limit cycles. Since normally we cannot solve the nonlinear equations which might have them it is important to be able to establish by indirect means whether a limit cycle is present in the phase diagram. If one exists, then various approximate methods can be used to locate it. In this section we give some simple indications and counter-indications, the question of existence being dealt with in more detail in Chapter 11.

The index of a limit cycle $\mathscr{C}$ is 1 since the vector (X, Y) is tangential to $\mathscr{C}$ at every point on it, and the change in ϕ around $\mathscr{C}$ is 2π. By Theorem 3.2, therefore, *a necessary condition for $\mathscr{C}$ to be a limit cycle is that the sum of the indices at the equilibrium points enclosed by $\mathscr{C}$ is 1.*

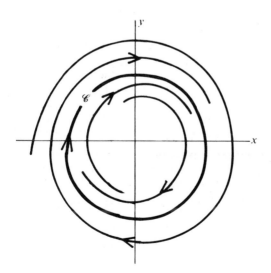

FIG. 3.17. An idealized limit cycle

This result applies equally to any type of periodic solution, but we can at least determine some cases when a limit cycle *cannot* exist. For example, a limit cycle cannot surround a region containing no equilibrium points, nor one containing only a saddle. However, consider the system

$$\dot{x} = y, \quad \dot{y} = (1 - x^2)x/(1 + x^2).$$

This has a saddle-point at the origin, index -1, and centres at $(0, 1)$, $(0, -1)$, with indices 1. The sum of the indices is 1, and it is therefore possible that there might be a limit cycle surrounding all three points. (In fact, however, the paths surrounding them are *all* closed.)

The following result is due to Bendixson (Cesari, 1971) and is called Bendixson's Negative Criterion:

THEOREM 3.5   *There are no closed paths in a simply-connected domain of the phase plane on which* $\partial X/\partial x + \partial Y/\partial y$ *is of one sign.*

Proof   We make the usual assumptions about the smoothness of the vector field (X, Y) necessary to justify the application of the divergence theorem. Suppose that there is a closed path $\mathscr{C}$ in the region $\mathscr{D}$ where $\partial X/\partial x + \partial Y/\partial y$ is of one sign; we shall show that this assumption leads to a contradiction. By the divergence theorem,

$$\int\int_{\mathscr{S}} \left(\frac{\partial X}{\partial x} + \frac{\partial Y}{\partial y} \right) \mathrm{d}x\,\mathrm{d}y = \int_{\mathscr{C}} (X, Y)\cdot\mathbf{n}\,\mathrm{d}s,$$

where $\mathscr{S}$ is the interior of $\mathscr{C}$, $\mathbf{n}$ is the unit outward normal, and $\mathrm{d}s$ is an undirected line element of $\mathscr{C}$ (Fig. 3.18). Since $\mathscr{C}$ is a path, (X, Y) is normal to $\mathbf{n}$, so the integral on the right is zero. But since the integrand on the left is of one sign its integral cannot be zero. Therefore $\mathscr{C}$ cannot be a path.

An extension of this result, called Dulac's test, is given in Exercise 23.

Example 3.7   *Show that the equation* $\ddot{x} + f(x)\dot{x} + g(x)x = 0$ *can have no periodic solution whose phase path lies in a region where f is of one sign. (Such regions have only 'positive damping' or 'negative damping'.)*

The equivalent system is $\dot{x} = y$, $\dot{y} = -f(x)y - g(x)x$, so that $(X, Y) = (y, -f(x)y - g(x)x)$, and

$$\frac{\partial X}{\partial x} + \frac{\partial Y}{\partial y} = -f(x),$$

which is of one sign whenever f is.

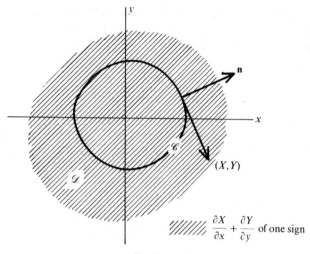

$$\frac{\partial X}{\partial x} + \frac{\partial Y}{\partial y} \text{ of one sign}$$

FIG. 3.18.

Example 3.8 The equations

$$\dot{V}_1 = -\sigma V_1 + f(E - V_2) = X(V_1, V_2), \text{ say};$$

$$\dot{V}_2 = -\sigma V_2 + f(E - V_1) = Y(V_1, V_2), \text{ say};$$

describe the voltages generated over the deflection plates in a simplified triggered sweeping circuit for an oscilloscope (Andronov and Chaikin, 1949). E and σ are positive constants, and f has the shape shown in Fig. 3.19. f is a continuous, odd function, tending to a limit as $x \to \infty$, and f' is positive and decreasing on $x > 0$. It can be shown (Chapter 2, Exercise 36) that there is an equilibrium point, at $V_1 = V_2, = V_0$, say. If, moreover, $f'(E - V_0) > \sigma$, then there

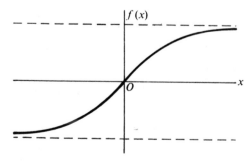

FIG. 3.19.

are two others, and in this case (V_0, V_0) is a saddle-point and the other two are stable nodes. The sum of the indices is 1, so periodic solutions with paths surrounding the three points are not ruled out by this test. However,

$$\frac{\partial X}{\partial V_1} + \frac{\partial Y}{\partial V_2} = -2\sigma,$$

which is of one sign everywhere, so in fact there are no periodic solutions.

3.5 Computation of the phase diagram

Detailed plotting of phase paths can give a much enhanced under-standing of the character of the solutions, even when we have obtained a general idea of the paths from analytical and geometrical arguments. We shall give no discussion of technical points in numerical analysis here (see, for example, Cohen, 1973), but simply indicate that very un-sophisticated methods can be made to work well so long as obvious precautions are taken and analytical reasoning is not lost sight of. We envisage calculations being done at a computer terminal, possibly with limited program and storage space, in an elementary language such as BASIC.

We consider autonomous equations of the form

$$\dot{x} = P(x, y), \quad \dot{y} = Q(x, y). \tag{3.17}$$

(P and Q are used since x and X are indistinguishable in computer programming.) In plotting a phase path we proceed step-by-step in some way from the initial point x_0, y_0:

$$x(t_0) = x_0, \quad y(t_0) = y_0. \tag{3.18}$$

The simplest step-by-step method is Euler's, using t as a supplementary variable, which we now describe and use as the basis of discussion.

Let the length of the time step be h, assumed small; and let $(x_0, y_0), (x_1, y_1), (x_2, y_2), \ldots$, be the points on the phase path at times $t_0, t_0 + h, t_0 + 2h, \ldots$. Assume that (x_n, y_n) is known. Then, writing $\dot{x}(t_0 + nh) \simeq (x_{n+1} - x_n)/h$, $\dot{y}(t_0 + nh) = (y_{n+1} - y_n)/h$, (3.17) becomes approximately

$$(x_{n+1} - x_n)/h = P(x_n, y_n), \quad (y_{n+1} - y_n)/h = Q(x_n, y_n),$$

or, after rearrangement,

$$x_{n+1} = x_n + hP(x_n, y_n), \quad y_{n+1} = y_n + hQ(x_n, y_n). \tag{3.19}$$

(x_{n+1}, y_{n+1}) is now recorded and we proceed to the next step. The danger of progressively increasing error is present, even when h is very small so

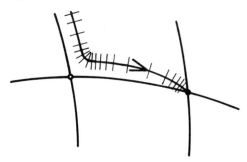

Fig. 3.20. Equal time steps near equilibrium points

that good accuracy is ostensibly being obtained: the reader should see, for example, Cohen (1973) for a discussion of error.

The point which concerns us here is the general unsuitability of t (time) as the parameter for the step-by-step construction of phase paths. The difficulty is shown schematically in Fig. 3.20, where a path is shown approaching a node, and passing near a saddle. Notionally equal time steps are marked. Since $\dot{x}$, $\dot{y}$, or P, Q, are small near an equilibrium point, progress is very uneven.

It is preferable to use arc length, s, as the parameter, which gives equally spaced points along the arc. Since $\delta s^2 = \delta x^2 + \delta y^2$, (3.17) with parameter s becomes

$$\frac{\mathrm{d}x}{\mathrm{d}s} = \frac{P}{\sqrt{(P^2 + Q^2)}} = U, \quad \frac{\mathrm{d}y}{\mathrm{d}s} = \frac{Q}{\sqrt{(P^2 + Q^2)}} = V \quad (3.20)$$

(s is measured increasing in the direction of t increasing). Letting h now represent the basic step length, (3.20) leads to the iterative scheme

$$x_{n+1} = x_n + hU(x_n, y_n), \quad y_{n+1} = y_n + hV(x_n, y_n) \quad (3.21)$$

with the initial values (x_0, y_0) given.

Improved processes, such as the Runge–Kutta method (Cohen, 1973) are generally to be preferred, and can easily be programmed. It is desirable to have a program which prints out only after a predetermined number of calculation steps. In this way it can be arranged, for example, to produce a point every centimetre of arc along a phase path, the interval being reduced if the path is turning rapidly.

In a practical plot it is helpful, if possible, to locate and classify the equilibrium points and to know at least whether limit cycles are likely to appear. In particular, saddle points are difficult to locate closely by

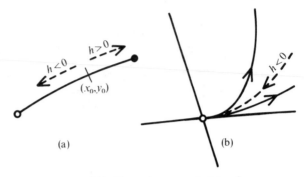

(a)   (b)

FIG. 3.21. Illustrating reversal of step sign

plotting paths. The location of equilibrium points may be obtained by
algebraic or numerical solution of the equations $P(x, y) = Q(x, y) = 0$.
Classification on the basis of linear approximation may not be possible,
for example when P or Q have a zero linear expansion, or when the linear
approximation predicts a centre (which may become a spiral in the
presence of nonlinear terms; end of Section 2.3 and Exercise 3, Chapter
2), or in certain other cases. In such cases many exploratory plots may
have to be made near the equilibrium point before an adequate
characterisation is obtained.

By reversing the sign of h in the input to suitable programs, a plot
'backwards' from the starting point is obtained. This is useful to
complete a path (Fig. 3.21a) or to get close to an unstable node (Fig.
3.21b).

It may be difficult to distinguish with confidence between certain
cases, for example between a node whose paths are seen to wind around

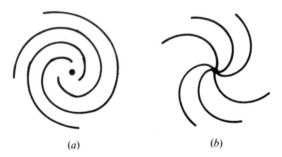

(a)   (b)

FIG. 3.22. (a) small-scale plot (spiral?)   (b) large-scale plot (node?)

it and a genuine spiral (Fig. 3.22). A display which appears to represent a spiral may, on scaling up the neighbourhood of the origin, begin to resemble a node, and conversely. This consideration applies, in principle, to all numerical and plotted data: it is impossible to deduce from the data alone that some qualitative deviation does not appear between the plotted points. However, in the case of equilibrium points and limit cycles there may be genuine uncertainty, and it is preferable to establish their nature analytically and to determine the directions of separatrices when this is possible.

The connections between equilibrium points are often unexpected and interesting: see, for example, Fig. 7.4, 7.11(b), 7.22(a) and 3.26.

A limit cycle is a feature requiring some care. The 'model' of a limit cycle which it is useful to have in mind is a pattern of spirals inside and outside the cycle as in Fig. 3.17. The figure shows a stable limit cycle, for which all nearby solutions drift towards the limit cycle. If they all recede

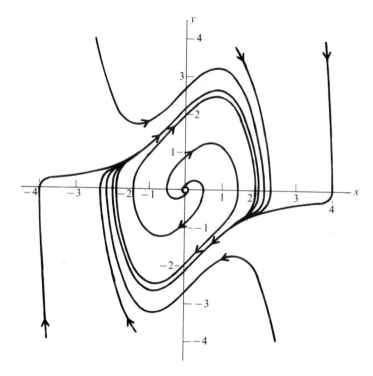

FIG. 3.23. Limit cycle for van der Pol's equation $\ddot{x} + (x^2 - 1)\dot{x} + x = 0$

(reverse the arrows) the limit cycle is unstable. A semi-stable limit cycle can also occur, in which the paths approach on one side and recede on the other.

Plotting a limit cycle may be difficult. Both the limit cycle and the neighbouring paths can diverge very considerably from the idealized pattern of Fig. 3.17; for example Fig. 3.23 shows the limit cycle for van der Pol's equation with a moderate value for the parameter. In general there is no way of finding a single point, exactly lying on the cycle, from which to start; the cycle has to be located by 'squeezing' it between inner and outer spirals. Clearly it is helpful to reverse the sign of t (or of h in the programme input) if necessary, so as to approach rather than recede from the cycle during the process of locating it.

Exercises

1. By considering the variation of path direction on closed curves round the equilibrium points, find the index in each case of Fig. 3.24.

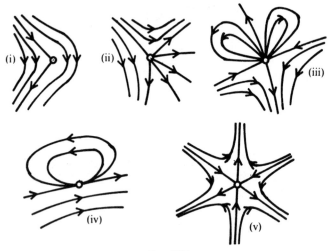

Fig. 3.24.

2. The motion of a damped pendulum is described by the equations

$$\dot{\theta} = \omega, \quad \dot{\omega} = -k\omega - v^2 \sin\theta,$$

where k (> 0) and v are constants. Find the indices of all the equilibrium states.

3. Find the index of the equilibrium points of the following systems: (i) $\dot{x} = 2xy$, $\dot{y} = 3x^2 - y^2$; (ii) $\dot{x} = y^2 - x^4$, $\dot{y} = x^3 y$; (iii) $\dot{x} = x - y$, $\dot{y} = x - y^2$.

4. For the linear system $\dot{x} = ax + by$, $\dot{y} = cx + dy$, where $ad - bc \neq 0$, obtain the index at the origin by evaluating equation (3.7), showing that it is equal to sgn($ad - bc$). (Hint: choose Γ to be the ellipse $(ax + by)^2 + (cx + dy)^2 = 1$.)

5. The equation of motion of a bar restrained by springs (see Fig. 3.25) and attracted by a parallel current-carrying conductor is

$$\ddot{x} + c\{x - \lambda/(a - x)\} = 0,$$

where c, a, and λ are positive constants. Sketch the phase paths for $-x_0 < x < a$, and find the indices of the equilibrium points for all $\lambda > 0$.

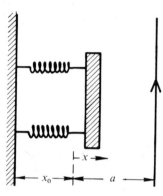

FIG. 3.25.

6. Show that the equation

$$\ddot{x} - \varepsilon(1 - x^2 - \dot{x}^2)\dot{x} + x = 0$$

has an equilibrium point of index 1 at the origin of the phase plane x, y with $\dot{x} = y$. (It also has a limit cycle, $x = \cos t$.) Use equation (3.7), with Γ a circle of radius a to show that, for all a,

$$\int_0^{2\pi} \frac{d\theta}{1 - 2\varepsilon(1 - a^2)\sin\theta\cos\theta + \varepsilon^2(1 - a^2)^2\sin^2\theta} = 2\pi.$$

7. A limit cycle encloses N nodes, F spirals, C centres, and S saddle points only, all of the type of Chapter 2. Show that $N + F + C - S = 1$.

8. Given the system

$$\dot{x} = X(x, y)\cos\alpha - Y(x, y)\sin\alpha, \quad \dot{y} = X(x, y)\sin\alpha + Y(x, y)\cos\alpha,$$

where α is a parameter, prove that the index of a simple closed curve which does not meet an equilibrium point is independent of α. (See also Exercise 45, Chapter 2.)

9. Suppose that the system $\dot{\mathbf{x}} = \mathbf{X}(\mathbf{x})$ has a finite number of equilibrium points, each of which is either a node, a centre, a spiral, or a saddle point, of the elementary types discussed in Chapter 2. Show that the total number of nodes, centres, and spirals is equal to the total number of saddle points plus two.

10. Obtain differential equations describing the behaviour of the linear system, $\dot{x} = ax + by$, $\dot{y} = cx + dy$, at infinity. Analyse the system $\dot{x} = 2x - y$, $\dot{y} = 3x - 2y$ near the horizon.

11. A certain system is known to have exactly two equilibrium points, both saddle points. Sketch phase diagrams in which
 (i) a separatrix connects the saddle points;
 (ii) no separatrix connects them.

12. Deduce the index at infinity for the system $\dot{x} = x - y$, $\dot{y} = x - y^2$ by calculating the indices of the equilibrium points.

13. Use the geometrical picture of the field (X, Y) in the neighbourhood of an ordinary point to confirm Theorem 3.1.

14. Suppose that, for the two plane systems $\dot{\mathbf{x}}_1 = \mathbf{X}_1(\mathbf{x}_1)$, $\dot{\mathbf{x}}_2 = \mathbf{X}_2(\mathbf{x}_2)$, and for a given closed curve Γ, there is no point on Γ at which $\mathbf{X}_1$ and $\mathbf{X}_2$ are opposite in direction. Show that the index of Γ is the same for both systems.
 The system $\dot{x} = y^2$, $\dot{y} = xy$ has a saddle point at the origin. Show that the index of the origin for the system $\dot{x} = y^2 + cx^2$, $\dot{y} = xy - cy^2$ is likewise -1 for $|c|$ small enough.

15. Use Exercise 14 to show that the index of the equilibrium point $x = 0, \dot{x} = 0$ for the equation $\ddot{x} + \sin x = 0$ on the usual phase plane has index 1, by comparing the equation $\ddot{x} + x = 0$.

16. The system
$$\dot{x} = ax + by + P(x, y), \quad \dot{y} = cx + dy + Q(x, y)$$
has an isolated equilibrium point at $(0, 0)$, and $P(x, y) = O(r^2)$, $Q(x, y) = O(r^2)$ as $r \to 0$, where $r^2 = x^2 + y^2$. Assuming that $ad - bc \neq 0$, show that the origin has the same index as its linear approximation.

17. Show that, on the phase plane with $\dot{x} = y$, $\dot{y} = Y(x, y)$, Y continuous, the index I_Γ of any simple closed curve Γ can only be 1, -1, or zero.
 Let $\dot{x} = y$, $\dot{y} = f(x, \lambda)$, f, f' continuous, represent a parameter-dependent system with parameter λ. Show that, at a bifurcation point (Section 1.7), where an equilibrium point divides as λ varies, the sum of the indices of the equilibrium points resulting from the splitting is unchanged. (Hint: the integrand in eqn (3.7) is continuous.)
 Deduce that the equilibrium points for the system $\dot{x} = y$, $\dot{y} = -\lambda x + x^3$ consist of a saddle point for $\lambda < 0$, and a centre and two saddles for $\lambda > 0$.

18. Prove a similar result to that of Exercise 17 for the system $\dot{x} = y$,

$\dot{y} = f(x, y, \lambda)$. Deduce that the system $\dot{x} = y$, $\dot{y} = -\lambda x - ky - x^3$, $(k > 0)$, has a saddle point at $(0,0)$ when $\lambda < 0$, which bifurcates into a stable spiral or node and two saddle points as λ becomes positive.

19. A system is known to have three closed paths, $\mathscr{C}_1$, $\mathscr{C}_2$, and $\mathscr{C}_3$, such that $\mathscr{C}_2$ and $\mathscr{C}_3$ are interior to $\mathscr{C}_1$ and such that $\mathscr{C}_2, \mathscr{C}_3$ have no interior points in common. Show that there must be at least one equilibrium point in the domain bounded by $\mathscr{C}_1$, $\mathscr{C}_2$, and $\mathscr{C}_3$.

20. For each of the following systems you are given some information about phase paths and equilibrium points. Sketch phase diagrams consistent with these requirements.
 (i) $x^2 + y^2 = 1$ is a phase path, $(0,0)$ a saddle point, $(\pm\frac{1}{2}, 0)$ centres.
 (ii) $x^2 + y^2 = 1$ is a phase path, $(-\frac{1}{2}, 0)$ a saddle point, $(0,0)$ and $(\frac{1}{2}, 0)$ centres.
 (iii) $x^2 + y^2 = 1$, $x^2 + y^2 = 2$ are phase paths, $(0, \pm\frac{3}{2})$ stable spirals, $(\pm\frac{3}{2}, 0)$ saddle points, $(0,0)$ a stable spiral.

21. Consider the system
$$\dot{x} = y(z - 2), \quad \dot{y} = x(2 - z) + 1, \quad x^2 + y^2 + z^2 = 1$$
(Exercise 29, Chapter 2), having exactly two equilibrium points, which lie on the unit sphere. Project the phase diagram on to the plane $z = -1$ through the point $(0, 0, 1)$. Deduce that $I_\infty = 0$ on this plane (consider the projection of a small circle on the sphere with its centre on the z axis). Explain, in general terms, why the sum of the indices of the equilibrium points on a sphere is two.

For a certain problem, the phase diagram on the sphere has centres and saddle points only, and it has exactly two saddle points. How many centres has the phase diagram?

22. Show that the following systems have no periodic solutions:
 (i) $\dot{x} = y + x^3$, $\dot{y} = x + y + y^3$;
 (ii) $\dot{x} = y$, $\dot{y} = -(1 + x^2 + x^4)y - x$.

23. (Dulac's Test) For the system $\dot{x} = X(x, y)$, $\dot{y} = Y(x, y)$, show that there are no closed paths in a simply-connected region in which $(\partial(\rho X)/\partial x) + (\partial(\rho Y)/\partial y)$ is of one sign, where $\rho(x, y)$ is any function having continuous first partial derivatives.

24. Explain in general terms how Dulac's test (Exercise 23) and Bendixson's test may be extended to cover the cases when $(\partial(\rho X)/\partial x) + (\partial(\rho Y)/\partial y)$ is of one sign except on isolated points or curves within a simply-connected region.

25. For the second-order system $\dot{x} = X(x)$, $\mathrm{div}(X) = 0$ and $\mathbf{curl}(X) = 0$ in a simply-connected region $\mathscr{D}$. Show that the system has no closed paths in $\mathscr{D}$. Deduce that
$$\dot{x} = y + 2xy, \quad \dot{y} = x + x^2 - y^2$$
has no periodic solutions.

26. In Exercise 25 show that $\operatorname{div}(\mathbf{X}) = 0$ may be replaced by $\operatorname{div}(\psi\mathbf{X}) = 0$, where $\psi(x, y)$ is of one sign in $\mathscr{D}$.

27. By using Dulac's test (Exercise 23) with $\rho = e^{-2x}$, show that

$$\dot{x} = y, \quad \dot{y} = -x - y + x^2 + y^2$$

has no periodic solutions.

28. Use Dulac's test (Exercise 23) to show that $\dot{x} = x(y-1)$, $\dot{y} = x + y - 2y^2$, has no periodic solutions.

29. Show that the following systems have no periodic solutions:
   (i) $\dot{x} = y$, $\dot{y} = 1 + x^2 - (1-x)y$,
   (ii) $\dot{x} = -(1-x)^3 + xy^2$, $\dot{y} = y + y^3$,
   (iii) $\dot{x} = 2xy + x^3$, $\dot{y} = -x^2 + y - y^2 + y^3$,
   (iv) $\dot{x} = x$, $\dot{y} = 1 + x + y^2$,
   (v) $\dot{x} = y$, $\dot{y} = -1 - x^2$,
   (vi) $\dot{x} = 1 - x^3 + y^2$, $\dot{y} = 2xy$,
   (vii) $\dot{x} = y$, $\dot{y} = (1 + x^2)y + x^3$.

30. Let $\mathscr{D}$ be a doubly-connected domain in the x, y plane. Show that, if $\rho(x, y)$ has continuous first partial derivatives and $\operatorname{div}(\rho\mathbf{X})$ is of constant sign in $\mathscr{D}$, then the system has not more than one closed path in $\mathscr{D}$. (An extension of Dulac's theorem, Exercise 23.)

31. A system has exactly two limit cycles with one lying interior to the other and with no equilibrium points between them. Can the limit cycles be described in opposite senses?
   Obtain the equations of the phase paths of the system

$$\dot{r} = \sin \pi r, \quad \dot{\theta} = \cos \pi r$$

as described in polar coordinates (r, θ). Sketch the phase diagram.

32. Using Bendixson's theorem (Section 3.4) show that the response amplitudes a, b for the van der Pol equation in the 'van der Pol plane' (Chapter 7), described by the equations

$$\dot{a} = \tfrac{1}{2}\varepsilon(1 - \tfrac{1}{4}r^2)a - \frac{\omega^2 - 1}{2\omega}b, \quad \dot{b} = \tfrac{1}{2}\varepsilon(1 - \tfrac{1}{4}r^2)b + \frac{\omega^2 - 1}{2\omega}a + \frac{\Gamma}{2\omega},$$

$$r = \sqrt{(a^2 + b^2)},$$

have no closed paths in the circle $r < \sqrt{2}$.

33. Let $\mathscr{C}$ be a closed path for the system $\dot{\mathbf{x}} = \mathbf{X}(\mathbf{x})$, having $\mathscr{D}$ as its interior. Show that

$$\int_{\mathscr{D}} \operatorname{div}(\mathbf{X}) \, \mathrm{d}x \, \mathrm{d}y = 0.$$

34. Assume that van der Pol's equation in the phase plane

$$\dot{x} = y, \quad \dot{y} = -\varepsilon(x^2-1)y-x$$

has a single closed path, which, for ε small, is approximately a circle of radius a. Use the result of Exercise 33 to show that approximately

$$\int_{-a}^{a} \int_{-\sqrt{(a^2-x^2)}}^{\sqrt{(a^2-x^2)}} (x^2-1) \, \mathrm{d}y \, \mathrm{d}x = 0,$$

and so deduce a.

35. Following Exercises 33 and 34, deduce a condition on the amplitudes of periodic solutions of

$$\ddot{x} + \varepsilon h(x, \dot{x})\dot{x} + x = 0, \quad |\varepsilon| \ll 1.$$

36. For the system

$$\ddot{x} + \varepsilon h(x, \dot{x})\dot{x} + g(x) = 0,$$

suppose that $g(0) = 0$ and $g'(x) > 0$. Let $\mathscr{C}$ be a closed path in the phase plane (as all paths must be) for the equation

$$\ddot{x} + g(x) = 0$$

having interior $\mathscr{D}$. Use the result of Exercise 33 to deduce that for small ε, $\mathscr{C}$ approximately satisfies

$$\int_{\mathscr{D}} \{h(x, y) + h_y(x, y)y\} \, \mathrm{d}x \, \mathrm{d}y = 0.$$

Adapt this result to the equation

$$\ddot{x} + \varepsilon(x^2-\alpha)\dot{x} + \sin x = 0,$$

with ε small, $0 < \alpha \ll 1$, and $|x| < \frac{1}{2}\pi$. Show that a closed path (a limit cycle) is given by

$$y^2 = 2A + 2\cos x$$

where A satisfies

$$\int_{-\cos^{-1}(-A)}^{\cos^{-1}(-A)} (x^2-\alpha)\sqrt{\{2A+2\cos x\}} \, \mathrm{d}x = 0.$$

37. For the system

$$\dot{\mathbf{x}} = \mathbf{X}_0 + \varepsilon\mathbf{X}_1$$

$\mathrm{div}(\mathbf{X}_0) = 0$, and ε is small. Suppose that the system has a centre at the origin when $\varepsilon = 0$. Adapt the method of Exercise 33 to find approximately the periodic solutions, if any, of the system. For the case when there are only limit-cycles, suggest a condition for stability of these. (The system $\dot{\mathbf{x}} = \mathbf{X}_0$ is a Hamiltonian system: for the condition for a centre see Exercise 30, Chapter 2.)

Examine the system

$$\dot{x} = -2y, \quad \dot{y} = x + \varepsilon(x^2 - 1)y$$

from this point of view.

38. The equation $\ddot{x} + F_0 \tanh k(\dot{x} - 1) + x = 0, F_0 > 0, k \gg 1$, can be thought of as a plausible continuous representation of the type of Coulomb friction problem of Section 1.6. Show, however, that the only equilibrium point is a stable spiral at the origin, and that there are no periodic solutions.

39. Show that the third-order system

$$\dot{x}_1 = x_2, \quad \dot{x}_2 = -x_1, \quad \dot{x}_3 = 1 - (x_1^2 + x_2^2)$$

has no equilibrium points but nevertheless has closed paths (periodic solutions).

40. Sketch the phase diagram for the quadratic system

$$\dot{x} = 2xy, \quad \dot{y} = y^2 - x^2.$$

41. Locate the equilibrium points of the system

$$\dot{x} = x(x^2 + y^2 - 1), \quad \dot{y} = y(x^2 + y^2 - 1),$$

and sketch the phase diagram.

42. In the system $\dot{x} = X(x, y), \dot{y} = Y(x, y)$, X and Y are relatively prime polynomials of degree two in x, y. Show that three equilibrium points can never be collinear.

43. In the general quadratic system of Exercise 42 show that any closed path Γ is convex (that is, any straight line cuts Γ at not more than two points).

The following exercises are to be carried out on a computer.

44. Calculate a phase diagram, showing the main features of the phase plane, for the equation $\ddot{x} + \varepsilon(1 - x^2 - \dot{x}^2)\dot{x} + x = 0, \varepsilon = 0.1$ and $\varepsilon = 5$.

45. Calculate a phase diagram for the damped pendulum equation $\ddot{x} + 0.15\dot{x} + \sin x = 0$. See Fig. 3.26.

46. The system

$$\dot{x} = -\frac{1}{2\omega}y\{(\omega^2 - 1) + \tfrac{1}{8}(x^2 + y^2)\}, \quad \dot{y} = \frac{1}{2\omega}x\{(\omega^2 - 1) + \tfrac{1}{8}(x^2 + y^2)\} + \frac{\Gamma}{2\omega}$$

occurs in the theory of the forced oscillations of a pendulum (eqns (7.16), (7.17)). Obtain the phase diagram when $\omega = 0.975, \Gamma = 0.005$.

47. A population of rabbits $R(t)$ and foxes $F(t)$ live together in a certain territory. The combined birth and death rate of the rabbits due to 'natural' causes is $\alpha_1 > 0$,

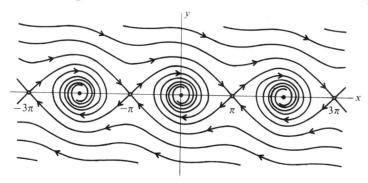

Fig. 3.26. Phase diagram for the pendulum equation with small damping

and the additional deaths due to their being eaten by foxes is introduced through an 'encounter factor' β_1, so that

$$\frac{dR}{dt} = \alpha_1 R - \beta_1 RF.$$

The foxes die of old age with death rate $\beta_2 > 0$, and the live birth rate is sustained through an encounter factor α_2, so that (compare Section 2.2)

$$\frac{dF}{dt} = \alpha_2 RF - \beta_2 F.$$

Plot the phase diagram, when $\alpha_1 = 10$, $\beta_1 = 0.2$, $\alpha_2 = 4 \times 10^{-5}$, $\beta_2 = 0.2$. Also plot the solution curves $R(t)$ and $F(t)$ (these are oscillatory, having the same period but different phase).

48. The system

$$\dot{x} = \tfrac{1}{2}e(1 - \tfrac{1}{4}r^2)x - \frac{\omega^2 - 1}{2\omega}y, \quad (r^2 = x^2 + y^2),$$

$$\dot{y} = \frac{\omega^2 - 1}{2\omega}x + \tfrac{1}{2}e(1 - \tfrac{1}{4}r^2)y + \frac{\Gamma}{2\omega},$$

occurs in the theory of forced oscillations of the van der Pol equation (eqns (7.11(b)), (7.13(b))). Plot phase diagrams for the cases
(i) $e = 1$, $\Gamma = 0.75$, $\omega = 1.2$;
(ii) $e = 1$, $\Gamma = 2.0$, $\omega = 1.6$.

49. The equation for a tidal bore on a shallow stream is

$$\varepsilon \frac{d^2\eta}{d\xi^2} - \frac{d\eta}{d\xi} + \eta^2 - \eta = 0$$

where (in appropriate dimensions), η is the height of the free surface, and $\xi = x - ct$ where c is the wave speed. For $0 < \varepsilon \ll 1$, find the equilibrium points of the equation and classify them according to their linear approximations.

Plot the phase paths in the plane of η, w, where

$$\frac{d\eta}{d\xi} = w, \quad \varepsilon \frac{dw}{d\xi} = \eta + w - \eta^2$$

and show that a separatrix from the saddle point at the origin reaches the other equilibrium point. Interpret this observation in terms of the shape of the wave.

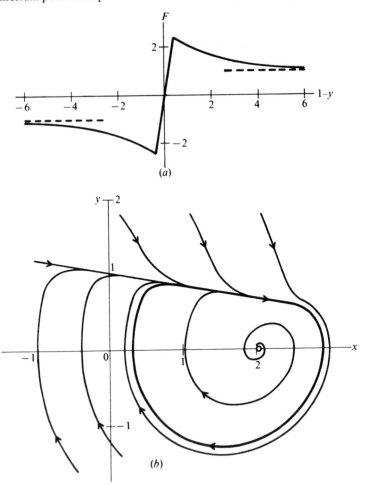

FIG. 3.27. Phase diagram for a Coulomb friction problem

50. Determine the nature of the equilibrium point, and compute the phase diagram for the Coulomb friction type problem

$$\ddot{x} + x = F(\dot{x})$$

where

$$F(y) = \begin{cases} -6 \cdot 0(y-1), & |y-1| \leqslant 0 \cdot 4 \\ -[1 + 1 \cdot 4 \exp\{-0 \cdot 5|y-1| + 0 \cdot 2\}] \operatorname{sgn}(y-1), & |y-1| \geqslant 0 \cdot 4. \end{cases}$$

(See Fig. 3.27, and compare the simpler case shown in Figs. 1.15(*b*) and 1.16.)

51. Compute the phase diagram for the system whose polar representation is

$$\dot{r} = r(1-r), \quad \dot{\theta} = \sin^2(\tfrac{1}{2}\theta).$$

52. Compute the phase diagrams for the following systems: (i) $\dot{x} = 2xy$, $\dot{y} = y^2 - x^2$; (ii) $\dot{x} = 2xy$, $\dot{y} = x^2 - y^2$; (iii) $\dot{x} = x^3 - 2x^2y$, $\dot{y} = 2xy^2 - y^3$.

4 Averaging methods

CONSIDER an equation of the form $\ddot{x} + \varepsilon h(x, \dot{x}) + x = 0$ where ε is small, and assume that it has at least one periodic solution, so that its phase diagram contains either a limit cycle or a centre. Such an equation is in a sense 'close' to the linear equation $\ddot{x} + x = 0$, and the paths will be close to circles. It should be possible to take advantage of this fact in constructing approximate solutions, in a way similar to that in which a linear approximation gives a guide to the nature of the phase paths near an equilibrium point. However, a solution may differ very little from some solution of the linear equation but it may still be sufficient to prevent the corresponding path's being closed (similarly, an equilibrium point which is a centre in the linear approximation may in fact be a spiral: see Chapter 2, Exercise 3). In order to determine periodic solutions we use a process which essentially assumes the existence of a closed path which is nearly circular but has unknown radius, then utilize the consistency conditions on the radius implied by the differential equation. Thus we make direct use of the term of order ε in the equation in order to establish the consistency. In doing this we simplify the function h by in effect considering only the first harmonic in the variation of $\dot{x}h(x, \dot{x})$ over the cycle; this can also be interpreted as approximation in the sense of taking the average value of $\dot{x}h(x, \dot{x})$ over a cycle.

Elaborations of the method permit the determination of the approximate period and stability of limit cycles, the shape of the spiral paths around limit cycles, and amplitude–frequency relations in the case of a centre. Similar estimates are successful even in very unpromising cases where ε is not small and the supporting arguments are no longer plausible (see Example 4.12 and Section 4.6). In such cases we can say that we are guided to the appropriate sense of how best to fit a circle to a non-circular path by arguments which are valid when ε is small. An error estimate is given for a class of such cases (Section 4.6).

4.1 An energy-balance method for limit-cycles

The nonlinear character of isolated periodic oscillations makes their detection and construction difficult. Here we discuss limit-cycles and other periodic solutions in the phase plane $\dot{x} = y$, $\dot{y} = Y(x, y)$, which allows the mechanical interpretation of Section 1.6 and elsewhere.

Consider the family of equations of the form

$$\ddot{x} + \varepsilon h(x, \dot{x}) + x = 0 \tag{4.1}$$

(note that the equation $\ddot{x} + \varepsilon h(x, \dot{x}) + \omega^2 x = 0$ can be put into this form by the change of variable $\tau = \omega t$). Then on the phase plane we have

$$\dot{x} = y, \quad \dot{y} = -\varepsilon h(x, y) - x. \tag{4.2}$$

Assume that $|\varepsilon| \ll 1$, so that the nonlinearity is small, and that $h(0,0) = 0$, so that the origin is an equilibrium point. Suppose we have reason to think that there is at least one periodic solution with phase path surrounding the origin: an appraisal of the regions of the phase plane in which energy loss and energy gain take place might give grounds for expecting a limit cycle.

When ε is zero, eqn (4.1) becomes $\ddot{x} + x = 0$, with solutions $x(t) = a\cos(t + \alpha)$ where a, α are constants. These solutions appear as circles of radius a on the phase plane, and the motion has period 2π. We do not lose any generality so far as the phase-plane representation is concerned if we limit consideration to the case

$$a > 0, \quad \alpha = 0,$$

since all other solutions simply correspond to a change of time origin.

Suppose now that ε is small but not zero. Then we should expect that *the limit cycle assumed above will be close to one of the circles*

$$x^2 + y^2 = a^2 \tag{4.3}$$

*generated by the **linearized equation** (with $\varepsilon = 0$), and that its period will be very nearly 2π.* In order to decide which circle most nearly satisfies the conditions for a limit cycle, we tighten up the argument of Section 1.5. Assume that the period of the postulated limit cycle is very nearly 2π, and that it is represented approximately (ignoring the unimportant phase angle) by

$$x(t) = a\cos t, \quad a > 0. \tag{4.4}$$

Then $\dot{x}(t) = y(t) = -a\sin t$. From (1.35) (with εh for h) the change in energy over a cycle is given approximately by

$$\mathscr{E}(2\pi) - \mathscr{E}(0) = -\varepsilon \int_0^{2\pi} (-a\sin t) h(a\cos t, -a\sin t)\, \mathrm{d}t. \tag{4.5}$$

(Obviously integration over any interval of length 2π gives the same result since the integrand is periodic.) But the change in energy must be zero over the cycle, since x and $\dot{x}$ return to their original values in the periodic motion. Therefore (4.5) implies that on a limit cycle, and, indeed, for any periodic solution,

$$\int_0^{2\pi} h(a\cos t, -a\sin t)\sin t\,dt = 0, \qquad (4.6)$$

an equation which, in principle, determines a in the case of a limit cycle, and is satisfied identically around a centre.

Example 4.1   *Find the approximate amplitude for the limit cycle of van der Pol's equation*

$$\ddot{x} + \varepsilon(x^2 - 1)\dot{x} + x = 0. \qquad (4.7)$$

Here,

$$h(x, \dot{x}) = (x^2 - 1)\dot{x}.$$

Assuming that $x \approx a\cos t$, the energy balance equation (4.6) gives

$$\int_0^{2\pi} (a^2\cos^2 t - 1)\sin t\sin t\,dt = 0, \qquad (4.8)$$

leading to $\frac{1}{4}a^2 - 1 = 0$, with positive solution $a = 2$. Figure 4.1 shows the calculated limit cycle for $\varepsilon = 0.1$.

By an extension of this argument the stability of the limit cycle can also be determined. Taking the model of a limit cycle as conforming to Fig. 3.17 we should expect that paths close enough to the limit cycle, though spiralling gradually, will also be given approximately by $x = a\cos t$, $y = -a\sin t$ where a is nearly constant. On a path segment corresponding to $0 \leqslant t \leqslant 2\pi$ the change in energy is approximately

$$\mathscr{E}(2\pi) - \mathscr{E}(0) = \varepsilon a \int_0^{2\pi} h(a\cos t, -a\sin t)\sin t\,dt = g(a), \qquad (4.9)$$

say. Let $a = a_0 > 0$ on the limit cycle, so that $g(a_0) = 0$. If the cycle is stable, then on interior segments $(a < a_0)$ energy must be gained, and on exterior segments $(a > a_0)$ energy must be lost: that is, there exists $\delta > 0$ such that

$$g(a) > 0 \quad \text{when} \quad a_0 - \delta < a < a_0$$

and

$$g(a) < 0 \quad \text{when} \quad a_0 < a < a_0 + \delta.$$

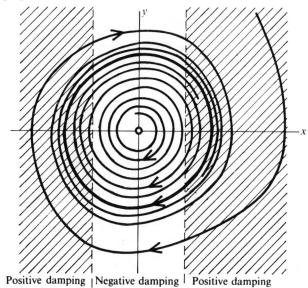

Positive damping | Negative damping | Positive damping

Fig. 4.1. Phase diagram, with limit-cycle, for $\ddot{x}+0\cdot1(x^2-1)\dot{x}+x=0$

Therefore

$$g'(a_0) < 0 \qquad (4.10)$$

is a sufficient condition for stability.

Example 4.2 *Check the stability of the limit cycle in Example 4.1.*

From eqn (4.14)

$$g(A) = -\varepsilon a^2 \int_0^{2\pi} (a^2\cos^2 t - 1)\sin^2 t \, dt = -\varepsilon a^2 \pi(\tfrac{1}{4}a^2 - 1).$$

Therefore

$$g'(a) = -\varepsilon \pi a(a^2 - 2)$$

and

$$g'(2) = -4\pi\varepsilon.$$

The cycle is therefore stable when $\varepsilon > 0$, unstable when $\varepsilon < 0$.

Direct differentiation of (4.9) gives an alternative criterion. From (4.9),

$$g'(a) = \varepsilon \int_0^{2\pi} h \sin t \, dt + \varepsilon a \int_0^{2\pi} \frac{dh}{da} \sin t \, dt$$

$$= \varepsilon[-h\cos t]_0^{2\pi} + \varepsilon \int_0^{2\pi} \frac{dh}{dt}\cos t \, dt + \varepsilon a \int_0^{2\pi} \frac{dh}{da}\sin t \, dt$$

(integrating by parts). But by the chain rule

$$\frac{dh}{dt} = -ah_1 \sin t - ah_2 \cos t$$

and

$$\frac{dh}{da} = h_1 \cos t - h_2 \sin t$$

(where $h_1(u,v) = \partial h(u,v)/\partial u$ and $h_2(u,v) = \partial h(u,v)/\partial v$). Therefore

$$g'(a) = -\varepsilon a \int_0^{2\pi} h_2(a\cos t, -a\sin t)\, dt.$$

For *stability* we require $g'(a_0) < 0$ which is equivalent to

$$\varepsilon \int_0^{2\pi} h_2(a_0 \cos t, -a_0 \sin t)\, dt > 0, \tag{4.11}$$

where a_0 is a positive solution of (4.6).

Example 4.3 *Check the stability of the limit cycle in Examples* 4.1.
  $h(x,y) = (x^2 - 1)y$ and $h_2(x,y) = x^2 - 1$. Equation (4.11), with $a_0 = 2$, gives

$$\varepsilon \int_0^{2\pi} (4\cos^2 t - 1)\, dt = 2\pi\varepsilon$$

and the cycle is stable if $\varepsilon > 0$.

4.2 Amplitude and frequency estimates

As described in the last section, a periodic solution of the equation

$$\ddot{x} + \varepsilon h(x, \dot{x}) + x = 0 \tag{4.12}$$

or of the system

$$\dot{x} = y, \quad \dot{y} = -\varepsilon h(x, y) - x, \tag{4.13}$$

with $h(0,0) = 0$ and $|\varepsilon| \ll 1$ is expected to appear in the phase plane as a small distortion, or perturbation, of one of the circular paths of the linearized equation $\ddot{x} + x = 0$.

In Fig. 4.2, P is a representative point for a phase path of the system and $a(t)\ (> 0)$, $\theta(t)$ are its polar coordinates. Note that the positive direction of θ is opposite to the direction of the path. Also $x = a\cos\theta$ and $y = a\sin\theta$ (not $(-a\sin\theta)$). Since $a^2 = x^2 + y^2$,

$$a\dot{a} = x\dot{x} + y\dot{y} = -\varepsilon yh(x, y)$$

so that

$$\dot{a} = -\varepsilon h(a\cos\theta, a\sin\theta)\sin\theta. \tag{4.14}$$

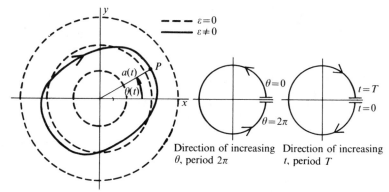

Fig. 4.2.

Similarly, since $\tan \theta = y/x$ we find that

$$\dot{\theta} = -1 - \frac{\varepsilon}{a} h(a \cos \theta, a \sin \theta) \cos \theta. \qquad (4.15)$$

These equations are exact. We also deduce that

$$\frac{\mathrm{d}a}{\mathrm{d}\theta} = \frac{\dot{a}}{\dot{\theta}} = \frac{-\varepsilon h(a \cos \theta, a \sin \theta) \sin \theta}{-1 - \dfrac{\varepsilon}{a} h(a \cos \theta, a \sin \theta) \cos \theta},$$

$$\approx \varepsilon h(a \cos \theta, a \sin \theta) \sin \theta \qquad (4.16)$$

accurate to order ε.

Since the motion is periodic, the interval $2\pi \geqslant \theta \geqslant 0$ corresponds to a single period (measured forwards in time: see Fig. 4.2). a returns to its original value in the period, so

$$\int_0^{2\pi} \frac{\mathrm{d}a}{\mathrm{d}\theta} \, \mathrm{d}\theta = 0 = \varepsilon \int_0^{2\pi} h(a \cos \theta, a \sin \theta) \sin \theta \, \mathrm{d}\theta. \qquad (4.17)$$

a in the integral is a function of θ through (4.16). Since $\mathrm{d}a/\mathrm{d}\theta = O(\varepsilon)$, then over one period

$$a(\theta) = a_0 + O(\varepsilon), \qquad (4.18)$$

where a_0 is the value, say, at the beginning of the period. In (4.17), therefore, by retaining only the dominant term we have

$$\int_0^{2\pi} h(a_0 \cos \theta, a_0 \sin \theta) \sin \theta \, \mathrm{d}\theta = 0. \qquad (4.19)$$

This is eqn (4.6) again and provides, in principle, a means for estimating a_0. The first approximation on the phase plane is then

$$x^2 + y^2 = a_0^2. \tag{4.20}$$

To find the perturbed *period*, T, we have

$$T = \int_0^T dt = \int_{2\pi}^0 \left(\frac{d\theta}{dt}\right)^{-1} d\theta.$$

(see Fig. 4.2 for the correspondence between the limits of integration). Therefore, from (4.15),

$$T = \int_0^{2\pi} \frac{d\theta}{1 + \dfrac{\varepsilon}{a} h(a\cos\theta, a\sin\theta)\cos\theta} \ ;$$

$$\approx 2\pi - (\varepsilon/a) \int_0^{2\pi} h(a\cos\theta, a\sin\theta)\cos\theta \, d\theta$$

(by expanding the integrand and retaining terms of order ε). Finally from (4.18)

$$T \approx 2\pi - \frac{\varepsilon}{a_0} \int_0^{2\pi} h(a_0\cos\theta, a_0\sin\theta)\cos\theta \, d\theta \tag{4.21}$$

to the same order, where a_0 is given by (4.19). The *frequency* ω is given to order ε by

$$\omega = 2\pi/T \approx 1 + \frac{\varepsilon}{2\pi a_0} \int_0^{2\pi} h(a_0\cos\theta, a_0\sin\theta)\cos\theta \, d\theta. \tag{4.22}$$

Example 4.4 *Obtain the frequency, to order ε, of the limit cycle of* $\ddot{x} + \varepsilon(x^2 - 1)\dot{x} + x = 0$.

Since $h(x, \dot{x}) = (x^2 - 1)\dot{x}$ and $a_0 = 2$ to order ε (Example 4.3), eqn (4.22) gives

$$\omega = 1 + \frac{\varepsilon}{4\pi} \int_0^{2\pi} (4\cos^2\theta - 1)(+2\sin\theta)\cos\theta \, d\theta = 1 + \text{zero}.$$

Therefore the frequency is 1 with error $O(\varepsilon^2)$.

Example 4.5 *Obtain the relation between frequency and amplitude for the pendulum equation* $\ddot{x} + \sin x = 0$.

Write the equation in the form $\ddot{x} + (\sin x - x) + x = 0$. Then for amplitudes of motion not too large we may write

$$\varepsilon h(x, \dot{x}) = \sin x - x, \quad 0 < \varepsilon \ll 1,$$

since $\sin x - x = -\frac{1}{6}x^3 + \ldots$, and $|\sin x - x| \ll 1$ so long as, say, $|x| < 1$. Equation (4.19) becomes

$$0 = \int_0^{2\pi} [\sin(a_0 \cos\theta) - a_0 \cos\theta] \sin\theta \, d\theta.$$

The integrand is anti-symmetric (odd) about $\theta = \pi$, so this equation is satisfied by all a_0, as expected.

For the frequency, (4.22) becomes

$$\omega = 1 + \frac{1}{2\pi a_0} \int_0^{2\pi} [\sin(a_0 \cos\theta) - a_0 \cos\theta] \cos\theta \, d\theta.$$

For the small amplitudes concerned, we may write $\sin x = x - \frac{1}{6}x^3$. Then

$$\omega = 1 - \frac{a_0^2}{12\pi} \int_0^{2\pi} \cos^4\theta \, d\theta = 1 - \frac{1}{16}a_0^2,$$

which is the required relationship for a_0 not too large. The frequency is seen to decrease with amplitude.

Example 4.6 *Find approximately the amplitude and frequency of the limit cycle for Rayleigh's equation $\ddot{x} + \varepsilon(\frac{1}{3}\dot{x}^3 - \dot{x}) + x = 0$.*

Here, $h(x, \dot{x}) = (\frac{1}{3}\dot{x}^2 - 1)\dot{x}$. Equation (4.19) becomes

$$\int_0^{2\pi} (\frac{1}{3}a_0^2 \sin^2\theta - 1)\sin^2\theta \, d\theta = 0,$$

which can be verified to have solution $a_0 = 2$. The frequency equation (4.15) becomes

$$\omega = 1 + \frac{\varepsilon}{2\pi} \int_0^{2\pi} (\frac{1}{3}a_0^2 \sin^2\theta - 1)\sin\theta \cos\theta \, d\theta.$$

The integral is zero, since the integrand is odd. The frequency is thus unchanged from unity to this order.

4.3 Slowly-varying amplitude: nearly periodic solutions

Equations (4.14), (4.15), and (4.16) apply to any solutions of $\ddot{x} + \varepsilon h(x, \dot{x}) + x = 0$ which are nearly circles in the phase plane, and not just to the periodic solutions considered in Section 4.1. However, they cannot in general be solved exactly, and we shall show how simpler, approximating, equations can be constructed.

We shall assume that a is nearly constant over a 'cycle', whether it is closed or not; specifically, that its rate of change with t or θ is of order ε. Then a is said to be a *slowly-varying amplitude*. The character of such solutions on the phase-plane is of slowly-winding, nearly circular spirals.

The right-hand side of (4.14) contains the function

$h(a \cos \theta, a \sin \theta) \sin \theta$, where a is a function of θ in any motion. We represent $h \sin \theta$ as a Fourier-type series over a 'cycle' $0 \leqslant \theta \leqslant 2\pi$ in the following way. Write

$$h[a(\theta) \cos \theta, a(\theta) \sin \theta] \sin \theta$$

$$= p_0[a(\theta)] + \sum_{n=1}^{\infty} \{p_n[a(\theta)] \cos n\theta + q_n[a(\theta)] \sin n\theta\}, \quad (4.23)$$

where $p_n(a)$ and $q_n(a)$ are obtained by treating a as constant in the usual definitions for Fourier series coefficients:

$$p_0[a(\theta)] = \frac{1}{2\pi} \int_0^{2\pi} h[a(\theta) \cos u, a(\theta) \sin u] \sin u \, du; \quad (4.24)$$

and for $n \geqslant 1$

$$p_n[a(\theta)] = \frac{1}{\pi} \int_0^{2\pi} h[a(\theta) \cos u, a(\theta) \sin u] \cos nu \sin u \, du,$$

$$q_n[a(\theta)] = \frac{1}{\pi} \int_0^{2\pi} h[a(\theta) \cos u, a(\theta) \sin u] \sin nu \cos u \, du.$$

Note that (4.23) is true exactly; the series represents the function on the left although the coefficients are not constant; also the equality is not restricted to the interval $0 \leqslant \theta \leqslant 2\pi$.

Equation (4.16) becomes, for all θ,

$$\frac{da}{d\theta} = \varepsilon p_0(a) + \varepsilon \sum_{n=1}^{\infty} [p_n(a) \cos n\theta + q_n(a) \sin n\theta]. \quad (4.25)$$

The terms under the summation sign can be looked on as nearly harmonic 'inputs' or forcing terms applied to a differential equation $(da/d\theta) - \varepsilon p_0(a) = 0$. We shall show that the *increment* in a over every 'cycle' $\theta_0 \leqslant \theta \leqslant \theta_0 + 2\pi$ is nearly independent of whether these terms are present or not.

By integrating (4.25) over this interval we have, for any θ,

$$a(\theta_0 + 2\pi) - a(\theta_0) = \varepsilon \int_{\theta_0}^{\theta_0 + 2\pi} p_0(a) \, d\theta + \varepsilon \sum_{n=1}^{\infty} \int_{\theta_0}^{\theta_0 + 2\pi} [p_n(a) \cos n\theta + q_n(a) \sin n\theta] \, d\theta.$$

Over such a cycle, a is constant to order ε (eqn (4.18)), so to this order the terms under the summation sign vanish, and

$$a(\theta_0 + 2\pi) - a(\theta_0) = 2\pi \varepsilon p_0(a)$$

where a is a representative value of $a(\theta)$ on the cycle. Therefore the simplified equation

$$\frac{da}{d\theta} = \varepsilon p_0(a) \qquad (4.26)$$

at least gives the increments in a per cycle correctly to order ε, though the actual shape of the spiral curves it predicts will lack contributions from the higher harmonics in (4.25). Note that $p_0(a)$ is approximately the mean value of $h[a(\theta) \cos \theta, a(\theta) \sin \theta] \sin \theta$ over a cycle, so we have replaced the right side of (4.16) by its mean value. This process characterizes a group of procedures called *averaging methods* (Krylov and Bogoliubov, 1949).

The simplified equation corresponding to (4.14) is obtained by writing

$$\frac{da}{dt} = \frac{da}{d\theta} \frac{d\theta}{dt} = -\frac{da}{d\theta} + O(\varepsilon^2)$$

from (4.15), so that, approximately,

$$\frac{da}{dt} = -\varepsilon p_0(a) \qquad (4.27)$$

where p_0 is given by (4.24). Similarly (4.15) becomes

$$\frac{d\theta}{dt} = -1 - \frac{\varepsilon}{a} r_0(a), \qquad (4.28)$$

where

$$r_0(a) = \frac{1}{2\pi} \int_0^{2\pi} h[a \cos u, a \sin u] \cos u \, du, \qquad (4.29)$$

and a is treated as constant for the integration.

Example 4.7 *Find approximate solutions to van der Pol's equation*

$$\ddot{x} + \varepsilon(x^2 - 1)\dot{x} + x = 0,$$

for small positive ε.

From (4.24),

$$p_0(a) = \frac{a}{2\pi} \int_0^{2\pi} (a^2 \cos^2 u - 1) \sin^2 u \, du = \tfrac{1}{2}a(\tfrac{1}{4}a^2 - 1).$$

We have the approximate equation, (4.27), for the radial coordinate:

$$\frac{da}{dt} = -\tfrac{1}{2}\varepsilon a(\tfrac{1}{4}a^2 - 1).$$

The constant solution $a = 2$ corresponds to the limit cycle. The equation separates, giving

$$\int \frac{\mathrm{d}a}{a(a^2-4)} = -\tfrac{1}{8}\varepsilon(t+C)$$

where C is a constant. Thus

$$-\tfrac{1}{4}\log a + \tfrac{1}{8}\log|a^2-4| = -\tfrac{1}{8}\varepsilon(t+C).$$

When $a(0) = a_0$, the solution is

$$a(t) = 2/\{1-(1-4/a_0^2)\mathrm{e}^{-\varepsilon t}\}^{1/2},$$

which tends to 2 as $t \to \infty$. It is easy to verify from eqn (4.29) that $r_0(a) = 0$. Equation (4.28) therefore gives $\theta(t) = -t+\theta_0$ where θ_0 is the initial polar angle. The frequency of the spiral motion is therefore the same as that of the limit cycle to our degree of approximation. Finally, the required approximate solutions are given by

$$x(t) = a(t)\cos\theta(t) = \frac{2\cos(t-\theta_0)}{\{1-(1-4/a_0^2)\mathrm{e}^{-\varepsilon t}\}^{1/2}}.$$

Example 4.8   *Find the approximate phase paths for the equation*

$$\ddot{x} + \varepsilon(|\dot{x}|-1)\dot{x} + x = 0.$$

We have $h(x,y) = (|y|-1)y$. Therefore, from (4.24),

$$p_0(a) = \frac{a}{2\pi}\int_0^{2\pi} (|a\sin\theta|-1)\sin^2\theta\,\mathrm{d}\theta$$

$$= \frac{a}{2\pi}\left(\int_0^{\pi} (a\sin\theta-1)\sin^2\theta\,\mathrm{d}\theta + \int_\pi^{2\pi} (-a\sin\theta-1)\sin^2\theta\,\mathrm{d}\theta\right)$$

$$= \frac{a}{2\pi}2\int_0^{\pi} (a\sin\theta-1)\sin^2\theta\,\mathrm{d}\theta = \frac{1}{\pi}a(\tfrac{4}{3}a-\tfrac{1}{2}\pi-\tfrac{1}{6}).$$

There is a limit cycle when $p_0(a) = 0$; that is, at $a = \tfrac{3}{8}(\pi+\tfrac{1}{3}) = a^*$, say. Equation (4.26) becomes

$$\frac{\mathrm{d}a}{\mathrm{d}\theta} = \frac{4\varepsilon}{3\pi}a(a-a^*).$$

The path satisfying $a = a_0$ at $\theta = \theta_0$ is given in polar coordinates by

$$a(\theta) = a^*\bigg/\left(1-(1-a^*/a_0)\exp\left[\varepsilon\frac{4a^*}{3\pi}(\theta-\theta_0)\right]\right).$$

Alternatively, eqn (4.27) gives a as a function of t.

4.4 Periodic solutions: harmonic balance

One of the easiest methods for estimating periodic solutions in simple cases is illustrated by the following examples.

Example 4.9 *Find an approximation to the amplitude and frequency of the limit cycle of van der Pol's equation* $\ddot{x} + \varepsilon(x^2 - 1)\dot{x} + x = 0$.

Assume an approximate solution $x = a\cos\omega t$. (The prescription of zero value for the phase is not a limitation; in the case of an autonomous equation we are, in effect, simply choosing the time origin so that $\dot{x}(0) = 0$.) Write the equation in the form

$$\ddot{x} + x = -\varepsilon(x^2 - 1)\dot{x}.$$

Upon substituting the assumed form of solution we obtain

$$(-\omega^2 + 1)a\cos\omega t = -\varepsilon(a^2\cos^2\omega t - 1)(-a\omega\sin\omega t)$$
$$= \varepsilon a\omega(\tfrac{1}{4}a^2 - 1)\sin\omega t + \tfrac{1}{4}\varepsilon a^3\omega\sin 3\omega t,$$

after some reduction. The right-hand side is just the Fourier series for $\varepsilon h(x, \dot{x})$: it is easier to get it this way than to work with the equations of the last section. Now, ignore the presence of the term in $\sin 3\omega t$, and match terms in $\cos\omega t$, $\sin\omega t$. We find

$$1 - \omega^2 = 0$$

from the $\cos\omega t$ term, and

$$\tfrac{1}{4}a^2 - 1 = 0$$

from the $\sin\omega t$ term. The second equation gives $a = \pm 2$, and the first $\omega = \pm 1$ (which are equivalent), as in several earlier examples.

Example 4.10 *Obtain the amplitude-frequency relation for the approximate pendulum equation* $\ddot{x} + x - \tfrac{1}{6}x^3 = 0$.

Assume an approximate solution of the form $x = a\cos\omega t$. We obtain, after substituting in the equation:

$$-\omega^2 a\cos\omega t + a\cos\omega t - \tfrac{1}{6}a^3(\tfrac{3}{4}\cos\omega t - \tfrac{1}{4}\cos 3\omega t) = 0.$$

Ignore the higher harmonic and collect the coefficients of $\cos\omega t$. We find that

$$a[(1 - \omega^2) - \tfrac{1}{8}a^2] = 0.$$

Therefore the frequency-amplitude relation becomes

$$\omega = \sqrt{(1 - \tfrac{1}{8}a^2)} \approx 1 - \tfrac{1}{16}a^2$$

for amplitude not too large (compare Example 4.5).

For the general equation

$$\ddot{x} + \varepsilon h(x, \dot{x}) + x = 0, \tag{4.30}$$

suppose that there is a periodic solution close to $a\cos\omega t$ and that h has a

Fourier series

$$h(x, \dot{x}) \approx h(a \cos \omega t, -a\omega \sin \omega t)$$
$$= A_1(a) \cos \omega t + B_1(a) \sin \omega t + \text{higher harmonics}, \quad (4.31)$$

with the constant term absent. Then (4.30) becomes

$$(1 - \omega^2)a \cos \omega t + \varepsilon A_1(a) \cos \omega t + \varepsilon B_1(a) \sin \omega t + \text{higher harmonics} = 0.$$
$$(4.32)$$

This equation can hold for all t only if

$$(1 - \omega^2)a + \varepsilon A_1(a) = 0, \quad B_1(a) = 0, \quad (4.33)$$

which determines a and ω.

It will not be possible in general to ensure that full matching is obtained in (4.32): since the solution is not exactly of the form $a \cos \omega t$ we should ideally have represented it, too, by a Fourier series and matching could then in principle be completed. A way of justifying the non-matching of the higher harmonics is to regard them as additional, neglected input (or forcing terms) to the linear equation $\ddot{x} + x = 0$. The periodic response of this equation to a force $K \cos n\omega t$ is equal to $-K \cos n\omega t/(n^2\omega^2 - 1)$, and is therefore of rapidly diminishing magnitude as n increases.

The method can also be applied to nonlinear equations with a periodic forcing term, and is used extensively in Chapter 7.

4.5   The equivalent linear equation by harmonic balance

The method of harmonic balance can be adapted to construct a linear substitute for the original nonlinear equation; the process can be described as psuedo-linearization. We illustrate the process by an example.

Example 4.11   *Obtain approximate solutions to the equation*

$$\ddot{x} + \varepsilon(x^2 - 1)\dot{x} + x = 0.$$

The phase diagram for this by now familiar equation (see Examples 4.1, 4.4, 4.7, 4.9) consists of a limit cycle and spirals slowly converging on it when ε is small and positive. We shall approximate to $\varepsilon(x^2 - 1)\dot{x} = \varepsilon h(x, \dot{x})$ on one 'cycle' of spiral: this nonlinear term is already small and a suitable approximation retaining the right characteristics should be sufficient. Suppose, then, that we assume (for the purpose of approximating h only) that $x = a \cos \omega t$ and $\dot{x} = -a\omega \sin \omega t$, where a, ω are considered constant on the 'cycle'. Then

$$\varepsilon(x^2-1)\dot{x} = -\varepsilon(a^2\cos^2\omega t-1)a\omega\sin\omega t$$

$$= -\varepsilon a\omega(\tfrac{1}{4}a^2-1)\sin\omega t - \tfrac{1}{4}\varepsilon a^3\sin 3\omega t. \qquad (4.34)$$

(see Example 4.9). As in the harmonic balance method, we neglect the effect of the higher harmonic, and see that since $-a\omega\sin\omega t = \dot{x}$ on the cycle, eqn (4.34) can be written

$$\varepsilon(x^2-1)\dot{x} \approx \varepsilon(\tfrac{1}{4}a^2-1)\dot{x}. \qquad (4.35)$$

We now replace this in the differential equation to give the linear equation

$$\ddot{x} + \varepsilon(\tfrac{1}{4}a^2-1)\dot{x} + x = 0. \qquad (4.36)$$

This equation is peculiar in that it contains a parameter of its own solutions, namely the amplitude a. If $a = 2$ the damping term vanishes and the solutions are periodic of the form $2\cos t$; hence $\omega = 1$. This is an approximation to the limit cycle.

The non-periodic solutions are spirals in the phase plane. Consider the motion for which $x(0) = a_0$, $\dot{x}(0) = 0$: for the next few 'cycles' a_0 will serve as the amplitude used in the above approximation, so (4.40) may be written

$$\ddot{x} + \varepsilon(\tfrac{1}{4}a_0^2-1)\dot{x} + x = 0.$$

With the initial conditions given, the solution is

$$x(t) = a_0\exp\left[-\tfrac{1}{2}\varepsilon(\tfrac{1}{4}a_0^2-1)t\right]\cos\left[1-\tfrac{1}{4}\varepsilon^2(\tfrac{1}{4}a_0^2-1)\right]^{1/2}t.$$

It can be confirmed, by expanding both in powers of εt, that so long as $\varepsilon t \ll 1$ this 'solution' agrees with the approximate solution found in Example 4.7. Damped or negatively damped behaviour of the solutions occurs according to whether $a_0 > 2$ or $a_0 < 2$; that is, according to whether we start inside or outside the limit cycle.

Example 4.12 *Find the frequency–amplitude relation for the equation $\ddot{x} + \mathrm{sgn}\,(x) = 0$ and compare it with the exact result.*

It can be shown that the system $\dot{x} = y$, $\dot{y} = -\mathrm{sgn}(x)$ has a centre at the origin: the behaviour is that of a particle on a spring, with a symmetrical restoring force of constant magnitude.

Suppose the oscillations have the form $a\cos\omega t$ approximately. Then $\mathrm{sgn}(x) = \mathrm{sgn}(a\cos\omega t)$. This has the shape shown in Fig. 4.3(b). The period is $2\pi/\omega$ and we shall approximate to $\mathrm{sgn}(a\cos\omega t)$ by the first term in its Fourier series on the interval $(0, 2\pi/\omega)$:

$$\mathrm{sgn}\,(a\cos\omega t) = A_1(a)\cos\omega t + \text{higher harmonics},$$

(there is no sine term, since the function is even), where

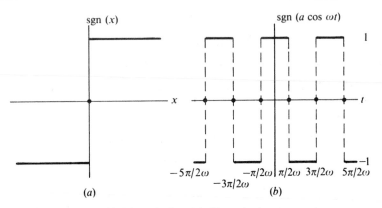

FIG. 4.3. (a) graph of sgn(x); (b) graph of sgn(a cos ωt)

$$A_1(a) = \frac{\omega}{\pi} \int_0^{2\pi/\omega} \text{sgn}\,(a\cos\omega t)\cos\omega t\,dt$$

$$= \frac{1}{\pi} \int_0^{2\pi} \text{sgn}\,(a\cos u)\cos u\,du$$

$$= \frac{1}{\pi}\left(\int_0^{\pi/2}\cos u\,du - \int_{\pi/2}^{3\pi/2}\cos u\,du + \int_{3\pi/2}^{2\pi}\cos u\,du\right) = \frac{4}{\pi}.$$

Therefore $\text{sgn}\,(x)$ is replaced by $(4/\pi)\cos\omega t$, which in turn is replaced by $(4/\pi)x/a$. The approximate equation is then

$$\ddot{x} + \frac{4}{\pi a}x = 0.$$

The solution, having amplitude a, of the form $a\cos\omega t$ is

$$x(t) = a\cos\left(\frac{4}{\pi a}\right)^{1/2} t.$$

Therefore

$$\omega = 2/\sqrt{(\pi a)}$$

and the period T is given by

$$T = 2\pi/\omega = \pi^{3/2}a^{1/2} \approx 5\cdot568\sqrt{a}.$$

The exact period can be found as follows. When $x > 0$, $\ddot{x} = -1$. The solution for $x > 0$ is therefore

$$x(t) = -\tfrac{1}{2}t^2 + \alpha t + \beta.$$

The solution for which $x(0) = 0$ and $\dot{x}(\tfrac{1}{4}T_e) = 0$, where T_e is the exact period, is

$-\frac{1}{2}t^2 + \frac{1}{4}T_e t$. Further, $x = a$ when $t = \frac{1}{4}T_e$, so that

$$T_e = 4\sqrt{(2a)} \approx 5 \cdot 657\sqrt{a},$$

which compares very favourably with the above estimate.

Example 4.13 *Obtain the period–amplitude relation for the Poisson–Boltzmann-type equation* $\ddot{x} + (e^x - e^{-x}) = 0$.

When $x = a\cos\omega t$,

$$e^x - e^{-x} = e^{a\cos\omega t} - e^{-a\cos\omega t} = A_1(a)\cos\omega t + \text{higher harmonics},$$

(there is no $\sin\omega t$ term since the function is even), where

$$A_1(a) = \frac{\omega}{\pi} \int_0^{2\pi/\omega} (e^{a\cos\omega t} - e^{-a\cos\omega t})\cos\omega t \, dt$$

$$= \frac{1}{\pi} \int_0^{2\pi} (e^{a\cos u} - e^{-a\cos u})\cos u \, du.$$

These integrals are expressible in terms of the modified Bessel function I_1 which has the integral representation (Abramowitz and Stegun, 1965)

$$I_1(z) = \frac{1}{\pi} \int_0^{\pi} e^{z\cos u}\cos u \, du,$$

and we find that

$$A_1(a) = 4I_1(a).$$

Neglecting the higher harmonics, we write the equation as

$$\ddot{x} + (4I_1(a)/a)x = 0.$$

The frequency ω is therefore given by

$$\omega = 2\sqrt{(I_1(a)/a)}$$

and the period T by

$$T = \pi\sqrt{(a/I_1(a))}.$$

Note that a is unrestricted in these equations—it can be verified that $x = 0$ is a centre for the original equation (see Exercise 21, Chapter 1).

4.6 Accuracy of a period estimate

Quantitative justification for the averaging methods is difficult to obtain, but the following treatment of a conservative system is suggestive. The method of Section 4.5 estimates the period of the conservative equation

$$\ddot{x} + g(x) = 0$$

(assuming that g is an odd function with $g(x) > 0$ for $x > 0$), to be given by

$$T \approx \frac{2\pi^{3/2}a^{1/2}}{\sqrt{\left(\int_0^{2\pi} g(a\cos v)\cos v\,dv\right)}} \tag{4.37}$$

where $a > 0$ is the amplitude. For example, when

$$g(x) = x - \tfrac{1}{6}x^3 \tag{4.38}$$

(as in Example 4.10) the period predicted is

$$T \approx 2\pi/\sqrt{(1 - \tfrac{1}{8}a^2)}. \tag{4.39}$$

For oscillations of amplitude a the phase paths are given by

$$\tfrac{1}{2}y^2 - \int_x^a g(u)\,du = 0,$$

so the correct period T^* is given by

$$T^* = 2\sqrt{2}\int_0^a \frac{dx}{\sqrt{\left(\int_x^a g(u)\,du\right)}} \tag{4.40}$$

(see eqn 1.11). Consider the particular case of eqn (4.38): we shall try to compare the exact period T^* with the harmonic balance estimate T. However, (4.40) cannot be evaluated in terms of elementary functions so we shall have to estimate T^* too: we shall be able to show that, at any rate, the harmonic balance estimate is as good as we can extract by simple methods from the exact expression (4.40).

Suppose we can find constants α, β, γ and δ (which in general will depend on a), such that

$$\alpha + \beta x \leqslant g(x) \leqslant \gamma + \delta x. \tag{4.41}$$

Then, from (4.40), provided $\alpha > -\tfrac{1}{2}\beta a$ so that the second integral has a real integrand,

$$2\sqrt{2}\int_0^a \frac{dx}{\sqrt{\{\gamma(a-x) + \tfrac{1}{2}\delta(a^2 - x^2)\}}} \leqslant T^*$$

$$\leqslant 2\sqrt{2}\int_0^a \frac{dx}{\sqrt{\{\alpha(a-x) + \tfrac{1}{2}\beta(a^2 - x^2)\}}}$$

or

$$\frac{4}{\sqrt{\delta}}\cos^{-1}\left(\frac{\gamma}{\gamma+\delta a}\right) \leqslant T^* \leqslant \frac{4}{\sqrt{\beta}}\cos^{-1}\left(\frac{\alpha}{\alpha+\beta a}\right). \quad (4.42)$$

Now consider the graph of $y = g(x) = x - \frac{1}{6}x^3$, our special case (4.38). Choose as the lower bounding line in (4.41) the chord joining the origin to the point $(a, g(a))$, and as the upper bounding line the tangent line at some point $(b, g(b))$ on the curve, $y = \frac{1}{3}b^3 + (1 - \frac{1}{2}b^2)x$. Then

$$\alpha = 0, \quad \beta = 1 - \frac{1}{6}a^2, \quad \gamma = \frac{1}{3}b^3, \quad \delta = 1 - \frac{1}{2}b^2.$$

From (4.42) we obtain as the upper bound on T^*:

$$T^* \leqslant 2\pi/\sqrt{(1 - \frac{1}{6}a^2)}.$$

Ideally the lower bound should be maximized with respect to b, but this is difficult. Instead we take as an example the special case $a = \sqrt{2}$ (the maximum on the graph $y = x - \frac{1}{6}x^3$), and $b = 1$. (4.42) then gives

$$7 \cdot 041 \leqslant T^* \leqslant 7 \cdot 695.$$

For the same amplitude, $a = \sqrt{2}$, the harmonic balance estimate (4.39) gives $T \approx 7 \cdot 255$.

Exercises

1. By transforming to polar coordinates, find the limit cycles of the systems
   (i) $\dot{x} = y + x(1 - x^2 - y^2), \quad \dot{y} = -x + y(1 - x^2 - y^2)$;
   (ii) $\dot{x} = (x^2 + y^2 - 1)x - y\sqrt{(x^2 + y^2)}, \quad \dot{y} = (x^2 + y^2 - 1)y + x\sqrt{(x^2 + y^2)}$;
   and investigate their stability.

2. Consider the system

$$\dot{x} = y + xf(r^2), \quad \dot{y} = -x + yf(r^2),$$

where $r^2 = x^2 + y^2$ and $f(u)$ is continuous on $u \geqslant 0$. Show that r satisfies

$$\frac{d(r^2)}{dt} = 2r^2 f(r^2).$$

If f has n zeros, at $r = r_k$, $k = 1, 2, \ldots, n$, how many periodic solutions has the system? Discuss their stability in terms of the sign of $f'(r_k^2)$.

3. Apply the energy balance method of Section 4.1 to each of the following equations, and find the amplitude and stability of any limit cycles:
   (i) $\ddot{x} + \varepsilon(x^2 + \dot{x}^2 - 1)\dot{x} + x = 0$;
   (ii) $\ddot{x} + \varepsilon(\frac{1}{3}\dot{x}^3 - \dot{x}) + x = 0$;
   (iii) $\ddot{x} + \varepsilon(x^4 - 1)\dot{x} + x = 0$;

(iv) $\ddot{x} + \varepsilon \sin(x^2 + \dot{x}^2) \operatorname{sgn}(\dot{x}) + x = 0$;

(v) $\ddot{x} + \varepsilon(|x| - 1)\dot{x} + x = 0$;

(vi) $\ddot{x} + \varepsilon(\dot{x} - 3)(\dot{x} + 1)\dot{x} + x = 0$;

(vii) $\ddot{x} + \varepsilon(x - 3)(x + 1)\dot{x} + x = 0$.

4. For the equation $\ddot{x} + \varepsilon(x^2 + \dot{x}^2 - 4)\dot{x} + x = 0$, the solution $x = 2\cos t$ is a limit cycle. Test its stability, using the method of Section 4.1, and obtain an approximation to the paths close to the limit cycle by the method of slowly-varying amplitude. Show that the period of the limit cycle is equal to 2π, with error of order ε^2.

5. For the equation $\ddot{x} + \varepsilon(|x| - 1)\dot{x} + x = 0$, find approximately the amplitude of the limit cycle and its period, and the polar equations for the phase paths near the limit cycle.

6. Repeat Exercise 5 with Rayleigh's equation, $\ddot{x} + \varepsilon(\tfrac{1}{3}\dot{x}^3 - \dot{x}) + x = 0$.

7. Find approximately the radius of the limit cycle, and its period, for the equation

$$\ddot{x} + \varepsilon(x^2 - 1)\dot{x} + x - \varepsilon x^3 = 0.$$

8. Show that the frequency–amplitude relation for the pendulum equation, $\ddot{x} + \sin x = 0$, is $\omega^2 = 2J_1(a)/a$, using the methods of Section 4.4 or 4.5. (J_1 is the Bessel function of order 1, with representations

$$J_1(a) = \frac{2}{\pi} \int_0^{\frac{1}{2}\pi} \sin(a\cos u)\cos u \, du = \sum_{n=0}^{\infty} \frac{(-1)^n (\tfrac{1}{2}a)^{2n+1}}{n!(n+1)!}.)$$

Show that, for small amplitudes,

$$\omega = 1 - \tfrac{1}{16}a^2.$$

9. In the equation

$$\ddot{x} + \varepsilon h(x, \dot{x}) + g(x) = 0$$

suppose that $g(0) = 0$, and that in some interval $|x| < \delta$, g is continuous and strictly increasing. Show that the origin for the equation $\ddot{x} + g(x) = 0$ is a centre. Let $\zeta(t, a)$ represent its periodic solutions near the origin, where a is a parameter which distinguishes the solutions, say the amplitude. Also, let $T(a)$ be the corresponding period.

By using an energy balance argument show that the periodic solutions of the original equation satisfy

$$\int_0^{T(a)} h(\zeta, \dot{\zeta})\dot{\zeta} \, dt = 0.$$

Apply this equation to obtain the amplitude of the limit cycle of the equation

$$\ddot{x} + \varepsilon(x^2 - 1)\dot{x} + \operatorname{sgn} x = 0.$$

(Confirm that the function ζ defined by

$$\zeta(t, a) = \begin{cases} \sqrt{(2a)}t + \tfrac{1}{2}t^2, & -2\sqrt{(2a)} \leqslant t \leqslant 0; \\ \sqrt{(2a)}t - \tfrac{1}{2}t^2, & 0 \leqslant t \leqslant 2\sqrt{(2a)}; \end{cases}$$

$$\zeta(t + 4\sqrt{(2a)}, a) = \zeta(t, a), \quad -\infty < t < \infty$$

is a periodic solution of $\ddot{x} + \operatorname{sgn} x = 0$ with amplitude a.)

10. For the following equations, show that, for small ε, the amplitude $a(t)$ satisfies approximately the equation given.
   (i) $\ddot{x} + \varepsilon(x^4 - 2)\dot{x} + x = 0$, $16\dot{a} = -\varepsilon a(a^4 - 16)$;
   (ii) $\ddot{x} + \varepsilon \sin(x^2 + \dot{x}^2)\operatorname{sgn}(\dot{x}) + x = 0$, $\pi\dot{a} = -\varepsilon 2a \sin(a^2)$;
   (iii) $\ddot{x} + \varepsilon(x^2 - 1)\dot{x}^3 + x = 0$, $16\dot{a} = -\varepsilon a^3(a^2 - 6)$.

11. Verify that the equation

$$\ddot{x} + \varepsilon h(x^2 + \dot{x}^2 - 1)\dot{x} + x = 0$$

where f is continuous, $h(u) < 0$ for $u < 0$, $h(0) = 0$, and $h(u) > 0$ for $u > 0$, has the periodic solutions $x = \cos(t + \alpha)$ for any α. Using the method of slowly varying amplitude show that this solution is a stable limit cycle when $\varepsilon > 0$.

12. Find, by the method of Section 4.5 the equivalent linear equation for

$$\ddot{x} + \varepsilon(x^2 + \dot{x}^2 - 1)\dot{x} + x = 0.$$

Show that it gives the limit cycle exactly. Obtain from the linear equation the equations of the nearby spiral paths, and compare these (to order ε) with those predicted by the method of slowly-varying amplitude.

13. Use the method of equivalent linearization to find the amplitude and frequency of the limit cycle of the equation

$$\ddot{x} + \varepsilon(|x| - 1)|\dot{x}|\dot{x} + x + \varepsilon x^3 = 0.$$

14. The equation $\ddot{x} + x^3 = 0$ has a centre at the origin of the phase plane.
(i) Substitute $x = a \cos \omega t$ to find by the harmonic balance method the frequency–amplitude relation $\omega = \sqrt{3a/2}$.
(ii) Construct, by the method of equivalent linearization, the associated linear equation, and show how the processes (i) and (ii) are equivalent.
(iii) Compare the linearization obtained in (ii) with that obtained by fitting x^3 on $[-a, a]$ by a linear function, $x^3 \approx \beta x$, by the method of least squares, and compare the amplitude–frequency relation. Explain why such a linearization is not likely to be so reliable as that obtained in (ii).

15. The displacement x of a relativistic oscillator satisfies

$$m_0\ddot{x} + k(1 - (\dot{x}/c)^2)^{3/2}x = 0.$$

Show that the equation becomes $\ddot{x} + (\alpha/a)x = 0$ when linearized with respect to the approximate solution $x = a \cos \omega t$ by the method of equivalent linearization, where

$$\alpha = \frac{1}{\pi} \int_0^{2\pi} \frac{ka}{m_0} \cos^2\theta \left(1 - \frac{a^2\omega^2}{c^2} \sin^2\theta\right)^{3/2} d\theta.$$

Confirm that, when $a^2\omega^2/c^2$ is small, the period of the oscillations is given approximately by

$$2\pi \sqrt{\left(\frac{m_0}{k}\right)} \left(1 + \frac{3a^2 k}{16 m_0 c^2}\right).$$

16. Show that the phase paths of the equation

$$\ddot{x} + (x^2 + \dot{x}^2)x = 0$$

are given by

$$e^{-x^2}(y^2 + x^2 - 1) = \text{constant}.$$

Show that the surface $e^{-x^2}(y^2 + x^2 - 1) = z$ has a maximum at the origin, and deduce that the origin is a centre.

Use the method of harmonic balance to obtain the frequency–amplitude relation $\omega^2 = 3a^2/(4 - a^2)$ for $a < 2$, assuming solutions of the approximate form $a \cos \omega t$. Verify that $\cos t$ is an exact solution, and that $\omega = 1$, $a = 1$ is predicted by harmonic balance.

Plot some exact phase paths to indicate where the harmonic balance method is likely to be unreliable.

17. Show, by the method of harmonic balance, that the frequency–amplitude relation for the periodic solutions of the approximate form $a \cos \omega t$, for

$$\ddot{x} - x + \alpha x^3 = 0, \quad \alpha > 0,$$

is $\omega^2 = 3\alpha a^2 - 1$.

By analysing the phase diagram, explain the lower bound $2/\sqrt{(3\alpha)}$ for the amplitude of periodic motion. Find the equation of the phase paths, and compare where the separatrix cuts the x-axis, with the amplitude $2/\sqrt{(3\alpha)}$.

18. Apply the method of harmonic balance to the equation

$$\ddot{x} + x - \alpha x^2 = 0, \quad \alpha > 0,$$

using the approximate form of solution $x = c + a \cos \omega t$ to show that, for α small,

$$\omega^2 = 1 - 2\alpha c, \quad c = \frac{1}{2\alpha} - \frac{1}{2\alpha}\sqrt{(1 - 2\alpha^2 a^2)}.$$

Deduce the frequency–amplitude relation

$$\omega = (1 - 2\alpha^2 a^2)^{1/4}, \quad a < 1/\sqrt{2}\alpha.$$

Explain, in general terms, why an upper bound on the amplitude is to be expected.

19. Apply the method of harmonic balance to the equation

$$\ddot{x} - x + x^3 = 0$$

in the neighbourhood of the centre at $x = 1$, using the approximate form of solution $x = 1 + c + a \cos \omega t$. Deduce that the mean displacement, frequency, and amplitude are related by

$$\omega^2 = 3c^2 + 6c + 2 + \tfrac{3}{4}a^2, \quad 2c^3 + 6c^2 + c(4 + 3a^2) + 3a^2 = 0.$$

20. Consider the van der Pol equation with nonlinear restoring force:

$$\ddot{x} + \varepsilon(x^2 - 1)\dot{x} + x - \alpha x^2 = 0,$$

where ε and α are small. By assuming solutions approximately of the form $x = c + a \cos \omega t$, show that the mean displacement, frequency, and amplitude are related by

$$c \approx 4\alpha, \quad \omega \approx 1 - 4\alpha^2, \quad a \approx 2(1 - 8\alpha^2).$$

21. Suppose that the nonlinear system

$$\dot{\mathbf{x}} = \mathbf{p}(\mathbf{x}), \quad \mathbf{x} = \begin{pmatrix} x \\ y \end{pmatrix},$$

has an isolated equilibrium point at $\mathbf{x} = \mathbf{0}$, and that solutions exist which are approximately of the form

$$\tilde{\mathbf{x}} = B \begin{pmatrix} \cos \omega t \\ \sin \omega t \end{pmatrix}, \quad B = \begin{pmatrix} a & b \\ c & d \end{pmatrix}.$$

Adapt the method of equivalent linearization to this problem by approximating $\mathbf{p}(\tilde{\mathbf{x}})$ by its first harmonic terms:

$$\mathbf{p}\{\tilde{\mathbf{x}}(t)\} = C \begin{pmatrix} \cos \omega t \\ \sin \omega t \end{pmatrix},$$

where C is a matrix of the Fourier coefficients. It is assumed that

$$\int_0^{2\pi/\omega} \mathbf{p}\{\tilde{\mathbf{x}}(t)\}\, dt = \mathbf{0}.$$

Substitute in the system to show that

$$BU = C, \quad U = \begin{pmatrix} 0 & -\omega \\ \omega & 0 \end{pmatrix}.$$

Deduce that the equivalent linear system is

$$\dot{\tilde{\mathbf{x}}} = BUB^{-1}\tilde{\mathbf{x}} = CUC^{-1}\tilde{\mathbf{x}},$$

when B and C are nonsingular.

22. Use the method of Exercise 21 to construct a linear system equivalent to the van der Pol equation

$$\dot{x} = y, \quad \dot{y} = -x - \varepsilon(x^2 - 1)y.$$

23. Apply the method of Exercise 21 to construct a linear system equivalent to

$$\begin{pmatrix} \dot{x} \\ \dot{y} \end{pmatrix} = \begin{pmatrix} \varepsilon & 1 \\ -1 & \varepsilon \end{pmatrix} \begin{pmatrix} x \\ y \end{pmatrix} + \begin{pmatrix} 0 \\ -\varepsilon x^2 y \end{pmatrix},$$

and show that if the amplitudes of x and y on the limit cycle are α and β respectively, then for small ε, $\alpha \approx \sqrt{8}, \beta \approx \alpha$.

24. Apply the method of Exercise 21 to the predator–prey equation (see Section 2.2)

$$\dot{x} = x - xy, \quad \dot{y} = -y + xy,$$

in the neighbourhood of the equilibrium point $(1, 1)$, by using the displaced solutions

$$x = 1 + \lambda + a\cos\omega t + b\sin\omega t, \quad y = 1 + \mu + c\cos\omega t + d\sin\omega t.$$

Show that, for small amplitudes,
  (i) the amplitudes of x and y are equal;
  (ii) $\lambda = \mu = \frac{1}{2}(-1 + \sqrt{(1 + 2\alpha^2)})$ where α is the amplitude;
  (iii) $\omega = (1 + 2\alpha^2)^{1/4}$.

25. Show that the amplitude–frequency relation for oscillations of the equation

$$\ddot{x} = x^2 - x^3$$

in the neighbourhood of $x = 1$ is $\omega^2 = 1 - \frac{13}{4}a^2$. (Take as the approximate solutions the form $x = c + a\cos\omega t$.)

26. Apply the method of Section 4.6 to the pendulum equation $\ddot{x} + \sin x = 0$ to show that for $0 \leqslant b \leqslant a \leqslant \frac{1}{2}\pi$:

$$\frac{4}{\sqrt{(\cos b)}} \cos^{-1}\left(\frac{\tan b - b}{\tan b - b + a}\right) \leqslant T^* \leqslant 2\pi\sqrt{(a/\sin a)}$$

where a is the amplitude and T^* is the period of oscillation with amplitude a. Confirm that when $a = \frac{1}{4}\pi$ and $b = 0.7$, this equation predicts $6.52 \leqslant T^* \leqslant 6.6$ and that the period estimate given by harmonic balance (Exercise 8) is $T \approx 6.56$. (Take $J_1(\frac{1}{4}\pi) = 0.36$.)

27. Use the method of Section 4.2 to obtain approximate solutions of the equation

$$\ddot{x} + \varepsilon\dot{x}^3 + x = 0, \quad |\varepsilon| \ll 1.$$

28. Suppose that the equation $\ddot{x} + f(x)\dot{x} + g(x) = 0$ has a periodic solution. Represent the equation in the phase plane given by

$$\dot{x} = y - F(x), \quad \dot{y} = -g(x)$$

where

$$F(x) = \int_0^x f(u)\,du$$

(the Liénard plane). Let

$$v(x, y) = \tfrac{1}{2}y^2 + \int_0^x g(u)\,du,$$

and by considering du/dt on the closed path $\mathscr{C}$ show that

$$\int_{\mathscr{C}} F(x)\,dy = 0$$

(Liénard's criterion).

Deduce that on the assumption that van der Pol's equation $\ddot{x} + \varepsilon(x^2 - 1)\dot{x} + x = 0$ has a periodic solution approximately of the form $x = A \cos t$, then $A \approx 2$.

5 Perturbation methods

IN THE PREVIOUS chapter approximations to periodic solutions of differential equations were achieved by in effect fitting a harmonic oscillation to the required solution. The methods of this chapter are for use when the nonlinearity is small, and generate—in principle—complete representations of periodic solutions of certain types of both autonomous and forced equations as power series in a small parameter. All the essentials are present in treating the pendulum equation, upon which much of this chapter is based.

5.1. Outline of the direct method

The equation for forced oscillations of a pendulum is

$$\ddot{x} + \omega_0^2 \sin x = F \cos \omega t, \qquad (5.1)$$

where x is the inclination, $\omega_0 > 0$ is the natural circular frequency for small-amplitude swings, and $F \cos \omega t$ is the forcing term. We can suppose without loss of generality that $\omega > 0$, also that $F > 0$, since $F < 0$ implies a phase difference which can be eliminated by change of time origin and an appropriate modification of initial conditions.

We put $\sin x \approx x - \frac{1}{6} x^3$ to allow for moderately large swings, which is good to 1% for $|x| < 1$ radian. Equation (5.1) then becomes

$$\ddot{x} + \omega_0^2 x - \frac{1}{6} \omega_0^2 x^3 = F \cos \omega t, \qquad (5.2)$$

which is nonlinear. We shall see that the effect of the nonlinearity is not confined to occasions when large swings actually occur.

Standardize the form of (5.2) by writing

$$\tau = \omega t, \quad \Omega^2 = \omega_0^2 / \omega^2 \quad (\Omega > 0), \quad \Gamma = F / \omega^2. \qquad (5.3)$$

Then

$$x'' + \Omega^2 x - \frac{1}{6} \Omega^2 x^3 = \Gamma \cos \tau, \qquad (5.4)$$

where the dashes represent differentiation with respect to τ. The argument depends on the nonlinear terms being in some sense small (which does not mean that the resulting phenomena are not spectacular): here we assume that $\frac{1}{6} \Omega^2$ is 'small' and write

$$\tfrac{1}{6} \Omega^2 = \varepsilon_0. \qquad (5.5)$$

The equation becomes

$$x'' + \Omega^2 x - \varepsilon_0 x^3 = \Gamma \cos \tau. \tag{5.6}$$

Instead of taking (5.6) as it stands, with ω, Γ, ε_0 as constants, we consider the *family of differential equations*

$$x'' + \Omega^2 x - \varepsilon x^3 = \Gamma \cos \tau, \tag{5.7}$$

where ε is a parameter occupying an interval I_ε which includes $\varepsilon = 0$. When $\varepsilon = \varepsilon_0$ we recover (5.6), *and when $\varepsilon = 0$ we obtain the* linearized equation *corresponding to the family* (5.7):

$$x'' + \Omega^2 x = \Gamma \cos \tau. \tag{5.8}$$

The solutions of (5.7) are now to be thought of as functions of both ε and τ and we will write $x(\varepsilon, \tau)$.

The most elementary version of the *perturbation method* is to attempt a representation of the solutions of (5.7) in the form of a power series in ε:

$$x(\varepsilon, \tau) = x_0(\tau) + \varepsilon x_1(\tau) + \varepsilon^2 x_2(\tau) + \dots, \tag{5.9}$$

whose coefficients $x_i(\tau)$ are functions only of τ. To form equations for $x_i(\tau)$, $i = 0, 1, 2, \dots$, substitute the series (5.9) into eqn (5.7):

$$(x_0'' + \varepsilon x_1'' + \dots) + \Omega^2(x_0 + \varepsilon x_1 + \dots) - \varepsilon(x_0 + \varepsilon x_1 + \dots)^3 = \Gamma \cos \tau.$$

Since this is assumed to hold for *every member of the family* (5.7), that is for *every* ε on I_ε the coefficients of powers of ε must balance and we obtain

$$x_0'' + \Omega^2 x_0 = \Gamma \cos \tau, \tag{5.10a}$$

$$x_1'' + \Omega^2 x_1 = x_0^3, \tag{5.10b}$$

$$x_2'' + \Omega^2 x_2 = 3x_0^2 x_1, \tag{5.10c}$$

and so on.

We shall be concerned only with *periodic solutions having the period, 2π, of the forcing term*. (There are periodic solutions having other periods: see Chapter 7; also, by a modification of the present procedure as in Section 5.3, we shall find yet more solutions having period 2π.) Then, for all ε on I_ε and for all τ,

$$x(\varepsilon, \tau + 2\pi) = x(\varepsilon, \tau) \tag{5.11}$$

which, by (5.9), implies that for all τ

$$x_i(\tau + 2\pi) = x_i(\tau), \quad i = 0, 1, 2, \dots. \tag{5.12}$$

Equations (5.10) and (5.12) provide the solutions required. The details will be worked out in Section 5.2; for the present, note that (5.10a) is the same as the 'linearized equation' (5.8): necessarily so, since putting $\varepsilon_0 = 0$ in (5.6) implies putting $\varepsilon = 0$ in (5.9). The major term in (5.9) is therefore a periodic solution of the linearized equation (5.8). It is therefore clear that this process restricts us to finding the solutions of the nonlinear equation which are close to (or branch from, or bifurcate from) the solution of the linearized equation. The method will not expose any other periodic solutions. The zero-order solution $x_0(\tau)$ is known as the *generating solution for the family of equations* (5.7).

5.2. Forced oscillation far from resonance

Suppose that

$$\Omega \neq \text{an odd integer.} \qquad (5.13)$$

We now solve (5.10) subject to the periodicity condition (5.12). The solutions of (5.10a) are

$$a_0 \cos \Omega\tau + b_0 \sin \Omega\tau + \frac{\Gamma}{\Omega^2 - 1} \cos \tau, \qquad (5.14)$$

where a_0, b_0 are constants. If Ω is not an integer *the only solution having period 2π is obtained by putting $a_0 = b_0 = 0$ in* (5.14); that is

$$x_0(\tau) = \frac{\Gamma}{\Omega^2 - 1} \cos \tau. \qquad (5.15)$$

(For Ω even the following argument must be modified.) Equation (5.10b) then becomes

$$x_1'' + \Omega^2 x_1 = \left(\frac{\Gamma}{\Omega^2 - 1}\right)^3 \cos^3 \tau = \frac{\Gamma^3}{(\Omega^2 - 1)^3} \left(\tfrac{3}{4}\cos \tau + \tfrac{1}{4}\cos 3\tau\right).$$

The only solution with period 2π is given by

$$x_1(\tau) = \frac{3}{4} \frac{\Gamma^3}{(\Omega^2 - 1)^4} \cos \tau + \frac{1}{4} \frac{\Gamma^3}{(\Omega^2 - 1)^3 (\Omega^2 - 9)} \cos 3\tau, \qquad (5.16)$$

for reasons similar to those above.

The first two terms of the expansion (5.9) give

$$x(\varepsilon, \tau) = \frac{\Gamma}{\Omega^2 - 1} \cos \tau + \varepsilon \left(\frac{3}{4} \frac{\Gamma^3}{(\Omega^2 - 1)^4} \cos \tau \right.$$
$$\left. + \frac{1}{4} \frac{\Gamma^3}{(\Omega^2 - 1)^3 (\Omega^2 - 9)} \cos 3\tau\right) + O(\varepsilon^2). \qquad (5.17)$$

The series continues with terms of order ε^2, ε^3 and involves the harmonics $\cos 5\tau$, $\cos 7\tau$ and so on. For the pendulum, $\varepsilon = \varepsilon_0 = \frac{1}{6}\Omega^2$, by (5.5).

The method obviously fails if Ω^2 takes one of the values $1, 9, 25, \ldots$, since certain terms would then be infinite, and this possibility is averted by condition (5.13). However, the series would not converge well if Ω^2 were even close to one of these values, and so a few terms of the series would describe $x(\varepsilon, \tau)$ only poorly. Such values of Ω correspond to conditions of *near-resonance*. $\Omega \approx 1$ is such a case; not surprisingly, since it is a value close to resonance for the linearized equation. The other critical values correspond to resonances peculiar to the nonlinear equation caused by the higher harmonics present in $x^3(\varepsilon, \tau)$, which can be regarded as 'feeding-back' into the linear equation like forcing terms. This idea is explored in Chapter 7.

Example 5.1 *Obtain an approximation to the forced response, of period 2π, for the equation* $x'' + \frac{1}{4}x + 0 \cdot 1x^3 = \cos \tau$.

Consider the family $x'' + \frac{1}{4}x + \varepsilon x^3 = \cos \tau$. Assume that $x(\varepsilon, \tau) = x_0(\tau) + \varepsilon x_1(\tau) + \ldots$. The periodicity requires $x_0(\tau), x_1(\tau), \ldots$, to have period 2π. The equations for x_0, x_1 are

$$x_0'' + \tfrac{1}{4}x_0 = \cos \tau,$$

$$x_1'' + \tfrac{1}{4}x_1 = -x_0^3.$$

The only periodic solution of the first equation is

$$x_0(\tau) = -\tfrac{4}{3}\cos \tau,$$

and the second becomes

$$x_1'' + \tfrac{1}{4}x_1 = \tfrac{16}{9}\cos \tau + \tfrac{16}{27}\cos 3\tau.$$

The periodic solution is

$$x_1(\tau) = -\tfrac{64}{27}\cos \tau - \tfrac{64}{945}\cos 3\tau.$$

Therefore

$$x(\varepsilon, \tau) = -\tfrac{4}{3}\cos \tau - \varepsilon(\tfrac{64}{27}\cos \tau + \tfrac{64}{945}\cos 3\tau) + O(\varepsilon^2).$$

With $\varepsilon = 0 \cdot 1$,

$$x(\varepsilon, \tau) \approx -1 \cdot 570 \cos \tau - 0 \cdot 006 \cos 3\tau.$$

5.3. Forced oscillations near resonance with weak excitation

Not only does the method of the previous sections fail to give a good representation near critical values of Ω, but it does not reveal all the solutions having the period 2π of the forcing term. The only solution it can detect is the one which bifurcates from the one-and-only solution having period 2π of the linear eqn (5.8). More solutions can be found by, so to speak, offering the nonlinear equation a greater range of choice of possible generating solutions by modifying the structure so that a different linearized equation is obtained.

We will generalize the pendulum equation to include small damping:

$$\ddot{x} + k\dot{x} + \omega_0^2 x - \tfrac{1}{6}\omega_0^2 x^3 = F \cos \omega t. \tag{5.18}$$

Corresponding to eqn (5.4) we have

$$x'' + Kx' + \Omega^2 x - \varepsilon_0 x^3 = \Gamma \cos \tau$$

where

$$\tau = \omega t, \quad \Omega^2 = \omega_0^2/\omega^2, \quad \varepsilon_0 = \tfrac{1}{6}\Omega^2, \quad K = k/\omega, \quad \Gamma = F/\omega^2 \tag{5.19}$$

Assume that Γ is small (weak excitation), and K is small (small damping) and therefore put

$$\Gamma = \varepsilon_0 \gamma, \quad K = \varepsilon_0 \kappa, \quad (\gamma, \kappa > 0). \tag{5.20}$$

Suppose also that Ω is close to one of the critical values $1, 3, 5, \ldots$ say that $\Omega \approx 1$ (near to resonance), so we write

$$\Omega^2 = 1 + \varepsilon_0 \beta. \tag{5.21}$$

Equation (5.18) becomes

$$x'' + x = \varepsilon_0(\gamma \cos \tau - \kappa x' - \beta x + x^3). \tag{5.22}$$

Now consider the family of equations

$$x'' + x = \varepsilon(\gamma \cos \tau - \kappa x' - \beta x + x^3), \tag{5.23}$$

with parameter ε, in which γ, κ, β retain the constant values given by (5.20) and (5.21). When $\varepsilon = \varepsilon_0$ we return to (5.22). When $\varepsilon = 0$ we obtain the new linearized equation $x'' + x = 0$. *This has infinitely many solutions with period 2π, offering a wide choice of generating solutions* (compare (5.8), in which only one presents itself).

Now assume that $x(\varepsilon, \tau)$ may be expanded in the form

$$x(\varepsilon, \tau) = x_0(\tau) + \varepsilon x_1(\tau) + \varepsilon^2 x_2(\tau) + \ldots, \tag{5.24}$$

where (by the same argument as led up to (5.12)), for all τ

$$x_i(\tau + 2\pi) = x_i(\tau), \quad i = 0, 1, 2, \ldots. \tag{5.25}$$

Substitute (5.24) into (5.23). By the argument leading to (5.10) we have

$$x_0'' + x_0 = 0, \tag{5.26a}$$

$$x_1'' + x_1 = \gamma \cos \tau - \kappa x_0' - \beta x_0 + x_0^3, \tag{5.26b}$$

$$x_2'' + x_2 = -x x_1' - \beta x_1 + 3x_0^2 x_1, \tag{5.26c}$$

and so on.

The solution of (5.26a) is

$$x_0(\tau) = a_0 \cos \tau + b_0 \sin \tau \tag{5.27}$$

for every a_0, b_0. Now put (5.27) into (5.26b). After writing

$$\cos^3 \tau = \tfrac{3}{4} \cos \tau + \tfrac{1}{4} \cos 3\tau, \quad \sin^3 \tau = \tfrac{3}{4} \sin \tau - \tfrac{1}{4} \sin 3\tau$$

and making other similar reductions† we have

$$\begin{aligned}
x_1'' + x_1 = &\{\gamma - \kappa b_0 + a_0[-\beta + \tfrac{3}{4}(a_0^2 + b_0^2)]\} \cos \tau \\
&+ \{\kappa a_0 + b_0[-\beta + \tfrac{3}{4}(a_0^2 + b_0^2)]\} \sin \tau \\
&+ \tfrac{1}{4} a_0(a_0^2 - 3b_0^2) \cos 3\tau + \tfrac{1}{4} b_0(3a_0^2 - b_0^2) \sin 3\tau.
\end{aligned} \tag{5.28}$$

This is the crucial stage: the solution $x_1(\tau)$ is required to have period 2π, but *unless the coefficients of* $\cos \tau$ *and* $\sin \tau$ *in (5.28) are zero, there are no periodic solutions*, since any solution would contain terms of the form $\tau \cos \tau$, $\tau \sin \tau$. Such non-periodic or otherwise undesirable constituents of a solution are usually called secular terms. We eliminate the secular terms by *requiring* the coefficients of $\cos \tau$ and $\sin \tau$ to be zero, which decides the acceptable values of a_0, b_0:

$$\kappa a_0 - b_0\{\beta - \tfrac{3}{4}(a_0^2 + b_0^2)\} = 0, \tag{5.29a}$$

$$\kappa b_0 + a_0\{\beta - \tfrac{3}{4}(a_0^2 + b_0^2)\} = \gamma. \tag{5.29b}$$

These equations settle the values of a_0 and b_0 as follows. Let r_0 be the amplitude of the generating solution:

$$r_0 = \sqrt{(a_0^2 + b_0^2)} > 0. \tag{5.30}$$

Square and add eqns (5.29a) and (5.29b), obtaining

$$r_0^2\{\kappa^2 + (\beta - \tfrac{3}{4} r_0^2)^2\} = \gamma^2. \tag{5.31}$$

---

† Often, the easiest way to carry out such calculations is to write the solution of (5.26a) as $A_0 e^{i\tau} + \bar{A}_0 e^{-i\tau}$. Then

$$x_0^3 = (A_0 e^{i\tau} + \bar{A}_0 e^{-i\tau})^3 = A_0^3 e^{3i\tau} + 3A_0^2 \bar{A}_0 e^{i\tau} + \text{complex conjugate}.$$

The substitution $A_0 = \tfrac{1}{2}(a_0 - ib_0)$, a_0 real, b_0 real, is left to the very end.

When (5.31) is solved for r_0, a_0 and b_0 can be obtained from (5.29) and (5.30).

Equation (5.31) will be analysed in the next section. For the present, note that there may be as many as three positive values of r_0^2 (hence of $r_0 > 0$) satisfying (5.31). This indicates that for certain values of κ, β, γ there may be three distinct solutions of (5.18), (5.22), or (5.23), each bifurcating from one of three distinct generating solutions.

Having selected a pair of values a_0 and b_0 satisfying (5.29), we solve (5.28) to give

$$x_1(\tau) = a_1 \cos \tau + b_1 \sin \tau - \tfrac{1}{32} a_0 (a_0^2 - 3b_0^2) \cos 3\tau - \tfrac{1}{32} b_0 (3a_0^2 - b_0^2) \sin 3\tau,$$

where a_1 and b_1 are any constants. This expression is substituted into eqn (5.26c); the requirement that $x_2(\tau)$ should have period 2π provides equations determining a_1, b_1, as before. In this and subsequent stages the determining equations for a_i, b_i are linear, so no further multiplicity of solutions is introduced.

5.4. The amplitude equation for the undamped pendulum

Suppose that the damping coefficient is zero: in (5.18), (5.19), (5.23)

$$k = K = \kappa = 0. \tag{5.32}$$

Instead of seeking r_0 through (5.31), the coefficients a_0, b_0 can be found directly from (5.29): the only solutions are given by

$$b_0 = 0 \tag{5.33a}$$

$$a_0(\beta - \tfrac{3}{4}a_0^2) = \gamma. \tag{5.33b}$$

We shall consider in detail only the pendulum case. The original parameters ω, ω_0, and F of eqn (5.18) can be restored through (5.3), (5.20), and (5.21). Equation (5.33b) becomes

$$a_0(\omega^2 - \omega_0^2 + \tfrac{1}{8}\omega_0^2 a_0^2) = -F. \tag{5.34}$$

The solutions a_0 can be obtained by drawing a cubic curve

$$z = a_0(\omega^2 - \omega_0^2 + \tfrac{1}{8}\omega_0^2 a_0^2) = f(a_0) \tag{5.35}$$

on a graph with axes a_0, z for each fixed value of ω and ω_0, then finding the intersections with the lines $z = -F$ ($F > 0$) for various F, as in Fig. 5.1.

The main features revealed are as follows.

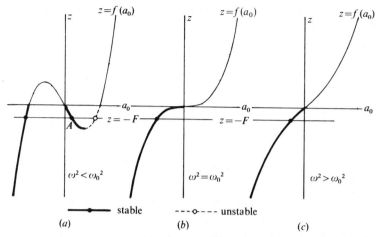

Fig. 5.1. Illustrating eqn (5.34)

(i) When $\omega^2 > \omega_0^2$ (Fig. 5.1(c)), there is exactly one periodic oscillation possible. When F is small, this approximates to the linear response (5.15), and to the 'corrected linear response' (5.17), unless $\omega^2 \approx \omega_0^2$ (very near to resonance) in which case it is considerably different (Fig. 5.1(b)). These responses are 180° out of phase with the forcing term.

(ii) When $\omega^2 < \omega_0^2$ (Fig. 5.1(a)), there is a single response, 180° out of phase, when F is relatively large. When F is smaller there are three distinct periodic responses, two in phase and one out of phase with the forcing term. The response marked 'A' in Fig. 5.1(a) corresponds to the response (5.15) of the linearized equation, and to the corrected linear response (5.17).

(iii) All three of the responses described in (ii) will be of small amplitude if the intersections of the curve $z = f(a_0)$ with the a_0 axis (Fig. 5.1(a)) occur close to the origin. These intersections are at $a_0 = 0$, $a_0 = \pm 2\sqrt{2}(1 - \omega^2/\omega_0^2)^{1/2}$. Three intersections are actually achieved provided $F < \frac{4}{3}\sqrt{\frac{2}{3}}\omega_0^2(1 - \omega^2/\omega_0^2)^{3/2}$ (recalling that F is positive). Therefore, by choosing ω^2/ω_0^2 sufficiently near to 1 (near to resonance) and F correspondingly small, the amplitudes of all the three responses can be made as small as we wish. In particular, they can all be confined in what we should normally regard as the linear range of amplitude for the pendulum.

(iv) Despite there being no damping, there are steady, bounded oscillations even when $\omega = \omega_0$ (unlike the linearized case). The non-linearity controls the amplitude in the following way. Amplitude increases indefinitely if the forcing term remains in step with the natural oscillation and reinforces it cycle by cycle. However (e.g. Example 4.10), the frequency of the natural oscillation varies with amplitude due to the nonlinearity and does not remain in step with the forcing term.

(v) Whether a steady oscillation is set up or approached at all, and if it is, which of the possible modes is adopted, depends on the initial conditions of the problem, which are not considered at all here (see Chapter 7).

(vi) Whether a particular mode can be sustained in practice depends on its stability, of which an indication can be got as follows. If, in the neighbourhood of amplitude a_0, the forcing amplitude required to sustain a_0 increases/decreases as a_0 increases/decreases, we expect a stable solution, in which an accidental small disturbance of amplitude cannot be sustained and amplified. If, however, F increases/decreases as a_0 decreases/increases, the conditions are right for growth of the disturbance and instability results. Further justification is given for this argument in Section 9.7. In anticipation of the analysis of Chapters 7 and 9, the stable and unstable branches are indicated in Figs. 5.1 and 5.2.

The nature of the solutions of (5.34) as a function of the parameters ω, ω_0, with F given, can be exhibited on a single 'response diagram', Fig. 5.2. The figure can be plotted directly by writing (5.34) in the form

$$\omega = \sqrt{\{\omega_0^2(1 - \tfrac{1}{8}a_0^2) - F/a_0\}}.$$

For each value of $F > 0$ the amplitudes lie on two smooth branches, a typical pair being shown as heavy lines on the figure. F increases on 'contours' increasingly distant from the curve $F = 0$ (which gives the amplitude–frequency curve for free oscillations) on both sides of it.

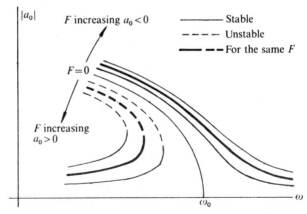

FIG. 5.2. Amplitude–frequency curves for the undamped pendulum, eqn (5.34)

Example 5.2 *Investigate the forced periodic solutions of the equation*

$$x'' + (9 + \varepsilon\beta)x - \varepsilon x^3 = \Gamma \cos \tau$$

where ε is small and β, Γ are not too large.

This is a case where $\Omega^2 = 9 + \varepsilon\beta$ (eqn (5.7)) has a value which causes the direct expansion (5.17) to fail. The given equation may be rewritten

$$x'' + 9x = \Gamma \cos \tau + \varepsilon(x^3 - \beta x).$$

Write $x(\varepsilon, \tau) = x_0(\tau) + \varepsilon x_1(\tau) + \dots$, where $x_0(\tau), x_1(\tau), \dots$ have period 2π. Then

$$x_0'' + 9x_0 = \Gamma \cos \tau,$$

$$x_1'' + 9x_1 = x_0^3 - \beta x_0,$$

and so on. These have the same form as (5.10), but since 9 is the square of an integer, the first equation has solutions of period 2π of the form

$$x_0(\tau) = a_0 \cos 3\tau + b_0 \sin 3\tau + \tfrac{1}{8}\Gamma \cos \tau.$$

When this is substituted into the equation for x_1, terms in $\cos 3\tau$, $\sin 3\tau$ emerge on the right-hand side, preventing periodic solutions unless their coefficients are zero. The simplest way to make the substitution is to write

$$x_0(\tau) = A_0 e^{3i\tau} + \bar{A}_0 e^{-3i\tau} + \tfrac{1}{16}\Gamma e^{i\tau} + \tfrac{1}{16}\Gamma e^{-i\tau},$$

where $A_0 = \tfrac{1}{2}a_0 - \tfrac{1}{2}ib_0$. We find

$$x_0^3 - \beta x_0 = \left[\left(\frac{\Gamma^3}{16^3} + \frac{6\Gamma^2}{16^2} A_0 + 3A_0^2 \bar{A}_0 - \beta A_0 \right) e^{3i\tau} + \text{complex conjugate} \right]$$

$$+ \text{other harmonics.}$$

Therefore we require

$$A_0 \left(3A_0 \bar{A}_0 - \beta + \frac{6\Gamma^2}{16^2} \right) = - \frac{\Gamma^3}{16^2}.$$

This implies that A_0 is real: $b_0 = 0$, $A = \tfrac{1}{2}a_0$, and the equation for a_0 is

$$a_0 \left(\tfrac{3}{4}a_0^2 - \beta + \frac{6\Gamma^2}{16^2} \right) + \frac{\Gamma^3}{16^3} = 0.$$

5.5. The amplitude equation for a damped pendulum

The amplitude equation (5.31) translated into the parameters of eqn (5.18) by (5.19), (5.20), and (5.21), becomes

$$r_0^2 \{ k^2\omega^2 + (\omega^2 - \omega_0^2 + \tfrac{1}{8}\omega_0^2 r_0^2)^2 \} = F^2. \tag{5.36}$$

Only solutions with $r_0 > 0$ are valid (see eqn (5.30)). Solving the quadratic equation for ω^2 given by (5.36) we find that

$$\omega^2 = \tfrac{1}{2}\{ 2\omega_0^2(1 - \tfrac{1}{8}r_0^2) - k^2 \pm \sqrt{\{ k^4 - 4\omega_0^2 k^2(1 - \tfrac{1}{8}r_0^2) + 4F^2/r_0^2 \}}.$$

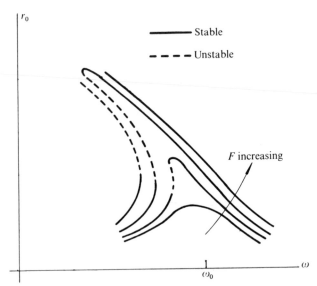

FIG. 5.3. Amplitude–frequency curves for the damped pendulum, eqn (5.36).
On each curve F is constant

Typical response curves are shown in Fig. 5.3 for fixed values of k and ω_0, and selected values of $F > 0$. k was chosen fairly 'small' so that comparison might be made with the undamped case shown in Fig. 5.2. The figure is similar to Fig. 5.2, with some important differences. There are no longer two distinct branches corresponding to each value of F: the branches join up to make continuous curves. In the near neighbourhood of the point $\omega = \omega_0$, $r_0 = 0$ the curves do not turn over: that is to say, a certain minimum value of F is required before it is possible to have three alternative forced responses (see Exercise 10). The response curves for small F represent the approximately linear response, as can be seen from (5.31).

The 'jump effect' associated with the response curves is presented in Section 7.3, and may be read at this stage.

5.6. Soft and hard springs

We have so far been concerned with equations of the type

$$\ddot{x} + k\dot{x} + cx + \varepsilon x^3 = F \cos \omega t,$$

which arise from an approximation to the pendulum equation. In the general case this is called a *Duffing's equation* with a forcing term. Now consider

$$\ddot{x} + k\dot{x} + cx + \varepsilon g(x) = F \cos \omega t, \tag{5.37}$$

where $k > 0, c > 0$, and $|\varepsilon| \ll 1$. This can be interpreted as the equation of forced motion with damping of a particle on a spring which provides a restoring force which is almost but not quite linear. This is a useful physical model, since it leads us to expect that the features of linear motion will be to some extent preserved; for example, if $k = 0$ and $F = 0$ we guess that solutions for small enough amplitude will be oscillatory, and that the effect of small damping is to reduce the oscillations steadily to zero.

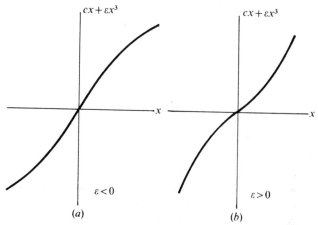

FIG. 5.4. Restoring force functions for (*a*) a soft spring; (*b*) a hard spring

Important classes of cases are represented by the example $g(x) = x^3$, with $\varepsilon > 0$ and $\varepsilon < 0$. The corresponding (symmetrical) restoring forces are illustrated in Figs. 5.4(*a*) and (*b*). When ε is negative the restoring force becomes progressively weaker in extension than for the linear spring: such a spring is called *soft*. A pendulum is modelled by a soft spring system. When ε is positive the spring becomes stiffer and is called a *hard spring*.

The nature of the response diagrams for the forced oscillations, period $2\pi/\omega$, of (5.37) in the case of a soft spring ($\varepsilon < 0$), a linear spring ($\varepsilon = 0$), and a hard spring ($\varepsilon > 0$) are shown in Fig. 5.5.

There are various ways in which the restoring force can be *unsymmetrical*: an important case is shown in Fig. 5.6, where $g(x) = -x^2$ and $\varepsilon > 0$.

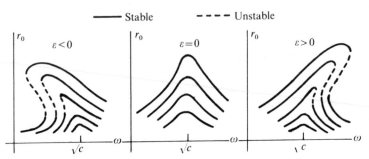

FIG. 5.5. Amplitude–frequency curves for Duffing's equation $\ddot{x} + k\dot{x} + cx + \varepsilon x^3 = F \cos \omega t$, $k > 0, c > 0$

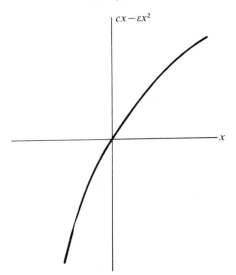

FIG. 5.6. Unsymmetrical restoring force

The spring is 'soft' for $x > 0$ and 'hard' for $x < 0$. Exercise 23 shows that for free oscillations ($F = 0$ in (5.37)) the effect is to shift the centre of oscillation to small positive x values, as the diagram suggests; the same occurs when the oscillation is forced as is shown in the following Example. The phenomenon is called *rectification*.

Example 5.3 *Obtain approximately the solutions of period 2π of the equation* $\ddot{x} + \Omega^2 x - \varepsilon x^2 = \Gamma \cos t, \varepsilon > 0$.

Firstly, suppose that Ω is not close to an integer (far from resonance). Writing

$x(\varepsilon, t) = x_0(t) + \varepsilon x_1(t) + \ldots$, the required sequence of equations is

$$\ddot{x}_0 + \Omega^2 x_0 = \Gamma \cos t,$$

$$\ddot{x}_1 + \Omega^2 x_1 = x_0^2,$$

and so on. The periodic solution of the first is

$$x_0(t) = \frac{\Gamma}{\Omega^2 - 1} \cos t,$$

and the second becomes

$$\ddot{x}_1 + \Omega^2 x_1 = \frac{\Gamma^2}{2(\Omega^2 - 1)^2} (1 + \cos 2t)$$

whose solutions are

$$x_1(t) = \frac{\Gamma^2}{2\Omega^2(\Omega^2 - 1)^2} + \frac{\Gamma^2}{2(\Omega^2 - 1)^2 (\Omega^2 - 4)} \cos 2t + a_1 \cos \Omega t + b_1 \sin \Omega t.$$

For $x_1(t)$ to have period 2π, $a_1 = b_1 = 0$: therefore

$$x(\varepsilon, t) = \frac{\Gamma}{\Omega^2 - 1} \cos t + \frac{\varepsilon \Gamma^2}{2\Omega^2(\Omega^2 - 1)^2} + \frac{\varepsilon \Gamma^2}{2(\Omega^2 - 1)^2 (\Omega^2 - 4)} \cos 2t.$$

Now suppose $\Omega \approx 1$, so put

$$\Omega^2 = 1 + \varepsilon \beta, \quad \Gamma = \varepsilon \gamma.$$

The equation becomes

$$\ddot{x} + x = \varepsilon(\gamma \cos t + x^2 - \beta x).$$

Assume that $x(\varepsilon, \tau) = x_0(\tau) + \varepsilon x_1(\tau) + \ldots$. The first two equations are

$$\ddot{x}_0 + x_0 = 0$$

$$\ddot{x}_1 + x_1 = \gamma \cos t + x_0^2 - \beta x_0.$$

The solutions to the first, of period 2π, are $a_0 \cos t + b_0 \sin t$, and the second becomes

$$\ddot{x}_1 + x_1 = \tfrac{1}{2}(a_0^2 + b_0^2) + (\gamma - \beta a_0) \cos t - \beta b_0 \sin t + \tfrac{1}{2}(a_0^2 - b_0^2) \cos 2t + a_0 b_0 \sin 2t.$$

Secular terms vanish only if

$$b_0 = 0, \quad a_0 = \gamma/\beta.$$

Then

$$x(\varepsilon, t) \approx \frac{\gamma}{\beta} \cos t + \varepsilon \left(\frac{1}{2} \frac{\gamma^2}{\beta^2} - \frac{1}{6} \frac{\gamma^2}{\beta^2} \cos 2t + a_1 \cos t + b_1 \sin t \right).$$

This is exactly the same as the earlier result: only one solution is revealed.

5.7. Amplitude-phase perturbation for the pendulum equation

In Section 5.3 we stopped at the first approximation to a solution because of the rapidly developing complexity of the algebra. The following method allows a higher approximation to be obtained rather more efficiently.

Consider again the family (5.23):

$$x'' + x = \varepsilon(\gamma \cos \tau - \kappa x' - \beta x + x^3). \tag{5.38}$$

Instead of seeking, as in Section 5.3, solutions in effect of the form

$$(a_0 + \varepsilon a_1 + \ldots) \cos \tau + (b_0 + \varepsilon b_1 + \ldots) \sin \tau + \text{higher harmonics}$$

we shall arrange for them to appear in the form

$$x(\varepsilon, \tau) = (r_0 + \varepsilon r_1 + \ldots) \cos(\tau + \alpha_0 + \varepsilon \alpha_1 + \ldots) + \text{higher harmonics}, \tag{5.39}$$

where

$$\alpha = \alpha_0 + \varepsilon \alpha_1 + \ldots$$

is the phase difference between the response and forcing term. α_0 is expected to be the phase of the generating solution and r_0 its amplitude.

It is more convenient for manipulative reasons to have the unknown phase, α, appear in the equation itself. Therefore we discuss instead the equation

$$X'' + X = \varepsilon(\gamma \cos(s - \alpha) - \kappa X' - \beta X + X^3) \tag{5.40}$$

where we have put $s = \tau + \alpha$ and $X(\varepsilon, s) = x(\varepsilon, \tau)$ into (5.38).

Assume that for all small enough ε

$$X(\varepsilon, s) = X_0(s) + \varepsilon X_1(s) + \ldots, \tag{5.41}$$
$$\alpha = \alpha_0 + \varepsilon \alpha_1 + \ldots.$$

We require solutions $X(\varepsilon, s)$ having the period, 2π, of the forcing term, which implies that for all s,

$$X_i(s + 2\pi) = X_i(s), \quad i = 1, 2, \ldots. \tag{5.42}$$

Finally we shall impose the extra condition $X'(\varepsilon, 0) = 0$. This is not a real restriction: we simply adjust the time origin, and hence the phase, so that it is so. Therefore

$$X_i'(0) = 0, \quad i = 0, 1, \ldots. \tag{5.43}$$

Substitute (5.41) into (5.40), writing

$$\cos(s - \alpha) = \cos(s - \alpha_0) + \varepsilon \alpha_1 \sin(s - \alpha_0) + \ldots.$$

By assembling powers of ε and equating their coefficients to zero we obtain

$$X_0'' + X_0 = 0, \tag{5.44a}$$

$$X_1'' + X_1 = \gamma \cos(s - \alpha_0) - \kappa X_0' - \beta X_0 + X_0^3, \tag{5.44b}$$

$$X_2'' + X_2 = \gamma \alpha_1 \sin(s - \alpha_0) - \kappa X_1' - \beta X_1 + 3X_0^2 X_1, \tag{5.44c}$$

and so on.

The periodic solutions of (5.44a) satisfying (5.43) are

$$X_0(s) = r_0 \cos s, \quad r_0 > 0 \tag{5.45}$$

where we choose $r_0 > 0$ by later adjusting the phase α_0. From (5.44b)

$$X_1'' + X_1 = (\gamma \cos \alpha_0 - \beta r_0 + \tfrac{3}{4} r_0^3) \cos s + (\kappa r_0 + \gamma \sin \alpha_0) \sin s + \tfrac{1}{4} r_0^3 \cos 3s. \tag{5.46}$$

For there to be a periodic solution the secular terms (in $\cos s$ and $\sin s$) must be eliminated, so

$$\beta r_0 - \tfrac{3}{4} r_0^3 = \gamma \cos \alpha_0 \tag{5.47a}$$

$$\kappa r_0 = -\gamma \sin \alpha_0 \tag{5.47b}$$

By squaring and adding we obtain eqn (5.31) again:

$$r_0^2 \{ \kappa^2 + (\beta - \tfrac{3}{4} r_0^2)^2 \} = \gamma^2. \tag{5.48}$$

α_0 is then obtainable from (5.47) (considering only $-\tfrac{1}{2}\pi \leqslant \alpha_0 \leqslant \tfrac{1}{2}\pi$):

$$\alpha_0 = -\sin^{-1}(\kappa r_0 / \gamma). \tag{5.49}$$

Equation (5.46) becomes

$$X_1'' + X_1 = \tfrac{1}{4} r_0^3 \cos 3s,$$

with solutions

$$X_1(s) = r_1 \cos s - \tfrac{1}{32} r_0^3 \cos 3s, \quad r_1 > 0, \tag{5.50}$$

satisfying (5.42) and (5.43). Substitute (5.50) into (5.44c):

$$X_2'' + X_2 = (-\gamma \alpha_1 \sin \alpha_0 - \beta r_1 + \tfrac{9}{4} r_0^2 r_1 - \tfrac{3}{128} r_0^5) \cos s$$
$$+ (\gamma \alpha_1 \cos \alpha_0 + \kappa r_1) \sin s$$
$$+ (\tfrac{3}{4} r_0^2 r_1 - \tfrac{3}{64} r_0^5 + \tfrac{1}{32} \beta r_0^3) \cos 3s - \tfrac{3}{32} \kappa r_0^3 \sin 3s - \tfrac{3}{128} r_0^5 \sin 5s. \tag{5.51}$$

Periodicity requires that the coefficients of $\cos s, \sin s$ should be zero, which leads to the result

$$r_1 = \tfrac{3}{128}r_0^5/(\tfrac{9}{4}r_0^2 - \beta + \kappa \tan \alpha_0), \tag{5.52a}$$

$$\alpha_1 = -\tfrac{3}{128}\frac{\kappa r_0^5}{\gamma \cos \alpha_0} \Big/ (\tfrac{9}{4}r_0^2 - \beta + \kappa \tan \alpha_0). \tag{5.52b}$$

The solution to eqn (5.38) is then

$$x(\varepsilon, \tau) = (r_0 + \varepsilon r_1)\cos(\tau + \alpha_0 + \varepsilon \alpha_1) - \varepsilon r_0^2 \cos 3(\tau + \alpha_0) + O(\varepsilon^2). \tag{5.53}$$

5.8. Periodic solutions of autonomous equations (Lindstedt's method)

Consider the oscillations of the pendulum-type equation

$$\frac{d^2 x}{dt^2} + x - \varepsilon x^3 = 0. \tag{5.54}$$

For a soft spring $\varepsilon > 0$, and for a hard spring $\varepsilon < 0$. The system is conservative, and the method of Section 1.3 can be used to show that all motions of small enough amplitude are periodic.

In this case of unforced vibration, *the frequency, ω, is not known in advance*, but depends on the amplitude. It reduces to 1 when $\varepsilon = 0$. Assume that

$$\omega = 1 + \varepsilon \omega_1 + \ldots, \tag{5.55}$$

$$x(\varepsilon, t) = x_0(t) + \varepsilon x_1(t) + \ldots. \tag{5.56}$$

We could substitute these into (5.54) and look for solutions of period $2\pi/\omega$, but it is mechanically simpler to cause ω to appear as a multiplier by writing

$$\omega t = \tau. \tag{5.57}$$

Then (5.54) becomes

$$\omega^2 x'' + x - \varepsilon x^3 = 0. \tag{5.58}$$

By this substitution we have replaced eqn (5.54), which has unknown period, by eqn (5.58) which has known period 2π. Therefore, as before, for all τ,

$$x_i(\tau + 2\pi) = x_i(\tau), \quad i = 0, 1, \ldots. \tag{5.59}$$

Equation (5.58) becomes

$$(1 + \varepsilon 2\omega_1 + \ldots)(x_0'' + \varepsilon x_1'' + \ldots) + (x_0 + \varepsilon x_1 + \ldots) = \varepsilon(x_0 + \varepsilon x_1 + \ldots)^3,$$

and by assembling powers of ε we obtain

$$x_0'' + x_0 = 0, \tag{5.60a}$$

$$x_1'' + x_1 = -2\omega_1 x_0'' + x_0^3, \tag{5.60b}$$

and so on.

To simplify the calculations we can impose the conditions

$$x(\varepsilon, 0) = a_0, \quad x'(\varepsilon, 0) = 0 \tag{5.61}$$

without loss of generality (only in the autonomous case!). This implies that

$$x_0(0) = a_0, \quad x_0'(0) = 0, \tag{5.62a}$$

and

$$x_i(0) = 0, \quad x_i'(0) = 0, \quad i = 1, 2, \ldots. \tag{5.62b}$$

The solution of (5.60a) satisfying (5.62a) is

$$x_0 = a_0 \cos \tau. \tag{5.63}$$

(5.60b) then becomes

$$x_1'' + x_1 = (2\omega_1 a_0 + \tfrac{3}{4} a_0^3) \cos \tau + \tfrac{1}{4} a_0^3 \cos 3\tau. \tag{5.64}$$

The solutions will be periodic only if

$$\omega_1 = -\tfrac{3}{8} a_0^2. \tag{5.65}$$

From (5.64), (5.65),

$$x_1(\tau) = a_1 \cos \tau + b_1 \sin \tau - \tfrac{1}{32} a_0^3 \cos 3\tau,$$

and (5.62b) implies

$$b_1 = 0, \quad a_1 = \tfrac{1}{32} a_0^3.$$

Therefore

$$x_1(\tau) = \tfrac{1}{32} a_0^3 (\cos \tau - \cos 3\tau). \tag{5.66}$$

Finally, from (5.63) and (5.66),

$$x(\varepsilon, \tau) \approx a_0 \cos \tau + \tfrac{1}{32} \varepsilon a_0^3 (\cos \tau - \cos 3\tau) + O(\varepsilon^2). \tag{5.67}$$

Returning to the variable t (eqn (5.57)), we have the approximation

$$x(\varepsilon, t) \approx a_0 \cos \omega t + \tfrac{1}{32} \varepsilon a_0^3 (\cos \omega t - \cos 3\omega t), \tag{5.68a}$$

where

$$\omega \approx 1 - \tfrac{3}{8} \varepsilon a_0^2; \tag{5.68b}$$

this gives the dependence of frequency on amplitude.

5.9. Forced oscillation of a self-excited equation

Consider the van der Pol equation with a forcing term:

$$\ddot{x} + \varepsilon(x^2 - 1)\dot{x} + x = F \cos \omega t. \qquad (5.69)$$

The unforced equation has a limit cycle with radius approximately 2 and period approximately 2π (see Sections 4.1 and 4.2). The limit cycle is generated by the balance between internal energy loss and energy generation (see Section 1.5), and the forcing term will alter this balance. If F is 'small' (weak excitation), its effect depends on whether or not ω is close to the natural frequency. If it is, it appears that an oscillation might be generated which is a perturbation of the limit cycle. If F is not small (hard excitation) or if the natural and imposed frequency are not closely similar, we should expect that the 'natural oscillation' will be extinguished, as occurs with the corresponding linear equation.

Firstly, write

$$\omega t = \tau \qquad (5.70)$$

then (5.69) becomes

$$\omega^2 x'' + \varepsilon\omega(x^2 - 1)x' + x = F \cos \tau, \qquad (5.71)$$

where the dashes signify differentiation with respect to τ.

Hard excitation, far from resonance. Assume that ω is not close to an integer. In (5.71), let

$$x(\varepsilon, \tau) = x_0(\tau) + \varepsilon x_1(\tau) + \dots. \qquad (5.72)$$

The sequence of equations for $x_0, x_1, \dots$ begins

$$\omega^2 x_0'' + x_0 = F \cos \tau, \qquad (5.73a)$$

$$\omega^2 x_1'' + x_1 = -(x_0^2 - 1)x_0', \qquad (5.73b)$$

$x_0(\tau), x_1(\tau)$ having period 2π.

The only solution of (5.73a) having period 2π is

$$x_0(\tau) = \frac{F}{1 - \omega^2} \cos \tau,$$

and therefore

$$x(\varepsilon, \tau) = \frac{F}{1 - \omega^2} \cos \tau + O(\varepsilon). \qquad (5.74)$$

The solution is therefore a perturbation of the ordinary linear response and the limit cycle is suppressed as expected.

Soft excitation, far from resonance. This case is similar to the last, and is left as Exercise 21. However, this solution is normally unstable (see Section 7.4).

Soft excitation near resonance. For soft excitation write in (5.71)

$$F = \varepsilon\gamma \tag{5.75}$$

and for near-resonance

$$\omega = 1 + \varepsilon\omega_1. \tag{5.76}$$

The expansion is assumed to be

$$x(\varepsilon, \tau) = x_0(\tau) + \varepsilon x_1(\tau) + \dots. \tag{5.77}$$

Equations (5.72), (5.75) and (5.76) lead to the sequence

$$x_0'' + x_0 = 0 \tag{5.78a}$$

$$x_1'' + x_1 = -2\omega_1 x_0'' - (x_0^2 - 1)x_0' + \gamma\cos\tau, \tag{5.78b}$$

and so on. We require solutions with period 2π. Equation (5.78a) has the solutions

$$x_0(\tau) = a_0\cos\tau + b_0\sin\tau. \tag{5.79}$$

After some manipulation, (5.78b) becomes

$$x_1'' + x_1 = \{\gamma + 2\omega_1 a_0 - b_0(\tfrac{1}{4}r_0^2 - 1)\}\cos\tau$$
$$+ \{2\omega_1 b_0 + a_0(\tfrac{1}{4}r_0^2 - 1)\}\sin\tau + \text{higher harmonics}, \tag{5.80}$$

where

$$r_0 = \sqrt{(a_0^2 + b_0^2)} > 0. \tag{5.81}$$

For a periodic solution we require

$$2\omega_1 a_0 - b_0(\tfrac{1}{4}r_0^2 - 1) = -\gamma, \tag{5.82a}$$

$$2\omega_1 b_0 + a_0(\tfrac{1}{4}r_0^2 - 1) = 0. \tag{5.82b}$$

By squaring and adding these two equations we obtain

$$r_0^2\{4\omega_1^2 + (\tfrac{1}{4}r_0^2 - 1)^2\} = \gamma^2 \tag{5.83}$$

which give the possible amplitudes r_0 of the response. The structure of this equation is examined in Chapter 7 in a different connection: it is sufficient to notice here its family resemblance to (5.31) for the pendulum equation. Like (5.31), it may have as many as three real solutions for $r_0 > 0$.

5.10. The perturbation method and Fourier series

In the examples given in Sections 5.2 and 5.3 the solutions emerge as series of sines and cosines with frequencies which are integer multiples of the forcing frequency. These appeared as a result of reorganising terms like x^3, but by making a direct attack using Fourier series we can show that this form always occurs.

Consider the more general forced equation

$$x'' + \Omega^2 x = F(\tau) - \varepsilon h(x, x') \tag{5.84}$$

where ε is a small parameter. Suppose that F is periodic, with the time variable already scaled to give it period 2π, and that its mean value is zero so that there is zero constant term in its Fourier series representation:

$$F(\tau) = \sum_{n=1}^{\infty} (A_n \cos n\tau + B_n \sin n\tau). \tag{5.85}$$

We shall allow the possibility that Ω *is close to some integer* N by writing

$$\Omega^2 = N^2 + \varepsilon \beta \tag{5.86}$$

($N = 1$ in Section 5.3). The perturbation method requires that the periodic solutions emerge from periodic solutions of some appropriate linear equation. If (5.85) has a non-zero term of order N then (5.84), with $\varepsilon = 0$, is clearly not an appropriate linearization, since the forcing term has a component equal to the natural frequency N and there will be no periodic solutions. However, if we write

$$A_N = \varepsilon A, \quad B_N = \varepsilon B, \tag{5.87}$$

the term in F giving resonance is removed from the *linearized* equation and we have a possible family of generating solutions. Now rearrange (5.84), isolating the troublesome term in F by writing

$$f(\tau) = F(\tau) - \varepsilon A \cos N\tau - \varepsilon B \sin N\tau = \sum_{n \neq N} (A_n \cos n\tau + B_n \sin n\tau). \tag{5.88}$$

Equation (5.84) becomes

$$x'' + N^2 x = f(\tau) + \varepsilon \{ -h(x, x') - \beta x + A \cos N\tau + B \sin N\tau \}. \tag{5.89}$$

The linearized equation is now $x'' + N^2 x = f(\tau)$, with no resonance.

Now write as usual

$$x(\varepsilon, \tau) = x_0(\tau) + \varepsilon x_1(\tau) + \dots, \tag{5.90}$$

where $x_0, x_1,...$ have period 2π. By expanding h in (5.84) in powers of ε we have

$$h(x, x') = h(x_0, x'_0) + \varepsilon h_1(x_0, x'_0, x_1, x'_1) + ..., \tag{5.91}$$

where h_1 can be calculated, and by substituting (5.90) and (5.91) into (5.89) we obtain the sequence

$$x''_0 + N^2 x_0 = \sum_{n \neq N} (A_n \cos n\tau + B_n \sin n\tau), \tag{5.92a}$$

$$x''_1 + N^2 x_1 = -h(x_0, x'_0) - \beta x_0 + A \cos N\tau + B \sin N\tau, \tag{5.92b}$$

$$x''_2 + N^2 x_2 = -h_1(x_0, x'_0, x_1, x'_1) - \beta x_1, \tag{5.92c}$$

and so on. The solution of (5.92a) is

$$x_0(\tau) = a_0 \cos N\tau + b_0 \sin N\tau + \sum_{n \neq N} \frac{A_n \cos n\tau + B_n \sin n\tau}{N^2 - n^2}$$

$$= a_0 \cos N\tau + b_0 \sin N\tau + \phi(\tau), \tag{5.93}$$

say; a_0, b_0 are constants to be determined at the next stage.

We require x_0, as determined by (5.93), to be such that (5.92b) has periodic solutions. This is equivalent to requiring that its right side has no Fourier term of order N, since such a term would lead to resonance. We require therefore that

$$\beta a_0 = -\frac{1}{\pi} \int_0^{2\pi} h(a_0 \cos N\tau + b_0 \sin N\tau + \phi(\tau),$$

$$-a_0 N \sin N\tau + b_0 N \cos N\tau + \phi'(\tau)) \cos N\tau \, d\tau + A, \tag{5.94a}$$

$$\beta b_0 = -\frac{1}{\pi} \int_0^{2\pi} h(a_0 \cos N\tau + b_0 \sin N\tau + \phi(\tau),$$

$$-a_0 N \sin N\tau + b_0 N \cos N\tau + \phi'(\tau)) \sin N\tau \, d\tau + B; \tag{5.94b}$$

which constitute two equations for the unknowns a_0, b_0. The reader should confirm that the resulting equations are the same as those for the first order approximation, with $N = 1$, obtained in Section 5.3.

Each equation in the sequence (5.92) has solutions containing constants $a_1, b_1; a_2, b_2$; whose values are similarly established at the succeeding step. However, the equations for subsequent constants are linear: the pair (5.94) are the only ones which may have several solutions.

Exercises

1. Find all the periodic solutions of

$$\ddot{x} + \Omega^2 x = \Gamma \cos t$$

for all values of Ω^2.

2. Find the first two harmonics of the solutions of period 2π of the following
  (i) $\ddot{x} - 0\!\cdot\!5x^3 + 0\!\cdot\!25x = \cos t$;
  (ii) $\ddot{x} - 0\!\cdot\!1x^3 + 0\!\cdot\!6x = \cos t$;
  (iii) $\ddot{x} - 0\!\cdot\!1\dot{x}^2 + 0\!\cdot\!5x = \cos t$.

3. Find a first approximation to the limit cycle for Rayleigh's equation

$$\ddot{x} + \varepsilon(\tfrac{1}{3}\dot{x}^3 - \dot{x}) + x = 0, \quad |\varepsilon| \ll 1,$$

using the method of Section 5.8.

4. Take the method of Section 5.8 to order ε to obtain solutions of period 2π, and the amplitude–frequency relation, for
  (i) $\ddot{x} - \varepsilon x \dot{x} + x = 0$;
  (ii) $(1 + \varepsilon \dot{x})\ddot{x} + x = 0$.

5. Apply the perturbation method to the equation $\ddot{x} + \Omega^2 \sin x = \cos t$ by considering $\ddot{x} + \varepsilon\Omega^2 x + \varepsilon\Omega^2(\sin x - x) = \cos t$, with $\varepsilon = 1$, and assuming that Ω is not close to an odd integer. Use the Fourier expansion

$$\sin(a \cos t) = 2 \sum_{n=0}^{\infty} (-1)^n J_{2n+1}(a) \cos\{(2n+1)t\},$$

where J_{2n+1} is the Bessel function of order $2n+1$. Confirm that the leading terms are given by

$$x = \frac{1}{\Omega^2 - 1}[1 + \Omega^2 - 2\Omega^2 J_1\{1/(\Omega^2 - 1)\}] \cos t + \frac{2}{\Omega^2 - 9} J_3\{1/(\Omega^2 - 1)\} \cos 3t.$$

Compute the coefficients for representative values of Ω^2.

6. For the equation $\ddot{x} + \Omega^2 x - 0\!\cdot\!1x^3 = \cos \tau$, where $\Omega^2 \not\approx 1$, sketch the variation in the ratio of the magnitudes of the first two harmonics.

7. In the equation

$$\ddot{x} + \Omega^2 x + \varepsilon f(x) = \Gamma \cos t$$

Ω is not close to an odd integer, and $f(x)$ is an odd function of x, with expansion

$$f(a \cos t) = -a_1(a) \cos t - a_3(a) \cos 3t - \ldots.$$

Derive a perturbation solution of period 2π, to order ε.

8. The Duffing equation near resonance at $\Omega = 3$, with weak excitation, is

$$\ddot{x} + 9x = \varepsilon(\gamma \cos t - \beta x + x^3).$$

Show that there are solutions of period 2π if the amplitude of the zero-order solution is 0 or $2\sqrt{(\beta/3)}$. Find the solution to order ε in the latter case.

9. From eqn (5.34), the amplitude equation for an undamped pendulum is

$$-F = a_0(\omega^2 - \omega_0^2 + \tfrac{1}{8}\omega_0^2 a_0^2).$$

When ω_0 is given, find for what values of ω there are three possible responses. (Find the stationary values of F with respect to a_0, with ω fixed. These are the points where the response curves of Fig. 5.2 turn over.)

10. From eqn (5.36), the amplitude equation for the damped pendulum is $F^2 = r_0^2\{k^2\omega^2 + (\omega^2 - \omega_0^2 + \tfrac{1}{8}\omega_0^2 r_0^2)^2\}$. By considering $d(F^2)/d(r_0^2)$, show that if $(\omega^2 - \omega_0^2)^2 \leqslant 3k^2\omega^2$, then the amplitude equation has one real root r_0 for all F.

11. Find the linear equivalent form (Section 4.5) of the expression $\ddot{x} + \Omega^2 x - \varepsilon x^3$, with respect to the periodic form $x = a\cos t$. Use the linear form to obtain the frequency–amplitude relation for the equation

$$\ddot{x} + \Omega^2 x - \varepsilon x^3 = \Gamma \cos t.$$

Solve the equation approximately by assuming that $a = a_0 + \varepsilon a_1$, and show that this agrees with the first harmonic in eqn (5.17). (Note that there may be three solutions, but that this method of solution shows only the one close to $\{\Gamma/(1 - \Omega^2)\} \cos t$.)

12. Generalize the method of Exercise 11 for the same equation by putting $x = x^{(0)} + x^{(1)} + \ldots$, where $x^{(0)}$ and $x^{(1)}$ are the first two harmonics to order ε, $a\cos t$ and $b\cos 3t$, say, in the expansion of the solution. Show that the linear form equivalent to x^3 is

$$(\tfrac{3}{4}a^2 + \tfrac{3}{4}ab + \tfrac{3}{2}b^2)x^{(0)} + (\tfrac{1}{4}a^3 + \tfrac{3}{2}a^2b + \tfrac{3}{4}b^3)x^{(1)}/b.$$

Split the pendulum equation into the two equations

$$\ddot{x}^{(0)} + \{\Omega^2 - \varepsilon(\tfrac{3}{4}a^2 + \tfrac{3}{4}ab + \tfrac{3}{2}b^2)\}x^{(0)} = \Gamma\cos t,$$
$$\ddot{x}^{(1)} + \{\Omega^2 - \varepsilon(\tfrac{1}{4}a^3 + \tfrac{3}{2}a^2b + \tfrac{3}{4}b^3)/b\}x^{(1)} = 0.$$

Deduce that a and b must satisfy

$$a\{\Omega^2 - 1 - \varepsilon(\tfrac{3}{4}a^2 + \tfrac{3}{4}ab + \tfrac{3}{2}b^2)\} = \Gamma$$
$$b\{\Omega^2 - 9 - \varepsilon(\tfrac{1}{4}a^3 + \tfrac{3}{2}a^2b + \tfrac{3}{4}b^3)\} = 0.$$

Assume that $a = a_0 + \varepsilon a_1$, $b = \varepsilon b_1$, and obtain a_0, a_1, and b_1 (giving the perturbation solution (5.17)).

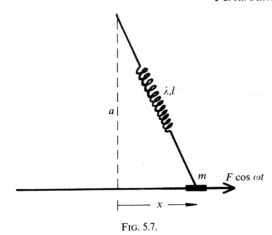

Fig. 5.7.

13. A sleeve of mass m slides on a smooth horizontal wire and is linked to a point height a above the wire by a linear spring of stiffness λ and natural length l, as shown in Fig. 5.7. Show that the approximate equation of motion is

$$\ddot{x} + \frac{\lambda}{m}\left(1 - \frac{l}{a}\right)x + \frac{\lambda l}{2a^3 m}x^3 = \frac{F}{m}\cos \omega t$$

for 'small' x when $l < a$, and a periodic horizontal force $F \cos \omega t$ is applied. For what parameter ranges does the method of Section 5.2 apply?

14. Apply the Lindstedt method, Section 5.8, to van der Pol's equation $\ddot{x} + \varepsilon(x^2 - 1)\dot{x} + x = 0$, $|\varepsilon| \ll 1$. Show that the frequency of the limit cycle is given by $\omega = 1 - \frac{1}{16}\varepsilon^2 + O(\varepsilon^3)$.

15. Investigate the forced periodic solutions of period $\frac{2}{3}\pi$ for the Duffing equation in the form $\ddot{x} + (1 + \varepsilon\beta)x - \varepsilon x^3 = \Gamma \cos 3t$.

16. For the equation $\ddot{x} + x + \varepsilon x^3 = 0$, $|\varepsilon| \ll 1$, with $x(0) = a$, $\dot{x}(0) = 0$, assume an expansion of the form $x(t) = x_0(t) + \varepsilon x_1(t) + \dots$, and carry out the perturbation process without assuming periodicity of the solution. Show that

$$x(t) = a \cos t + \varepsilon a^3 \{-\tfrac{3}{8}t \sin t + \tfrac{1}{32}(\cos 3t - \cos t)\} + O(\varepsilon^2).$$

(This expansion is valid, so far as it goes. Why is it not so suitable as those already obtained for describing the solutions?)

17. Find the first few harmonics in the solution, period 2π, of $\ddot{x} + \Omega^2 x + \varepsilon x^2 = \Gamma \cos t$, by the direct method of Section 5.2. Explain the presence of a constant term in the expansion.

For what values of Ω does the expansion fail? Show how, for small values of Γ, an expansion valid near $\Omega = 1$ can be obtained.

18. Use the method of amplitude-phase perturbation (Section 5.7) to approximate to the solutions, period 2π, of $\ddot{x} + x = \varepsilon(\gamma \cos t - x\dot{x} - \beta x)$.

19. Investigate the solutions, period 2π, of $\ddot{x} + 9x + \varepsilon x^2 = \Gamma \cos t$.

20. For the damped pendulum equation with a forcing term:

$$\ddot{x} + k\dot{x} + \omega_0^2 x - \tfrac{1}{6}\omega_0^2 x^3 = F \cos \omega t,$$

show that the amplitude–frequency curves have their maxima on

$$\omega^2 = \omega_0^2(1 - \tfrac{1}{8}r_0^2) - \tfrac{1}{2}k^2.$$

21. Show that the first harmonic for the forced van der Pol equation $\ddot{x} + \varepsilon(x^2 - 1)\dot{x} + x = F \cos \omega t$ is the same for both weak and hard excitation, far from resonance.

22. The orbital equation of a planet about the sun is

$$\frac{d^2 u}{d\theta^2} + u = k(1 + \varepsilon u^2),$$

where $u = r^{-1}$ and r, θ are polar coordinates, $k = \gamma m/h^2$, γ is the gravitational constant, m is the mass of the planet and h is its moment of momentum, a constant. $\varepsilon k u^2$ is the relativistic correction term.

Obtain a perturbation expansion for the solution with initial conditions $u(0) = k(e+1), \dot{u}(0) = 0$ (e is the eccentricity of the unperturbed orbit, and these are initial conditions at the perihelion: the nearest point to the sun on the unperturbed orbit.) Note that the solution of the perturbed equation is not periodic, and that 'secular' terms cannot be eliminated. Show that the expansion to order ε predicts that in each orbit the perihelion advances by $2k^2\pi\varepsilon$.

23. Use the Lindstedt procedure to find the first few terms in the expansion of the periodic solutions of

$$\ddot{x} + x + \varepsilon x^2 = 0.$$

Explain the presence of a constant term in the expansion.

24. Investigate the forced periodic solutions of period 2π of the equation

$$\ddot{x} + (4 + \varepsilon\beta)x - \varepsilon x^3 = \Gamma \cos t$$

where ε is small and β and Γ are not too large. Confirm that there is always a periodic solution of the form

$$a_0 \cos 2t + b_0 \sin 2t + \tfrac{1}{3}\Gamma \cos t$$

where

$$a_0(3r_0^2 + \tfrac{1}{6}\Gamma^2 - \beta) = b_0(3r_0^2 + \tfrac{1}{6}\Gamma^2 - \beta) = 0.$$

6 Singular perturbation methods

THIS CHAPTER is concerned with approximating to the solutions of differential equations containing a small parameter ε in what might be called difficult cases, where, for one reason or another, a straightforward expansion of the solution in powers of ε is unobtainable or unusable. Often in such cases it is possible to modify the method so that the essential features of such expansions are redeemed. Ideally, we wish to be able to take a few terms of an expansion, and to be able to say that for some small fixed numerical value of ε supplied in a practical problem the truncated series is close to the required solution *for the whole range of the independent variable* in the differential equation. Failure of this condition is the rule rather than the exception. In fact the examples of Chapter 5 do not give useful approximations if we work solely from the initial conditions (see Chapter 5, Exercise 16): for satisfactory approximation we must use the technically redundant information that the solutions are periodic.

This chapter illustrates several other methods which have been used in such cases. If one method fails, another may work, or perhaps a combination of methods may work, but generally speaking the approach is tentative, and considerable skill and intuition is necessary to get the desired result. For a treatment of the whole topic, not restricted to ordinary differential equations, see for example Nayfeh (1973), van Dyke (1964), and O'Malley (1974).

6.1. Non-uniform approximations to functions on an interval

The solutions of the differential equations we are considering,

$$\ddot{x} = f(x, \dot{x}, t, \varepsilon),$$

are functions of t and ε. Some of the problems which may arise in approximating to them can be illustrated by using functions which are not necessarily the solutions of any particular differential equation. For example, consider the function

$$x(\varepsilon, t) = e^{-\varepsilon t} \tag{6.1}$$

on $t \geqslant 0$, where ε lies in a neighbourhood of zero. The first few terms of the Taylor expansion in powers of ε are

$$1 - \varepsilon t + \tfrac{1}{2}\varepsilon^2 t^2, \tag{6.2}$$

where the error is

$$O(\varepsilon^3). \tag{6.3}$$

The error estimate (6.3) implies that for any *fixed t*, however large, we can choose ε small enough for the error to be as small as we please, and, indeed, smaller than the smallest term in the approximation, $\tfrac{1}{2}\varepsilon^2 t^2$. However, the trend of the terms shows clearly that *if ε is fixed*, at however small a value (which is the usual situation), t may be chosen large enough to destroy the approximation completely (see Fig. 6.1).

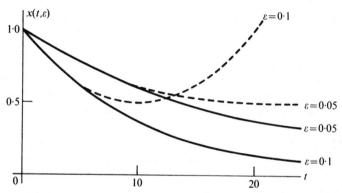

FIG. 6.1. Equations (6.1) and (6.2) compared, for $\varepsilon = 0.05$ and $\varepsilon = 0.1$. ———— $e^{-\varepsilon t}$, ———— $1 - \varepsilon t + \tfrac{1}{2}\varepsilon t^2$

Consider now a function which appears in Section 6.4 as part of the solution to a differential equation:

$$x(\varepsilon, t) = \cos\{(1 - \varepsilon)^{1/2} t\}, \quad 0 \leqslant t < \infty, \tag{6.4}$$

where ε lies in a neighbourhood of zero. The first three terms of the Taylor series for small ε gives

$$\cos t + \tfrac{1}{2}\varepsilon t \sin t + \tfrac{1}{8}\varepsilon^2 (t \sin t - t^2 \cos t) \tag{6.5}$$

with error $O(\varepsilon^3)$. Once again, for t large enough the approximation fails. We can see that it fails when t is so large that εt ceases to be small. The condition that (6.5) should be an approximation *for fixed ε* is that

$$t \ll \varepsilon^{-1}. \tag{6.6}$$

If this is satisfied, then the error, which is dominated by a term like $\varepsilon^3 t^3$, is small. It is also small compared with each term in (6.5), unless t has a fortuitous value making one of these terms very small or zero.† We say that (6.5) *does not provide an approximation uniform on $t \geqslant 0$*. Failure of the ordinary perturbation method to produce a uniform approximation to the solution of a differential equation is very common, and the occasions of failure are called *singular perturbation problems.*

Formally, we require an approximation $\bar{x}(\varepsilon, t)$ to a function $x(\varepsilon, t)$, usually to be valid for an infinite range of t:

$$x(\varepsilon, t) = \bar{x}(\varepsilon, t) + E(\varepsilon, t), \quad t_0 \leqslant t < \infty,$$

with error $E(\varepsilon, t)$, where $\lim_{\varepsilon \to 0} E(\varepsilon, t) = 0$ *uniformly* on $t_0 \leqslant t < \infty$. That is to say, given any $\delta > 0$, there exists $\eta > 0$ independent of t such that

$$|\varepsilon| < \eta \Rightarrow |E(\varepsilon, t)| < \delta.$$

The general question of approximation to functions is further illustrated in Section 6.5.

6.2. Coordinate perturbation (renormalization)

Consider the equation

$$\ddot{x} + x = \varepsilon x^3. \tag{6.7}$$

The expansion

$$x(\varepsilon, t) = x_0(t) + \varepsilon x_1(t) + \dots$$

leads to

$$\ddot{x}_0 + x_0 = 0,$$
$$\ddot{x}_1 + x_1 = x_0^3,$$

and so on. The solution of the first equation is given by

$$x_0(t) = A\,\mathrm{e}^{it} + \bar{A}\,\mathrm{e}^{-it},$$

where A is a complex constant. Therefore

$$\ddot{x}_1 + x_1 = A^3\,\mathrm{e}^{3it} + 3A^2\bar{A}\,\mathrm{e}^{it} + 3\bar{A}^2 A\,\mathrm{e}^{-it} + \bar{A}^3\,\mathrm{e}^{-3it}, \tag{6.8}$$

and secular terms begin to appear. They cannot be eliminated: only if $A = 0$ are they absent. Therefore a series similar in form to (6.5) emerges,

---

† The same conclusion is reached no matter how many terms of the Taylor series we take: if more terms are taken the approximation may be better while it lasts, but it still fails when t becomes comparable with ε^{-1}. Note that if the series is asymptotic rather than convergent as $\varepsilon \to 0$, the approximation does not necessarily improve by taking more terms (Copson, 1965).

and the truncated series does not approximate $x(\varepsilon, t)$ uniformly on $t \geqslant 0$.

This problem was treated in Section 5.8. There, the difficulty was avoided by anticipating a *periodic solution* of period $2\pi/\omega$ in t, ω being unknown; we put

$$\omega = 1 + \varepsilon\omega_1 + \varepsilon^2\omega_2 + \dots$$

where $\omega_1, \omega_2, \dots$, are unknown constants, and then change the variable from t to τ:

$$\tau = \omega t, \tag{6.9}$$

so that the equation in τ has known period 2π (the Lindstedt procedure). Equation (6.9) introduces sufficient free constants ω_i to allow the secular terms to be eliminated. This method can be looked on as a device for obtaining uniform approximations by adopting prior knowledge of the periodicity of the solutions.

We can look at this procedure in another way. Write (6.9) in the form

$$t = \tau/\omega = \tau/(1 + \varepsilon\omega_1 + \varepsilon^2\omega_2 + \dots) = \tau(1 + \varepsilon\tau_1 + \varepsilon^2\tau_2 + \dots), \tag{6.10}$$

where $\tau_1, \tau_2, \dots$, are unknown constant coefficients. Also put, as in the earlier method,

$$x(\varepsilon, t) = X(\varepsilon, \tau) = X_0(\tau) + \varepsilon X_1(\tau) + \varepsilon^2 X_2(\tau) + \dots, \tag{6.11}$$

and substitute (6.10) and (6.11) into (6.7). We know this leads to an expansion uniform on $t \geqslant 0$ since it is equivalent to the Lindstedt procedure. However, the interpretation is now different: it appears that by (6.10) and (6.11) we have made *a simultaneous expansion in powers of ε of both the dependent and the independent variables*, generating an implicit relation between x and t through a parameter τ. The coefficients can be adjusted to eliminate terms which would give non-uniformity (in fact, the 'secular terms'), and so a uniform (and, as it turns out, a periodic) solution is obtained.

We were guided to the form (6.10) by the prior success of the Lindstedt procedure, which ensures its appropriateness. If, however, we assume no prior experience, we have to abandon the assumption of constant coefficients in (6.10). Therefore let us take, in stead of the constants $\tau_1, \tau_2, \dots$, a set of unknown functions $T_1, T_2, \dots$, of τ, and see what happens. We have

$$x(\varepsilon, t) = X(\varepsilon, \tau) = X_0(\tau) + \varepsilon X_1(\tau) + \varepsilon^2 X_2(\tau) + \dots \tag{6.12a}$$

and

$$t = T(\varepsilon, \tau) = \tau + \varepsilon T_1(\tau) + \varepsilon^2 T_2(\tau) + \dots. \tag{6.12b}$$

The first term in the expansion of t remains as τ, since this is appropriate when $\varepsilon \to 0$: τ is called a *strained coordinate*, or a perturbed coordinate.

Equations (6.12a) and (6.12b) are the basis of Lighthill's method (see Section 6.3), in which the expansions are substituted directly into the differential equation. Here we show a different approach. The ordinary perturbation process is fairly easy to apply, and gives a series of the form

$$x(\varepsilon, t) = x_0(t) + \varepsilon x_1(t) + \varepsilon^2 x_2(t) + \dots . \tag{6.13}$$

A finite number of terms of this series will not generally give an approximation to $x(\varepsilon, t)$ holding uniformly for all t, but when (6.12b) is substituted into (6.13), it may be possible to choose $T_1(\tau), T_2(\tau)$, so as to force (6.13) into a form which gives a uniform approximation. This process is called coordinate perturbation (Crocco, 1972), and will be carried out in the following two examples.

Example 6.1 *Obtain an approximate solution of the equation*

$$\ddot{x} + x = \varepsilon x^3,$$

with $x(\varepsilon, 0) = 1$, $\dot{x}(\varepsilon, 0) = 0$, and error $O(\varepsilon^3)$ uniformly on $t \geqslant 0$, by the method of coordinate perturbation.

The expansion (6.13), together with the initial conditions, gives

$$\ddot{x}_0 + x_0 = 0, \qquad x_0(0) = 1, \quad \dot{x}_0(0) = 0;$$
$$\ddot{x}_1 + x_1 = x_0^3, \qquad x_1(0) = 0, \quad \dot{x}_1(0) = 0;$$
$$\ddot{x}_2 + x_2 = 3x_0^2 x_1, \qquad x_2(0) = 0, \quad \dot{x}_2(0) = 0;$$

and so on. Then

$$x(\varepsilon, t) = \cos t + \varepsilon(\tfrac{1}{32} \cos t + \tfrac{3}{8} t \sin t - \tfrac{1}{32} \cos 3t)$$
$$+ \varepsilon^2 (\tfrac{23}{1024} \cos t + \tfrac{3}{32} t \sin t - \tfrac{9}{128} t^2 \cos t - \tfrac{3}{128} \cos 3t - \tfrac{9}{256} t \sin 3t$$
$$+ \tfrac{1}{1024} \cos 5t) + O(\varepsilon^3). \tag{6.14}$$

The expansion is clearly non-uniform on $t \geqslant 0$.

Now put

$$t = \tau + \varepsilon T_1(\tau) + \varepsilon^2 T_2(\tau) + \dots \tag{6.15}$$

into (6.14), expand the terms in powers of ε, and rearrange:

$$X(\varepsilon, \tau) = \cos \tau + \varepsilon(\tfrac{1}{32} \cos \tau - T_1 \sin \tau + \tfrac{3}{8} \tau \sin \tau - \tfrac{1}{32} \cos 3\tau)$$
$$+ \varepsilon^2 (\tfrac{23}{1024} \cos \tau - \tfrac{1}{2} T_1^2 \cos \tau + \tfrac{11}{32} T_1 \sin \tau - T_2 \sin \tau + \tfrac{3}{8} \tau T_1 \cos \tau$$
$$+ \tfrac{3}{32} \tau \sin \tau - \tfrac{9}{128} \tau^2 \cos \tau - \tfrac{3}{128} \cos 3\tau + \tfrac{3}{32} T_1 \sin \tau - \tfrac{9}{128} \tau^2 \cos \tau$$
$$- \tfrac{3}{128} \cos 3\tau + \tfrac{3}{32} T_1 \sin 3\tau - \tfrac{9}{256} \tau \sin 3\tau + \tfrac{1}{1024} \cos 5\tau) + O(\varepsilon^3).$$

To avoid the form $\tau \sin \tau$ in the ε term, which would give an obvious non-uniformity, define T_1 by

$$T_1(\tau) = \tfrac{3}{8}\tau. \tag{6.16}$$

The ε^2 coefficient then becomes

$$\tfrac{23}{1024}\cos \tau - T_2 \sin \tau + \tfrac{57}{256}\tau \sin \tau - \tfrac{3}{128}\cos 3\tau + \tfrac{1}{1024}\cos 5\tau,$$

and we must have

$$T_2(\tau) = \tfrac{57}{256}\tau \tag{6.17}$$

eliminating the non-uniformity $\tau \sin \tau$. On the assumption that this process could continue indefinitely we have finally

$$x = X(\varepsilon, \tau) = \cos \tau + \tfrac{1}{32}\varepsilon(\cos \tau - \cos 3\tau)$$
$$+ \varepsilon^2(\tfrac{23}{1024}\cos \tau - \tfrac{3}{128}\cos 3\tau + \tfrac{1}{1024}\cos 5\tau) + O(\varepsilon^3) \tag{6.18a}$$

where

$$t = \tau(1 + \tfrac{3}{8}\varepsilon + \tfrac{57}{256}\varepsilon^2) + O(\varepsilon^3). \tag{6.18b}$$

Example 6.2 (*Lighthill's equation.*) *Find an approximation, uniform on* $0 \leqslant t \leqslant 1$, *to the solution of*

$$(\varepsilon x + t)\dot{x} + (2 + t)x = 0, \quad \varepsilon \geqslant 0, \tag{6.19}$$

satisfying $x(\varepsilon, 1) = \mathrm{e}^{-1}$.

Begin by noticing the general character of the solution curve for $\varepsilon > 0$.

$$\frac{\mathrm{d}x}{\mathrm{d}t} = -\frac{(2 + t)x}{\varepsilon x + t},$$

so $x > 0$ and $\dot{x} < 0$ at $t = 1$. x therefore increases as $t = 0$ is approached from $t = 1$. It is still finite at $t = 0$, since $\dot{x}$ is infinite only on $\varepsilon x + t = 0$, an asymptote to our solution curve shown on Fig. 6.2. Obviously $x(0)$ is large, but still finite, when $0 < \varepsilon \ll 1$.

The direct approach, writing

$$x(\varepsilon, t) = x_0(t) + \varepsilon x_1(t) + \varepsilon^2 x_2(t) + \ldots \tag{6.20}$$

leads to the sequence of linear equations

$$t\dot{x}_0 + (2 + t)x_0 = 0, \quad x_0(1) = \mathrm{e}^{-1}; \tag{6.21a}$$

$$t\dot{x}_1 + (2 + t)x_1 = -x_0\dot{x}_0, \quad x_1(1) = 0; \tag{6.21b}$$

$$t\dot{x}_2 + (2 + t)x_2 = -\dot{x}_0 x_1 - x_0 \dot{x}_1, \quad x_2(1) = 0; \tag{6.21c}$$

and so on. The solution of (6.21a) is

$$x_0(t) = \mathrm{e}^{-t}/t^2 \tag{6.22}$$

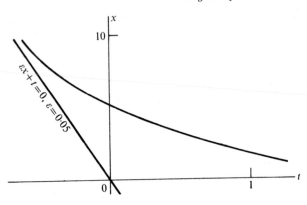

FIG. 6.2.

(predicting, incorrectly, that $x(0) = \infty$). Equation (6.21b) becomes

$$\dot{x}_1 + \left(\frac{2}{t} + 1\right)x_1 = e^{-2t}\left(\frac{2}{t^6} + \frac{1}{t^5}\right).$$

Therefore

$$x_1(t) = \frac{e^{-t}}{t^2} \int_1^t e^{-u}\left(\frac{2}{u^4} + \frac{1}{u^3}\right)du. \qquad (6.23)$$

This is of order $1/t^5$ at $t = 0$, and is even more singular than (6.22). A similar treatment of (6.21c) produces a singularity $O(1/t^8)$ as $t \to 0$. The approximation (6.20):

$$x(\varepsilon, t) \approx x_0(t) + \varepsilon x_1(t),$$

with error $O(\varepsilon^2)$ for fixed t, is clearly not uniform on $0 < t \leqslant 1$ and breaks down completely at $t = 0$.

As in the last example, write the near-identity transformation

$$t = T(\varepsilon, \tau) = \tau + \varepsilon T_1(\tau) + \varepsilon^2 T_2(\tau) + \dots. \qquad (6.24)$$

Then

$$\frac{e^{-t}}{t^2} = \frac{e^{-\tau}}{\tau^2}\left\{1 - \varepsilon T_1(\tau)\left(\frac{2}{\tau} + 1\right)\right\} + O(\varepsilon^2) \qquad (6.25)$$

($O(\varepsilon^2)$ referring to fixed τ, not fixed t) and

$$\int_1^t e^{-u}\left(\frac{2}{u^4} + \frac{1}{u^3}\right)du = \int_1^\tau e^{-u}\left(\frac{2}{u^4} + \frac{1}{u^3}\right)du + \int_\tau^{\tau + \varepsilon T_1(\tau) + \dots} e^{-u}\left(\frac{2}{u^4} + \frac{1}{u^3}\right)du$$

$$= \int_1^\tau e^{-u}\left(\frac{2}{u^4} + \frac{1}{u^3}\right)du + O(\varepsilon). \qquad (6.26)$$

From (6.22), (6.23), (6.25), and (6.26) we find, to $O(\varepsilon^2)$ (referring to fixed τ),

$$x = \frac{e^{-\tau}}{\tau^2}\left\{1 + \varepsilon\left[\int_1^\tau e^{-u}\left(\frac{2}{u^4} + \frac{1}{u^3}\right)du - T_1(\tau)\left(\frac{2}{\tau} + 1\right)\right]\right\}. \quad (6.27)$$

The best thing we can do is to choose T_1 to eliminate the worst singularity for small τ arising from the integral. This is $-\frac{2}{3}\tau^{-3}$; so

$$T_1(\tau) = -\frac{1}{3\tau^2} \quad (6.28)$$

eliminates this singularity (of course, we are left with one of order τ^{-2}).

Finally, from (6.27), (6.24), and (6.28)

$$x = e^{-\tau}/\tau^2 \quad (6.29a)$$

together with

$$t = \tau - \frac{\varepsilon}{3\tau^2} \quad (6.29b)$$

gives a uniform approximation for $0 \leqslant t \leqslant 1$.

Note that there is now no infinity predicted at $t = 0$: from (6.29b), $t = 0$ corresponds to $\tau = (\varepsilon/3)^{1/3}$, and $x(0)$ therefore has the value approximately $(\varepsilon/3)^{-2/3}$. The error in (6.29a, b) is $O(\varepsilon^2)$ for fixed τ, but only $O(\varepsilon^{1/3})$ for fixed t.

6.3. Lighthill's method

We reconsider the equation

$$(\varepsilon x + t)\dot{x} + (2 + t)x = 0, \quad (6.30a)$$

$$x(\varepsilon, 1) = e^{-1}, \quad (6.30b)$$

on $0 \leqslant t \leqslant 1$, using the direct attack suggested in the previous section. We substitute

$$x = X(\varepsilon, \tau) = X_0(\tau) + \varepsilon X_1(\tau) + \dots \quad (6.31a)$$

and

$$t = T(\varepsilon, \tau) = \tau + \varepsilon T_1(\tau) + \dots \quad (6.31b)$$

into the differential equation and boundary condition.

First, we do not expect that $t = 1$ will correspond with $\tau = 1$. Suppose that $t = 1$ corresponds to $\tau = \tau^*(\varepsilon)$. Then we must solve (6.31b) for τ^*; that is

$$1 = \tau^* + \varepsilon T_1(\tau^*) + \dots. \quad (6.32)$$

Then (6.30b), (6.31a) give the transformed boundary condition

$$e^{-1} = X_0(\tau^*) + \varepsilon X_1(\tau^*) + \dots. \quad (6.33)$$

To solve (6.32), assume τ^* is close to 1 for ε small and has the expansion
$$\tau^* = 1 + \varepsilon\tau_1 + \dots, \tag{6.34}$$

where $\tau_1, \dots$ are constants. Equation (6.32) becomes (writing $T_1(\tau^*)$ $= T_1(1) + \varepsilon\tau_1 T_1'(1) + \dots$)

$$1 = 1 + \varepsilon(\tau_1 + T_1(1)) + \dots$$

so $\tau_1 = -T_1(1)$, and from (6.34), the boundary $t = 1$ corresponds to $\tau = \tau^*$, where

$$\tau^* = 1 - \varepsilon T_1(1) + \dots.$$

Therefore, by expanding $X_0(\tau^*)$ and $X_1(\tau^*)$ in (6.33), the boundary condition (6.33) becomes

$$e^{-1} = X_0(1) + \varepsilon(X_1(1) - X_0'(1)T_1(1)) + \dots. \tag{6.35}$$

Next, the derivative in (6.30a) is transformed by

$$\frac{dx}{dt} = \frac{dX}{d\tau} \bigg/ \frac{dT}{d\tau} = \frac{X_0' + \varepsilon X_1' + \dots}{1 + \varepsilon T_1' + \dots} = X_0' + \varepsilon(X_1' - X_0' T_1') + \dots. \tag{6.36}$$

Thus eqn (6.30a) becomes, to order ε,

$$(\varepsilon X_0 + \tau + \varepsilon T_1)(X_0' + \varepsilon\{X_1' - X_0' T_1'\}) + (2 + \tau + \varepsilon T_1)(X_0 + \varepsilon X_1) = 0. \tag{6.37}$$

Equations (6.35) and (6.37) give

$$\tau X_0' + (2 + \tau)X_0 = 0, \quad X_0(1) = e^{-1}; \tag{6.38a}$$

$$\tau X_1' + (2 + \tau)X_1 = -T_1(X_0' + X_0) + \tau X_0' T_1' - X_0 X_0',$$
$$X_1(1) = X_0'(1)T_1(1). \tag{6.38b}$$

From (6.38a) the zero-order approximation is

$$X_0(\tau) = e^{-\tau}/\tau^2. \tag{6.39}$$

Then (6.38b) becomes

$$\tau X_1' + (2 + \tau)X_1 = \frac{2e^{-\tau}}{\tau^3}T_1 - e^{-\tau}\left(\frac{2}{\tau^2} - \frac{1}{\tau}\right)T_1' + e^{-2\tau}\left(\frac{2}{\tau^5} - \frac{1}{\tau^4}\right), \tag{6.40}$$

with initial condition

$$X_1(1) = -3e^{-1} T_1(1).$$

We now have a free choice of $T_1(\tau)$: it could be chosen to make the right-hand side of (6.40) zero, for example; this would lead to a solution for (6.40) of the type (6.39) again, but is in any case impracticable. We shall choose T_1 to nullify the worst visible singularity on the right of (6.40), which is of order $1/\tau^5$. We attempt this by writing $e^{-\tau}, e^{-2\tau} \approx 1$ for small τ and solving

$$\frac{2}{\tau^3} T_1 - \frac{2}{\tau^2} T_1' + \frac{2}{\tau^5} = 0.$$

The simplest solution is

$$T_1(\tau) = -1/3\tau^2 \tag{6.41}$$

(compare (6.28)). We have therefore achieved the same result as in the last section; though, in this example, at considerably more effort. Note that throughout the argument we have regarded the equation as a member of a family of equations with parameter ε, as in Chapter 5.

6.4. Multiple time scales (two-timing)

This method can be used to extend the interval of validity of the simple perturbation method. We shall illustrate it with the linear equation

$$\ddot{x} + x = \varepsilon x, \quad t \geqslant 0. \tag{6.42}$$

This has the general solution which it is convenient to write in the form (*cf.* eqn (6.4))

$$x(\varepsilon, t) = a \cos \{(1-\varepsilon)^{1/2} t + \alpha\} \tag{6.43}$$

where a, α are arbitrary. We can expand this in powers of ε:

$$x(\varepsilon, t) = x_0(t) + \varepsilon x_1(t) + \ldots + \varepsilon^{n-1} x_{n-1}(t) + R_n(\varepsilon, t),$$

where R_n is the remainder after n terms. (This is also the expansion given by the ordinary perturbation method.) To three terms we have the obviously non-uniform approximation

$$x(\varepsilon, t) \approx a \cos (t + \alpha) + \tfrac{1}{2}\varepsilon a t \sin (t + \alpha)$$
$$+ \tfrac{1}{8}\varepsilon^2 a \{t \sin (t + \alpha) - t^2 \cos (t + \alpha)\}. \tag{6.44}$$

If any fixed number of terms is taken the approximation eventually starts to fail as t increases since the remainder will cease to be small compared with the terms retained. For example, in (6.44) the terms retained consist of bounded sines and cosines, together with the combinations

$$\varepsilon t, \quad \varepsilon^2 t, \quad \varepsilon^2 t^2,$$

and $R_3 = O(\varepsilon^3 t^3)$. Therefore, so long as $t = O(1)$, the terms in (6.44) are successively $O(1), O(\varepsilon), O(\varepsilon^2)$, and the remainder is $O(\varepsilon^3)$. However, when t becomes comparable with $1/\varepsilon$, say

$$t = O(\varepsilon^{-1}),$$

the terms are all of order 1 and so is the remainder. The approximation then fails. The failure will occur for $t = O(\varepsilon^{-1})$ no matter how many terms are taken. This is the situation described in Section 6.1.

Now, other series representations, still retaining the general power series form, but satisfactory over a larger range of t, are possible. For instance, if we pick out the first two terms in the Taylor series for $(1-\varepsilon)^{1/2}$ by writing

$$(1-\varepsilon)^{1/2}t \equiv (1-\tfrac{1}{2}\varepsilon)t + \{(1-\varepsilon)^{1/2} - (1-\tfrac{1}{2}\varepsilon)\}t, \tag{6.45}$$

and correspondingly expand (6.43) over the first couple of terms, we find that

$$x(\varepsilon, t) = a\cos\left[(t - \tfrac{1}{2}\varepsilon t) + \alpha\right] + \tfrac{1}{8}\varepsilon^2 at \sin\left[(t - \tfrac{1}{2}\varepsilon t) + \alpha\right] + O(\varepsilon^4 t^2). \tag{6.46}$$

In general, the remainder after n terms is of $O(\varepsilon^{2n}t^n)$. This expansion is well-behaved for t even as large as

$$t = O(\varepsilon^{-1})$$

the successive terms being $O(1), O(\varepsilon), \ldots, O(\varepsilon^{n-1})$ and the remainder $O(\varepsilon^n)$. To obtain an expansion valid for t as large as $O(\varepsilon^{-2})$ we should have to write

$$(1-\varepsilon)^{1/2}t \equiv (1 - \tfrac{1}{2}\varepsilon - \tfrac{1}{8}\varepsilon^2)t + O(\varepsilon^3 t),$$

recast (6.45) and (6.46) accordingly, and so on for further extensions of the range of t (Nayfeh, 1973).

The procedure is possible when we already know the exact solution; the problem is to achieve an expansion of this type starting only with a differential equation. In the following process we aim at attaining an approximation with an error of $O(\varepsilon^2)$, at the cost of accepting a limitation on the range of t in non-periodic cases:

$$t = O(\varepsilon^{-1}). \tag{6.47}$$

When $x(\varepsilon, t)$ is periodic, however, the approximation will be uniform for $t \geqslant 0$.

Introduce a new time-like variable η ('slow time') defined by

$$\eta = \varepsilon t. \tag{6.48}$$

Then for $t = O(\varepsilon^{-1})$,

$$\eta = O(1). \tag{6.49}$$

The expansion (6.46) has the form

$$a \cos (t - \tfrac{1}{2}\eta + \alpha) + \tfrac{1}{8}\varepsilon a\eta \sin (t - \tfrac{1}{2}\eta + \alpha) + O(\varepsilon^2), \tag{6.50}$$

in which the function $x(\varepsilon, t)$ begins to emerge in the form

$$x(\varepsilon, t) = X(\varepsilon, t, \eta). \tag{6.51}$$

Now t and η are not in fact independent, being connected by (6.48), but X is certainly *a solution of some equation in which t and η figure as independent variables*. We obtain the required equation in the following way.

When $\eta = \varepsilon t$

$$\frac{dx}{dt} = \frac{d}{dt} X(\varepsilon, t, \varepsilon t) = \frac{\partial X}{\partial t} + \varepsilon \frac{\partial X}{\partial \eta}, \tag{6.52}$$

and

$$\frac{d^2 x}{dt^2} = \frac{\partial^2 X}{\partial t^2} + 2\varepsilon \frac{\partial^2 X}{\partial \eta \partial t} + \varepsilon^2 \frac{\partial^2 X}{\partial \eta^2}. \tag{6.53}$$

Since x satisfies (6.42), X must satisfy

$$\frac{\partial^2 X}{\partial t^2} + 2\varepsilon \frac{\partial^2 X}{\partial t \partial \eta} + \varepsilon^2 \frac{\partial^2 X}{\partial \eta^2} + X = \varepsilon X. \tag{6.54}$$

We now carry out a straightforward perturbation expansion by writing

$$X(\varepsilon, t, \eta) = X_0(t, \eta) + \varepsilon X_1(t, \eta) + \ldots, \tag{6.55}$$

expecting the remainder, or error, after the nth term to be $O(\varepsilon^n)$. Matching the powers of ε in (6.54) we obtain

$$\frac{\partial^2 X_0}{\partial t^2} + X_0 = 0, \tag{6.56a}$$

$$\frac{\partial^2 X_1}{\partial t^2} + X_1 = X_0 - 2\frac{\partial^2 X_0}{\partial t \partial \eta}, \tag{6.56b}$$

$$\frac{\partial^2 X_2}{\partial t^2} + X_2 = X_1 - 2\frac{\partial^2 X_1}{\partial t \partial \eta} - \frac{\partial^2 X_0}{\partial \eta^2}, \tag{6.56c}$$

and so on.

The general solution of (6.56a) is

$$X_0(t, \eta) = A_0(\eta) e^{it} + \bar{A}_0(\eta) e^{-it}, \tag{6.57}$$

where A_0 is an arbitrary function. When (6.56a) is substituted into (6.56b) we have

$$\frac{\partial^2 X_1}{\partial t^2} + X_1 = e^{it}(A_0 - 2iA_0') + \text{complex conjugate.} \tag{6.58}$$

The secular terms must be removed; since otherwise X_1 will be $O(t)$, and εX_1 of order εt, instead of $\varepsilon^2 t$. Therefore we require

$$A_0 - 2i\frac{dA_0}{d\eta} = 0$$

so that

$$A_0(\eta) = A e^{-\frac{1}{2}i\eta} \tag{6.59}$$

where A is any constant. From (6.57) we obtain

$$X_0(t, \eta) = A e^{i(t - \frac{1}{2}\eta)} + \text{complex conjugate} = a\cos(t - \tfrac{1}{2}\eta + \alpha), \tag{6.60}$$

where $A = \frac{1}{2}a e^{i\alpha}$. Thus we have recovered the first term of (6.50).

The right-hand side of (6.56b) is now zero, so

$$X_1(t, \eta) = A_1(\eta) e^{it} + \bar{A}_1(\eta) e^{-it} \tag{6.61}$$

is its general solution. Putting (6.60) and (6.61) into (6.56c) we find

$$\frac{\partial^2 X_2}{\partial t^2} + X_2 = (A_1 - 2iA_1' + \tfrac{1}{4}A e^{-\frac{1}{2}i\eta}) e^{it} + \text{complex conjugate.} \tag{6.62}$$

Again it is necessary to remove the secular terms by equating the bracket to zero. We find that

$$\begin{aligned} X_1(t, \eta) &= -\tfrac{1}{8}A\eta i e^{i(t - \frac{1}{2}\eta)} + \text{complex conjugate} \\ &= \tfrac{1}{8}a\eta \sin(t - \tfrac{1}{2}\eta + \alpha), \end{aligned} \tag{6.63}$$

and we have recovered the second term of (6.50).

Example 6.3 *Find solutions of Rayleigh's equation*

$$\ddot{x} + \varepsilon(\tfrac{1}{3}\dot{x}^3 - \dot{x}) + x = 0, \tag{6.64}$$

correct to $O(\varepsilon)$ so long as $t = O(\varepsilon^{-1})$.

We are not concerned only with the limit cycle. Suppose that x has the expansion

$$x(\varepsilon, t) = X(\varepsilon, t, \eta) = X_0(t, \eta) + \varepsilon X_1(t, \eta) + O(\varepsilon^2) \tag{6.65}$$

at least for

$$t = O(\varepsilon^{-1}) \tag{6.66}$$

By substituting (6.52) and (6.53) into (6.64), retaining terms up to order ε, and matching the coefficients of ε we obtain

$$\frac{\partial^2 X_0}{\partial t^2} + X_0 = 0, \tag{6.67}$$

$$\frac{\partial^2 X_1}{\partial t^2} + X_1 = \frac{\partial X_0}{\partial t} - \frac{1}{3}\left(\frac{\partial X_0}{\partial t}\right)^3 - 2\frac{\partial^2 X_0}{\partial t \partial \eta}. \tag{6.68}$$

Equation (6.67) has the general solution

$$X_0(\eta, t) = A_0(\eta)\,\mathrm{e}^{\mathrm{i}t} + \bar{A}_0(\eta)\,\mathrm{e}^{-\mathrm{i}t}. \tag{6.69}$$

Equation (6.68) becomes

$$\frac{\partial^2 X_1}{\partial t^2} + X_1 = \mathrm{i}(A_0 - A_0^2\bar{A}_0 - 2A_0')\,\mathrm{e}^{\mathrm{i}t} - \tfrac{1}{3}A_0^3\,\mathrm{e}^{3\mathrm{i}t}$$

$$+ \text{complex conjugate.} \tag{6.70}$$

The secular terms arise from the presence of $\mathrm{e}^{\mathrm{i}t}, \mathrm{e}^{-\mathrm{i}t}$ in (6.70), and vanish when

$$A_0' - \tfrac{1}{2}A_0 + \tfrac{1}{2}A_0^2\bar{A}_0 = 0. \tag{6.71}$$

The conjugate equation gives nothing new. To solve (6.71), write

$$A_0(\eta) = \rho(\eta)\exp(\mathrm{i}\alpha(\eta)), \tag{6.72}$$

where ρ and α are real functions. We obtain

$$\rho' - \tfrac{1}{2}\rho + \tfrac{1}{2}\rho^3 = 0,$$

$$\alpha' = 0;$$

from which

$$\rho(\eta) = (1 + A\,\mathrm{e}^{-\eta})^{-1/2}, \quad A \text{ real,}$$
$$\alpha(\eta) = \text{real constant.} \tag{6.73}$$

From (6.72) and (6.73)

$$A_0(\eta) = (1 + A\,\mathrm{e}^{-\eta})^{-1/2}\,\mathrm{e}^{\mathrm{i}\alpha},$$

where A, α are real.

Equation (6.69) becomes

$$X_0(t, \eta) = 2\mathrm{Re}\{A_0(\eta)\,\mathrm{e}^{\mathrm{i}t}\} = 2(1 + A\,\mathrm{e}^{-\eta})^{-1/2}\cos(t + \alpha);$$

or

$$X_0(t, \eta) = 2(1 + A\,\mathrm{e}^{-\varepsilon t})^{-1/2}\cos(t + \alpha),$$

after writing $\eta = \varepsilon t$. From the initial conditions

$$x(0) = a, \quad \dot{x}(0) = 0$$

we obtain

$$a = 2(1+A)^{-1/2}\cos\alpha, \quad 0 = -2(1+A)^{-1/2}\sin\alpha + O(\varepsilon).$$

Therefore, to order ε, $\alpha = 0$ and $A = -1 + 4/a^2$. Finally

$$x(t) = 2[1 - (1 - 4/a^2)e^{-\varepsilon t}]^{-1/2}\cos t \qquad (6.74)$$

to order ε, at least when $t = O(\varepsilon^{-1})$. The limit cycle is given by $a = 2$.

6.5. Matching approximations on an interval

In this section we shall mainly be concerned with boundary-value problems, and we conform with the literature by using y as the dependent and x as the independent variable. For fuller information on the methods of this section the reader is referred to Nayfeh (1973) and O'Malley (1974).

As in Section 6.1, we shall illustrate the problem by looking at the approximation to a particular function y:

$$y(\varepsilon, x) = e^{-\frac{1}{2}x} - e^{\frac{1}{2}x}e^{-2x/\varepsilon}, \quad 0 \leqslant x \leqslant 1, \quad \varepsilon \geqslant 0. \qquad (6.75)$$

(The particular form chosen appears in the solution to a differential equation later on.)

Note that, for $x > 0$,

$$\lim_{\varepsilon \to 0+} (e^{-2x/\varepsilon}/\varepsilon^n) = 0$$

for every positive n: the function tends to zero very rapidly as $\varepsilon \to 0+$. Therefore, for every *fixed* $x > 0$

$$y(\varepsilon, x) \approx e^{-\frac{1}{2}x} \qquad (6.76)$$

with error $O(\varepsilon^n)$ for every $n > 0$. But by looking back at (6.75) it can be seen that as x takes smaller and smaller values, smaller and smaller values of ε are required before (6.76) becomes an acceptable approximation. It fails altogether at $x = 0$, where (6.75) gives zero and (6.76) gives 1. Therefore (6.76) is not a uniform approximation on $0 \leqslant x \leqslant 1$, and another form is needed near $x = 0$. Figure 6.3 shows the nature of the approximation for some particular values of ε; as ε decreases the interval of good fit becomes extended, but it is always poor near $x = 0$. The approximation begins to fail when x is comparable with ε in magnitude. The region of failure is called a *boundary layer* (from a hydrodynamical analogy), having in this case *thickness of order ε*. The function (6.76) is called the *outer approximation* to y.

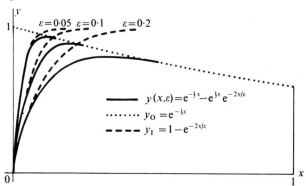

FIG. 6.3. Inner approximation, y_I, and outer approximation, y_O, to $y(\varepsilon, x)$ (eqn 6.75)

To get an approximation near $x = 0$ it is no use trying to work with x fixed: this is covered by (6.76). We therefore consider x tending to zero with ε by putting, say

$$x(\varepsilon) = \xi\varepsilon, \tag{6.77}$$

where ξ may take any value. ξ is called a *stretched variable*: from one point of view we are magnifying the boundary layer to thickness $O(1)$. Then

$$y(\varepsilon, x) \approx 1 - e^{-2\xi}. \tag{6.78}$$

The error is $O(\varepsilon)$ for every fixed ξ, though naturally it works better when ξ is not too large. We can express this idea alternatively by saying

$$y(\varepsilon, x) \approx 1 - e^{-2x/\varepsilon} \tag{6.79}$$

with error $O(\varepsilon)$ so long as $x = O(\varepsilon)$ (which includes the case $x = o(\varepsilon)$ too). This is rather like the situation described for $t > 0$ at the beginning of Section 6.1. The approximation (6.79) is shown in Fig. 6.3. The function (6.79) is called the *inner approximation* to y.

If our information about y consisted only of the approximations (6.76) and (6.79), together with the associated error estimates, it would be pure guesswork, given a value of x and a value of ε, to decide which to use, since an error of low *order* in ε is not necessarily small for any *given* small value of ε. We can see what the regions of validity look like in a general way by experimenting and displaying the results in the plane of ε, x. The regions in which (6.76) and (6.79) give errors of less than 0.05 and 0.01 are shown in Fig. 6.4.

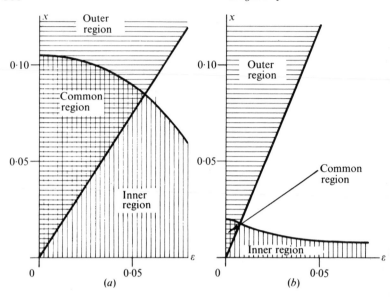

FIG. 6.4. Regions of validity for the approximations (6.76) (outer approximation) and (6.79) (inner approximation) with error E: (a) $E = 0.05$; (b) $E = 0.01$

The figures show that there is a region on the ε, x plane where (6.76) and (6.79) both give a small error. We therefore expect that there may be cases 'between' the cases $x = $ constant and $x = O(\varepsilon)$, in which both approximations have a small remainder. For example, if

$$x = \eta\sqrt{\varepsilon}, \quad \eta \text{ constant}, \tag{6.80}$$

$y(\varepsilon, x)$ becomes

$$e^{-\frac{1}{2}\eta\sqrt{\varepsilon}} - e^{\frac{1}{2}\eta\sqrt{\varepsilon}} e^{-2\eta/\sqrt{\varepsilon}} = 1 + O(\sqrt{\varepsilon}); \tag{6.81}$$

also (6.76) becomes

$$e^{-\frac{1}{2}\eta\sqrt{\varepsilon}} = 1 + O(\sqrt{\varepsilon}) \tag{6.82}$$

and (6.79) becomes

$$1 - e^{-2\eta/\sqrt{\varepsilon}} = 1 + O(\sqrt{\varepsilon}); \tag{6.83}$$

(in fact, $1 + o(\sqrt{\varepsilon})$). Thus the original function and both approximations have an error tending to zero with ε when $x = \eta\sqrt{\varepsilon}, \eta$ fixed. We say that the functions 'match' to $o(1)$ in the 'overlap region'.

Figure 6.5 indicates the progress of a point $(\varepsilon, \eta\sqrt{\varepsilon}) = (\varepsilon, x)$ as it moves into regions where the two approximations (6.76) and (6.79) have in common an error diminishing to zero with ε. It is desirable to show that there are no 'gaps' in the postulated 'common region'. Instead of (6.80),

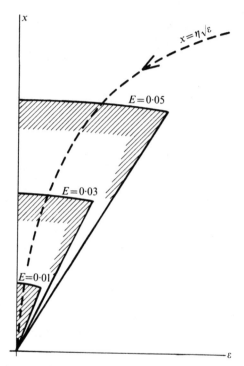

FIG. 6.5. As $\varepsilon \to 0$, the point $(\varepsilon, \eta\sqrt{\varepsilon})$ (η constant) always lies in the 'overlap region' for the approximations (6.76), (6.79). Error E.

therefore, consider the more general case

$$x(\varepsilon) = \zeta\psi(\varepsilon),$$

where ζ is any constant and ψ tends to zero, but more slowly than ε, so that

$$\lim_{\varepsilon \to 0} \varepsilon/\psi(\varepsilon) = 0.$$

Then y, from (6.75), becomes

$$y(\varepsilon, x) = e^{-\frac{1}{2}\zeta\psi(\varepsilon)} - e^{\frac{1}{2}\zeta\psi(\varepsilon)}e^{-2\zeta\psi(\varepsilon)/\varepsilon} = 1 + o(1).$$

The outer approximation (6.76) gives

$$e^{-\frac{1}{2}x} = e^{-\frac{1}{2}\zeta\psi(\varepsilon)} = 1 + o(1),$$

and the inner approximation (6.79) gives

$$1 - e^{-2x/\varepsilon} = 1 - e^{-2\zeta\psi(\varepsilon)/\varepsilon} = 1 + o(1).$$

These are in agreement to $o(1)$.

The following example shows how the assumption of a 'common region' is used in connection with differential equations to establish an unknown constant. In this case we have put

$$\psi(\varepsilon) = \varepsilon^{1-\delta}, \quad 0 < \delta < 1,$$

to give a certain generality.

Example 6.4 *A function y has the two approximations on $0 \leqslant x \leqslant 1$: an inner approximation*

$$y(\varepsilon, x) \approx A + (1 - A)e^{-x/\varepsilon} = y_{\mathrm{I}}, \text{ say,} \qquad (6.84)$$

with error $O(\varepsilon)$ when $x = O(\varepsilon)$, and an outer approximation

$$y(\varepsilon, x) \approx e^{1-x} = y_{\mathrm{O}}, \text{ say,} \qquad (6.85)$$

with error $O(\varepsilon)$ for x constant. Find the value of A.

Assuming that both approximations are valid (though possibly with a larger error) for

$$x = \eta\varepsilon^{1-\delta}, \quad 0 < \delta < 1, \quad \eta \text{ constant,}$$

we must have at least

$$\lim_{\substack{\varepsilon \to 0 \\ (\eta \text{ constant})}} \left[A + (1 - A)e^{-\eta/\varepsilon^{\delta}} \right] = \lim_{\substack{\varepsilon \to 0 \\ (\eta \text{ constant})}} e^{1 - \eta\varepsilon^{1-\delta}}$$

Therefore $A = e^{1}$.

Example 6.5 *From (6.84) and (6.85) make up a single approximate expression applying uniformly on $0 \leqslant x \leqslant 1$.*

Write, (with $A = e^{1}$),

$$y_{\mathrm{I}} + y_{\mathrm{O}} = e^{1} + (1 - e^{1})e^{-x/\varepsilon} + e^{1-x}.$$

When x is constant and $\varepsilon \to 0$, we obtain

$$e^{1} + e^{1-x}$$

instead of e^{1-x} required by (6.85); and when $x = \xi\varepsilon$ and $\varepsilon \to 0$ (ξ constant), we obtain

$$2e^1 + (1 - e^1)e^{-\xi}$$

instead of $e^1 + (1 - e^1)e^{-\xi}$ required by (6.84). Therefore the required uniform expansion is given by

$$y_1 + y_0 - e^1 = (1 - e^1)e^{-x/\varepsilon} + e^{1-x}.$$

6.6. A matching technique for differential equations

In the following problems the solution has different approximations in different (x, ε) regions as described in Section 6.5. Characteristic of these is *the presence of ε as the coefficient of the highest derivative appearing in the equation*. The term in the highest derivative is, then, negligible except where the derivative itself is correspondingly large. Thus, over most of the range the equation is effectively of lower order, and can satisfy fewer boundary conditions. However, in certain intervals (boundary layers) the derivative may be large enough for the term to be significant. In such intervals, y will change very rapidly, and may be caused to satisfy another boundary condition.

Example 6.6 (An initial value problem) *Find a first-order approximation to the solution of*

$$\varepsilon \frac{dy}{dx} + y = x, \quad x > 0, \quad \varepsilon > 0, \tag{6.86}$$

subject to

$$y(\varepsilon, 0) = 1. \tag{6.87}$$

The obvious first step is to put $\varepsilon = 0$ into (6.86): this gives the 'outer approximation'

$$y \approx y_0 = x \tag{6.88}$$

The error is $O(\varepsilon)$, as can be seen by viewing it as the first step in a perturbation process. It is not clear where this holds, but assume we have numerical or other indications that it is right except near $x = 0$, where there is a boundary layer. We do not know how thick this is, so we allow some freedom in choosing a new, stretched variable, ξ, by writing

$$x = \xi\phi(\varepsilon) \tag{6.89}$$

where

$$\lim_{\varepsilon \to 0} \phi(\varepsilon) = 0.$$

Equation (6.86) becomes

$$\frac{\varepsilon}{\phi(\varepsilon)} \frac{dy}{d\xi} + y = \xi\phi(\varepsilon). \tag{6.90}$$

The choice of $\phi(\varepsilon)$ which retains $dy/d\xi$ as a leading term in order of magnitude, and which gives an equation which can describe a boundary layer is

$$\phi(\varepsilon) = \varepsilon; \tag{6.91}$$

and (6.90) becomes

$$\frac{dy}{d\xi} + y = \varepsilon\xi. \tag{6.92}$$

The first approximation to the solution fitting the initial condition (6.87) (the inner approximation), is given by

$$y \approx y_1 = e^{-\xi} \tag{6.93}$$

with error $O(\varepsilon)$ as before. This may be interpreted

$$y_1 = e^{-x/\varepsilon}, \quad x = O(\varepsilon). \tag{6.94}$$

There is no scope for 'matching' y_0 and y_1 here: either they are approximations to the same solution or they are not; there are no arbitrary constants left to be settled. We note, however, that in an 'intermediate region' such as

$$x = \eta\sqrt{\varepsilon}$$

for η constant, both approximations do agree to order $\sqrt{\varepsilon}$.

To construct a uniform approximation, start with the form $y_1 + y_0$ as in Example 6.5:

$$y_1 + y_0 = e^{-x/\varepsilon} + x. \tag{6.95}$$

This agrees to order ε with eqns (6.88) and (6.93) for $x = $ constant and $x = \xi\varepsilon$ respectively, so that eqn (6.95) is already a uniform approximation. (The reader should compare the exact solution:

$$y(\varepsilon, x) = x - \varepsilon + e^{-x/\varepsilon} + \varepsilon e^{-x/\varepsilon}.)$$

Example 6.7 (A boundary-value problem) *Find a first approximation to the equation*

$$\varepsilon\frac{d^2y}{dx^2} + 2\frac{dy}{dx} + y = 0, \quad 0 < x < 1 \tag{6.96}$$

subject to

$$y(0) = 0, \quad y(1) = 1. \tag{6.97}$$

Putting $\varepsilon = 0$ in (6.96) (or finding the first term in an ordinary perturbation process), gives the differential equation for the outer approximation $y_0(\varepsilon, x)$:

$$2\frac{dy_0}{dx} + y_0 = 0. \tag{6.98}$$

Since this is first order, only one boundary condition can be satisfied. We shall assume (if necessary by showing that the contrary assumption leads to failure of the method) that (6.98) holds approximately at $x = 1$. We require $y_0(\varepsilon, 1) = 1$, so

$$y \approx y_0(x, \varepsilon) = e^{1/2} e^{-\frac{1}{2}x} \tag{6.99}$$

with error $O(\varepsilon)$. The non-uniformity is not self-evident, but certainly the boundary condition at $x = 0$ is not satisfied. Assuming that the condition is attained by a sudden change in the nature of the solution near $x = 0$, introduce a new, stretched, variable ξ, where

$$x = \xi\phi(\varepsilon), \tag{6.100}$$

ξ fixed, where

$$\lim_{\varepsilon \to 0} \phi(\varepsilon) = 0. \tag{6.101}$$

Equation (6.96) becomes

$$\frac{\varepsilon}{\phi^2(\varepsilon)} \frac{d^2y}{d\xi^2} + \frac{2}{\phi(\varepsilon)} \frac{dy}{d\xi} + y = 0.$$

The choice

$$\phi(\varepsilon) = \varepsilon \tag{6.102}$$

yields the equation

$$\frac{d^2y}{d\xi^2} + 2\frac{dy}{d\xi} + \varepsilon y = 0 . \tag{6.103}$$

which simplifies to the equation for the inner approximation y_1:

$$\frac{d^2y_1}{d\xi^2} + 2\frac{dy_1}{d\xi} = 0 \tag{6.104}$$

with boundary condition

$$y_1(\varepsilon, 0) = 0. \tag{6.105}$$

Therefore

$$y \approx y_1(\varepsilon, x) = A(1 - e^{-2\xi}) \tag{6.106}$$

where ξ is a constant, the error being $O(\varepsilon)$.

The value of A must be determined by the condition that (6.99) and (6.106) should both approximate to the same function (the solution): if there is an 'overlapping region' of approximation to some order, the approximations themselves must agree to this order, and there is only one choice of A making this agreement possible.

Allow the behaviour of x to be given by

$$x = \eta\psi(\varepsilon) \tag{6.107}$$

where η is any constant, and

$$\lim_{\varepsilon \to 0} \psi(\varepsilon) = 0 \tag{6.108}$$

but where ψ does not tend to zero so fast as ϕ, so that

$$\lim_{\varepsilon \to 0} \phi(\varepsilon)/\psi(\varepsilon) = \lim_{\varepsilon \to 0} \varepsilon/\psi(\varepsilon) = 0. \tag{6.109}$$

(Previously we put $\psi(\varepsilon) = \sqrt{\varepsilon}$, which satisfies these conditions.) Then

$$y_0(\varepsilon, x) = e^{1/2} e^{-\frac{1}{2}\eta\psi(\varepsilon)} = e^{\frac{1}{2}} + O(\psi) \tag{6.110}$$

by (6.108), and

$$y_1(\varepsilon, x) = A(1 - e^{-2\eta\psi/\varepsilon}) = A + o(1) \tag{6.111}$$

by (6.109). Agreement is only possible if $A = e^{1/2}$, so finally

and
$$y \approx y_0(\varepsilon, x) = e^{\frac{1}{2}} e^{-\frac{1}{2}x} \quad \text{for} \quad x = O(1) \tag{6.112}$$

$$y \approx y_1(\varepsilon, x) = e^{1/2}(1 - e^{-2x/\varepsilon}) \quad \text{for} \quad x = O(\varepsilon). \tag{6.113}$$

To find an approximation uniform on $0 \leqslant x \leqslant 1$, form

$$y_0 + y_1 = e^{1/2} e^{-\frac{1}{2}x} + e^{1/2}(1 - e^{-2x/\varepsilon}).$$

Then for $x = O(1)$ this becomes $e^{1/2} e^{-\frac{1}{2}x} + e^{1/2}$ as $\varepsilon \to 0$; that is it has the unwanted term $e^{1/2}$ in it. Similarly, when $x = \xi\varepsilon$ it becomes $e^{1/2} + e^{1/2}(1 - e^{-2\xi})$ as $\varepsilon \to 0$, which again contains an unwanted $e^{1/2}$. Therefore

$$y \approx y_0 + y_1 - e^{1/2} = e^{1/2}(e^{-\frac{1}{2}x} - e^{-2x/\varepsilon}) \tag{6.114}$$

is a uniform approximation on $0 \leqslant x \leqslant 1$.

The reader should compare the exact solution

$$y(\varepsilon, x) = (e^{\lambda_1 x} - e^{\lambda_2 x})/(e^{\lambda_1} - e^{\lambda_2}) \tag{6.115}$$

where λ_1, λ_2 are the roots of $\varepsilon\lambda^2 + 2\lambda + 1 = 0$: the expansions of the roots are

$$\lambda_1 = -\tfrac{1}{2} - \tfrac{1}{8}\varepsilon - \dots,$$

$$\lambda_2 = -\frac{2}{\varepsilon} + \tfrac{1}{2} + \tfrac{1}{8}\varepsilon + \dots.$$

The expansion and matching process may be extended to take in terms beyond the first-order terms we have considered here. Details are given by Nayfeh (1973).

Exercises

1. Work through the details of Example 6.1 to obtain an approximate solution of $\ddot{x} + x = \varepsilon x^3$, with $x(\varepsilon, 0) = 1$, $\dot{x}(\varepsilon, 0) = 0$, with error $O(\varepsilon^3)$ uniformly on $t \geqslant 0$.

Obtain the coefficients of $\cos t$ and $\cos 3t$ to order ε^3 for the approximate pendulum equation.

2. How does the period obtained by the method of Exercise 1 compare with that derived in Exercise 34, Chapter 1?

3. Apply the method of Exercise 1 to the equation

$$\ddot{x} + x = \varepsilon x^3 + \varepsilon^2 \alpha x^5$$

with $x(\varepsilon, 0) = 1$, $\dot{x}(\varepsilon, 0) = 0$. Obtain the period to order ε^3, and confirm that the period is correct for the pendulum when the right-hand side is the first two terms in the expansion of $x - \sin x$. (Compare the result of Exercise 33, Chapter 1. To obtain the required equations simply add the appropriate term to the right-hand side of the equation for x_2 in Example 6.1.)

4. Use a substitution to show that the case considered in Exercise 1 covers all boundary conditions $x(\varepsilon, 0) = a$, $\dot{x}(\varepsilon, 0) = 0$.

5. The equation for the relativistic perturbation of a planetary orbit is

$$\frac{d^2 u}{d\theta^2} + u = k(1 + \varepsilon u^2)$$

(see Exercise 22, Chapter 5). Apply the coordinate perturbation technique to eliminate the secular term in $u_1(\theta)$ in the expansion $u(\varepsilon, \theta) = u_0(\theta) + \varepsilon u_1(\theta) + \dots$, with $\theta = \phi + \varepsilon T_1(\phi) + \dots$. Confirm that the perihelion of the orbit advances by approximately $2\pi\varepsilon k^2$ in each planetary year.

6. Apply the multiple scale method to van der Pol's equation $\ddot{x} + \varepsilon(x^2 - 1)\dot{x} + x = 0$. Show that, for $x(0) = a$, $\dot{x}(0) = 0$, and for $t = O(\varepsilon^{-1})$,

$$x = 2\{1 + (4/a^2 - 1)e^{-\varepsilon t}\}^{-1/2} \cos t.$$

7. Apply the multiple-scale method to the equation $\ddot{x} + x - \varepsilon x^3 = 0$, with initial conditions $x(0) = a$, $\dot{x}(0) = 0$. Show that, for $t = O(\varepsilon^{-1})$,

$$x(t) = a \cos \{t(1 - \tfrac{3}{8}\varepsilon a^2)\}.$$

8. Obtain the exact solution of Example 6.7, and show that it has the first approximation equal to that obtained by the matching method.

9. Consider the problem

$$\varepsilon y'' + y' + y = 0, \quad y(\varepsilon, 0) = 0, \quad y(\varepsilon, 1) = 1,$$

on $0 \leqslant x \leqslant 1$, where ε is small and positive.
  (i) Obtain the outer approximation

$$y(\varepsilon, x) \approx y_O = e^{1-x}, \quad x \text{ fixed}, \quad \varepsilon \to 0+ ;$$

and the inner approximation

$$y(\varepsilon, x) \approx y_I = C(1 - e^{-x/\varepsilon}), \quad x = O(\varepsilon), \quad \varepsilon \to 0+,$$

where C is a constant.

(ii) Obtain the value of C by matching y_0 and y_1 in the intermediate region.

(iii) Construct from y_0 and y_1 a first approximation to the solution which is uniform on $0 \leqslant x \leqslant 1$.

10. Repeat the procedure of Exercise 9 for the problem

$$\varepsilon y'' + y' + xy = 0, \quad y(0) = 0, \quad y(1) = 1$$

on $0 \leqslant x \leqslant 1$.

11. Explain why the procedure of Exercise 9 fails for the van der Pol equation $\varepsilon \ddot{x} + (x^2 - 1)\dot{x} + x = 0, x(\varepsilon, 0) = 2, x(\varepsilon, 1) = 1$.

12. By using the method of multiple scales, with variables x and $\xi = x/\varepsilon$, obtain a first approximation uniformly valid on $0 \leqslant x \leqslant 1$ to the solution of

$$\varepsilon y'' + y' + xy = 0, \quad y(\varepsilon, 0) = 0, \quad y(\varepsilon, 1) = 1,$$

on $0 \leqslant x \leqslant 1$, with $\varepsilon > 0$. Show that the result agrees to order ε with that of Exercise 10.

13. The steady flow of a conducting liquid between insulated parallel plates under the influence of a transverse magnetic field satisfies

$$w'' + Mh' = -1, \quad h'' + Mw' = 0, \quad w = (\pm 1) = h(\pm 1) = 0,$$

where, in dimensionless form, w is the fluid velocity, h the induced magnetic field, and M is the Hartmann number. By putting $p = w + h$ and $q = w - h$, find the exact solution. Alternatively, solve the equations as a singular perturbation problem for large M.

14. Obtain an approximation, to order ε and for $t = O(\varepsilon^{-1})$, to the solutions of $\ddot{x} + 2\varepsilon \dot{x} + x = 0$, by using the method of multiple scales with the variables t and $\eta = \varepsilon t$.

15. Use the method of multiple scales to obtain a uniform approximation to the solutions of the equation $\ddot{x} + \omega^2 x + \varepsilon x^3 = 0$, in the form

$$x(\varepsilon, t) \approx a_0 \cos\left[\{\omega_0 + (3\varepsilon a_0^2/8\omega_0)\}t + \alpha\right],$$

α constant. Explain why the approximation is uniform, and not merely valid for $t = O(\varepsilon^{-1})$.

16. Use the coordinate perturbation technique to obtain the first approximation

$$x = \tau^{-1}, \quad t = \tau + \tfrac{1}{2}\varepsilon\tau(1 - \tau^{-2})$$

to the solution of

$$(t + \varepsilon x)\dot{x} + x = 0, \quad x(\varepsilon, 1) = 1, \quad 0 \leqslant x \leqslant 1.$$

Confirm that the approximation is, in fact, the exact solution, and that an alternative approximation

$$x = \tau^{-1} + \tfrac{1}{2}\varepsilon\tau^{-1}, \quad t = \tau - \tfrac{1}{2}\varepsilon\tau^{-1}$$

is correct to order ε, for fixed τ.

17. Apply the method of multiple scales, with variables t and $\eta = \varepsilon t$, to van der Pol's equation $\ddot{x} + \varepsilon(x^2 - 1)\dot{x} + x = 0$. Show that, for $t = O(\varepsilon^{-1})$,

$$x(\varepsilon, t) = \frac{2a_0^{\frac{1}{2}} e^{\frac{1}{2}\varepsilon t}}{\sqrt{(1 + a_0 e^{\varepsilon t})}} \cos(t + \alpha) + O(\varepsilon),$$

where a_0 is the initial radius.

18. Use the method of matched approximations to obtain a uniform approximation to the solution of

$$\varepsilon\left(y'' + \frac{2}{x}y'\right) - y = 0, \quad y(\varepsilon, 0) = 0, \quad y'(\varepsilon, 1) = 1,$$

$(\varepsilon > 0)$ on $0 \leqslant x \leqslant 1$. Show that there is a boundary layer of thickness $O(\varepsilon^{1/2})$ near $x = 1$, by putting $1 - x = \xi\phi(\varepsilon)$.

19. Use the method of matched approximations to obtain a uniform approximation to the solution of the problem

$$\varepsilon(y'' + y') - y = 0, \quad y(\varepsilon, 0) = 1, \quad y(\varepsilon, 1) = 1,$$

$(\varepsilon > 0)$ given that there are boundary layers at $x = 0$ and $x = 1$. Show that both boundary layers have thickness $O(\varepsilon^{1/2})$. Compare with the exact solution.

20. Obtain a first approximation, uniformly valid on $0 \leqslant x \leqslant 1$, to the equation

$$\varepsilon y'' + \frac{1}{1+x}y' + \varepsilon y = 0,$$

with $y(\varepsilon, 0) = 0$, $y(\varepsilon, 1) = 1$.

21. Apply the Lighthill technique to obtain a uniform approximation to the solution of

$$(t + \varepsilon x)\dot{x} + x = 0, \quad x(\varepsilon, 1) = 1, \quad 0 \leqslant x \leqslant 1.$$

(Compare Exercise 16.)

22. Obtain a first approximation, uniform on $0 \leqslant x \leqslant 1$, to the solution of

$$\varepsilon y' + y = x, \quad y(\varepsilon, 0) = 1,$$

using inner and outer approximations. Compare the exact solution and explain geometrically why the outer approximation is independent of the boundary condition.

23. Use the method of multiple scales with variables t and $\eta = \varepsilon t$ to show that, to a first approximation, the response of the van der Pol equation to a 'soft' forcing term described by

$$\ddot{x} + x = \varepsilon\{(1 - x^2)\dot{x} + \gamma \cos \omega t\}, \quad \varepsilon > 0,$$

is the same as the unforced response. (Consider $|\omega|$ not near to 1; that is, the non-resonant case.)

24. Repeat Exercise 23 for

$$\ddot{x} + x = \varepsilon(1 - x^2)\dot{x} + \Gamma \cos \omega t, \quad \varepsilon > 0,$$

where $\Gamma = O(1)$ and $|\omega|$ is not near 1. Show that

$$x(\varepsilon, t) = \frac{\Gamma}{1 - \omega^2} \cos \omega t + O(\varepsilon), \quad \Gamma^2 \geqslant 2(1 - \omega^2)^2;$$

and that for $\Gamma^2 < 2(1 - \omega^2)^2$

$$x(\varepsilon, t) = 2\{1 - \Gamma^2/2(1 - \omega^2)^2\}^{1/2} \cos t + \frac{\Gamma}{1 - \omega^2} \cos \omega t + O(\varepsilon).$$

25. Apply the matching technique to the damped pendulum equation

$$\varepsilon\ddot{x} + \dot{x} + \sin x = 0, \quad x(\varepsilon, 0) = 1, \quad \dot{x}(\varepsilon, 0) = 1,$$

for ε small and positive. Show that the inner and outer approximations are given by

$$x_{\text{I}} = 1, \quad x_{\text{O}} = 2 \tan^{-1}\{e^{-t} \tan \tfrac{1}{2}\}.$$

(The pendulum has strong damping and strong restoring action, but the damping dominates.)

26. The equation for a tidal bore on a shallow stream is

$$\varepsilon \frac{d^2\eta}{d\zeta^2} - \frac{d\eta}{d\zeta} - \eta + \eta^2 = 0;$$

where (in appropriate dimensions) η is the height of the free surface, and $\zeta = x - ct$, where c is the wave speed. For $0 < \varepsilon \ll 1$, find the equilibrium points for the equation and classify them according to their linear approximations. Apply the coordinate perturbation method to the equation for the phase paths,

$$\varepsilon \frac{dw}{d\eta} = \frac{w^2 + \eta - \eta^2}{w},$$

and show that

$$w = -\zeta + \zeta^2 + O(\varepsilon^2), \quad \eta = \zeta - \varepsilon(-\zeta + \zeta^2) + O(\varepsilon^2).$$

Confirm that, to this degree of approximation, a separatrix from the origin reaches the other equilibrium point. Interpret the result in terms of the shape of the bore.

27. Following Exercise 35, Chapter 2, on the motion of a satellite, show that the equation can be written in the form

$$\frac{d^2 z}{d\tau^2} = -\frac{1-\mu}{z^2} + \frac{\mu}{(1-z)^2},$$

where

$$\tau = \gamma^{1/2}(m_1 + m_2)^{1/2} a^{-3/2} t, \quad \mu = m_2/(m_1 + m_2), \quad z = x/a.$$

Show that, on the path with zero energy,

$$\tfrac{1}{2} z'^2 = \frac{1-\mu}{z} + \frac{\mu}{1-z}.$$

Construct the first two terms of a perturbation solution of the form

$$\tau(\mu, z) = T_0(z) + \mu T_1(z) + \dots, \quad \tau(\mu, 0) = 0,$$

valid for small μ, and indicate for what values of z it is valid.

28. Replace the time, t, in van der Pol's equation by the 'slow' variable $\tau = \varepsilon t$, so that $\ddot{x} + \varepsilon(x^2 - 1)\dot{x} + x = 0$ becomes

$$\varepsilon^2 x'' + \varepsilon^2(x^2 - 1)x' + x = 0$$

where the dashes represent derivatives with respect to τ.

Attempt an asymptotic solution

$$x \sim u_0(\varepsilon, t, \tau) + \varepsilon u_1(\varepsilon, t, \tau) + \varepsilon^2 u_2(\varepsilon, t, \tau) + \dots$$

for $\varepsilon \to 0$. Show that u_0 must satisfy $(\partial^2 u_0/\partial t^2) + u_0 = 0$. Write the solution in the form

$$u_0 = a(\varepsilon, \tau) \cos(t + b(\varepsilon, \tau))$$

and assume expansions of the form

$$a(\varepsilon, \tau) \sim a_0(\tau) + \varepsilon a_1(\tau) + \dots, \quad b(\varepsilon, \tau) \sim b_0(\tau) + \varepsilon b_1(\tau) + \dots.$$

Show that u_1 must satisfy

$$u_{1_{tt}} + u_1 = (2a_0' - a_0 + \tfrac{1}{4}a_0^3) \sin(t + B) + 2a_0 b_0' \cos(t + B) + \tfrac{1}{4}a_0^3 \sin 3(t + B),$$

and that the secular terms vanish if

$$2a_0' - a_0 + \tfrac{1}{4}a_0^3 = 0, \quad b_0' = 0.$$

Assume the initial conditions $x(\varepsilon, 0) = 2$, $\dot{x}(\varepsilon, 0) = 0$, and deduce that

$$x \sim 2\cos(t - \tfrac{1}{16}\varepsilon^2 t) + \tfrac{3}{4}\varepsilon \sin(t - \tfrac{1}{16}\varepsilon^2 t) - \tfrac{1}{4}\varepsilon \sin(3t - \tfrac{3}{16}\varepsilon^2 t).$$

7 Forced oscillations: harmonic and subharmonic response, stability, entrainment

WE CONSIDER second-order differential equations of the general form $\ddot{x} + f(x, \dot{x}) = F \cos \omega t$, which have a periodic forcing term. When the equation is linear the structure of its solutions is very simple. There are two parts combined additively. One part, the 'free oscillation' is a linear combination of the solutions of the homogeneous equation and involves the initial conditions. The second part, the 'forced solution', is proportional to F and is independent of initial conditions. When damping is present the free oscillation dies away in time. This independence of the free and the forced components, and the ultimate independence of initial conditions when there is damping allows a very restricted range of phenomena: typically the only things to look at are the amplitude and phase of the ultimate response, the rate of decay of the free oscillation and the possibility of resonance. When the equation is nonlinear, on the other hand, there is no such simple separation possible, and the resulting interaction between the free and forced terms, especially when a self-excited motion is possible, and the enduring importance of the initial conditions in some cases, generates a range of entirely new phenomena. The present chapter is concerned with using the method of harmonic balance (Chapter 4) and the perturbation method (Chapter 5) to obtain approximate solutions which clearly show these new features.

7.1. General forced periodic solutions

Consider, for definiteness, Duffing's equation

$$\ddot{x} + k\dot{x} + \alpha x + \beta x^3 = \Gamma \cos \omega t, \qquad (7.1)$$

of which various special cases have appeared in earlier chapters. Suppose that $x(t)$ is a periodic solution with period $2\pi/\lambda$. Then $x(t)$ can be represented by a Fourier series for all t:

$$x(t) = a_0 + a_1 \cos \lambda t + b_1 \sin \lambda t + a_2 \cos 2\lambda t + \dots. \qquad (7.2)$$

If this series is substituted into (7.1) the nonlinear term x^3 is periodic, and so generates a series of similar type. When the contributions are assembled the equation (7.1) takes the form

$$A_0 + A_1 \cos \lambda t + B_1 \sin \lambda t + A_2 \cos 2\lambda t + \ldots = \Gamma \cos \omega t \quad (7.3)$$

for all t, the coefficients being functions of $a_0, a_1, b_1, a_2 \ldots$. Matching the two sides gives, in principle, an infinite set of equations for $a_0, a_1, a_2, \ldots$ and enables λ to be determined. For example, the obvious matching

$$\lambda = \omega$$

and

$$A_1 = \Gamma, \quad A_0 = B_1 = A_2 = \ldots = 0$$

resembles the harmonic balance method of Section 4.4; though in the harmonic balance method a drastically truncated series (7.2), consisting of only the single term $a_1 \cos \lambda t$, is employed. The above matching leads to a harmonic (or autoperiodic) response to frequency ω.

A less obvious matching can sometimes be achieved when

$$\lambda = \omega/n \quad (n \text{ an integer});$$

$$A_n = \Gamma;$$

$$A_i = 0 \ (i \neq n); \quad B_i = 0 \ (\text{all } i).$$

If solutions of this set exist, the system will deliver responses of period $2\pi n/\omega$ and frequency ω/n. These are called *subharmonics* of order $\frac{1}{2}, \frac{1}{3}, \ldots$. Not all of these actually occur (see Section 7.7 and Exercises 16 and 17).

There will never be pure 'superharmonics' (of frequency $2\omega, 3\omega, \ldots$), though the higher harmonics in (7.2) are always present, and may have large amplitude.

It is important to realize where the terms in (7.3) come from: there is 'throwback' from terms of high order in (7.2) to contribute to terms of low order in (7.3). For example, in the expansion of x^3:

$$x^3(t) = (a_0 + \ldots + a_{11} \cos 11\omega t + \ldots + a_{21} \cos 21\omega t + \ldots)^3,$$

one of the terms is

$$3a_{11}^2 a_{21} \cos^2 11\omega t \cos 21\omega t = 3a_{11}^2 a_{21} (\tfrac{1}{2} \cos 21\omega t + \tfrac{1}{4} \cos 43\omega t + \tfrac{1}{4} \cos \omega t),$$

which includes a term in $\cos \omega t$. Normally we assume that terms in (7.2) above a certain small order are negligible, that is, that the coefficients are small. Hopefully, then, the combined throwback as measured by the modification of coefficients by high-order terms will be small.

An alternative way of looking at the effect of a nonlinear term is as a feedback. Consider the undamped form of (7.1) (with $k = 0$) written as

$$\mathscr{L}(x) \equiv \left(\frac{d^2}{dt^2} + \alpha\right)x = -\beta x^3 + \Gamma \cos \omega t.$$

We can represent this equation by the block diagram shown in Fig. 7.1.

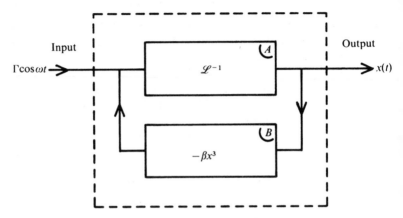

FIG. 7.1.   Feedback diagram for $\ddot{x} + \alpha x + \beta x^3 = \Gamma \cos \omega t$

Regard $\Gamma \cos \omega t$ as the input to the system and $x(t)$ as the output. The box A represents the operator $\mathscr{L}^{-1}$ which solves the equation $\ddot{x} + \alpha x = f(t)$ for a given input f and for assigned initial conditions. Here, the input to A is equal to the current value of $-\beta x^3 + \Gamma \cos \omega t$. Its output is $x(t)$. Suppose the output from A is assumed to be simple, containing only a few harmonic components. Then B generates a shower of harmonics of higher and possibly of lower orders which are fed back into A. The higher harmonics are the most attenuated on passing through A (roughly like n^{-2} where n is the order). It is therefore to be expected that a satisfactory consistency between the inputs might be obtained by a representation of in terms only of the lowest harmonics present. The low-order approximation should be most adequate when the lowest harmonic present is *amplified* by $\mathscr{L}^{-1}$; that is, when $\omega^2 \simeq n^2\alpha$, being a state of near-resonance.

7.2. Harmonic solutions, transients, and stability for Duffing's equation

As a direct illustration of the method described in Section 7.1, consider the approximate equation for the forced undamped pendulum in the standardized form

$$\ddot{x} + x - \tfrac{1}{6}x^3 = \Gamma \cos \omega t, \qquad (7.4)$$

where $\Gamma > 0$. As an approximation to the solution we use the truncated Fourier series

$$x(t) = a \cos \omega t + b \sin \omega t. \qquad (7.5)$$

This form allows for a possible phase difference between the forcing term and the solution. The omission of the constant term involves fore-knowledge that the only relevant solutions of reasonably small amplitude will give this constant as zero. Then

$$\ddot{x}(t) = -a\omega^2 \cos \omega t - b\omega^2 \sin \omega t \qquad (7.6)$$

and

$$x^3(t) = a^3 \cos^3 \omega t + 3a^2 b \cos^2 \omega t \sin \omega t + 3ab^2 \cos \omega t \sin^2 \omega t + b^3 \sin^3 \omega t$$
$$= \tfrac{3}{4}a(a^2 + b^2) \cos \omega t + \tfrac{3}{4}b(a^2 + b^2) \sin \omega t$$
$$+ \tfrac{1}{4}a(a^2 - 3b^2) \cos 3\omega t + \tfrac{1}{4}b(3a^2 - b^2) \sin 3\omega t. \qquad (7.7)$$

We disregard the terms in $\cos 3\omega t$, $\sin 3\omega t$ on the grounds that, regarded as feedback input to the differential equation, the attenuation will be large compared with that of terms having the fundamental frequency. When (7.5), (7.6), (7.7) are substituted into (7.4) and the coefficients of $\cos \omega t$, $\sin \omega t$ are matched we obtain

$$b\{(\omega^2 - 1) + \tfrac{1}{8}(a^2 + b^2)\} = 0, \qquad (7.8)$$

$$a\{(\omega^2 - 1) + \tfrac{1}{8}(a^2 + b^2)\} = -\Gamma. \qquad (7.9)$$

The only admissible solution of (7.8) and (7.9) requires $b = 0$, and the corresponding values of a are the solutions of (7.9):

$$\tfrac{1}{8}a^3 + (\omega^2 - 1)a + \Gamma = 0. \qquad (7.10)$$

The roots are given by the intersections, in the plane of a and z, of the lines

$$z = -\Gamma, \quad z = \tfrac{1}{8}a^3 + (\omega^2 - 1)a.$$

From Fig. 7.2 it can be seen that there are three solutions for Γ small and one for Γ larger when $\omega^2 < 1$, and that when $\omega^2 > 1$ there is one solution only. This reproduces the results of Sections 5.4 and 5.5. The oscillations are in phase with the forcing term when the critical value of a is positive and out of phase by a half cycle when a is negative.

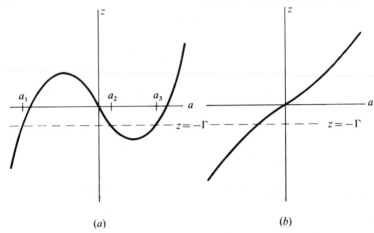

FIG. 7.2. Graph of $z = \frac{1}{8}a^3 + (\omega^2 - 1)a$ for: (a) $\omega^2 < 1$; (b) $\omega^2 > 1$

We do not yet know the stability of these various oscillations and it is necessary to decide this matter since an unstable oscillation will not occur in practice. To investigate stability we shall look at the 'transient' conditions, which lead to or away from periodic states, by supposing the coefficients a and b to be slowly-varying functions of time, at any rate near to periodic states. Assume that

$$x(t) = a(t)\cos \omega t + b(t)\sin \omega t, \qquad (7.11)$$

where a and b are *slowly-varying amplitudes* (compared with $\cos \omega t$ and $\sin \omega t$).

Then

$$\dot{x}(t) = (\dot{a} + \omega b)\cos \omega t + (-\omega a + \dot{b})\sin \omega t, \qquad (7.12)$$

and (neglecting $\ddot{a}, \ddot{b}$)

$$\ddot{x}(t) = (-\omega^2 a + 2\omega \dot{b})\cos \omega t + (-2\omega \dot{a} - \omega^2 b)\sin \omega t. \qquad (7.13)$$

Also, as in (7.7),

$$x^3(t) = \tfrac{3}{4}a(a^2 + b^2)\cos \omega t + \tfrac{3}{4}b(a^2 + b^2)\sin \omega t$$
$$+ \text{harmonics in } \cos 3\omega t, \sin 3\omega t. \qquad (7.14)$$

As before we shall ignore the terms in $\cos 3\omega t, \sin 3\omega t$.

When (7.12), (7.13), (7.14) are substituted into the differential equation

(7.4), and the terms are rearranged, we have

$$[2\omega\dot{b} - a\{(\omega^2 - 1) + \tfrac{1}{8}(a^2 + b^2)\}]\cos\omega t$$
$$+ [-2\omega\dot{a} - b\{(\omega^2 - 1) + \tfrac{1}{8}(a^2 + b^2)\}]\sin\omega t = \Gamma\cos\omega t. \quad (7.15)$$

Appealing again to the supposition that a and b are slowly-varying we may approximately match the coefficients of $\cos\omega t$ and $\sin\omega t$, giving the *autonomous system*

$$\dot{a} = -\frac{1}{2\omega}b\{(\omega^2 - 1) + \tfrac{1}{8}(a^2 + b^2)\} \equiv A(a, b), \quad \text{say}; \quad (7.16)$$

$$\dot{b} = \frac{1}{2\omega}a\{(\omega^2 - 1) + \tfrac{1}{8}(a^2 + b^2)\} + \frac{\Gamma}{2\omega} \equiv B(a, b), \quad \text{say}. \quad (7.17)$$

Initial conditions are given in terms of those for the original equation, (7.4), by

$$a(0) = x(0), \quad b(0) = \dot{x}(0)/\omega. \quad (7.18)$$

The phase plane for a, b in the system above is called the *van der Pol plane*. The equilibrium points, given by $A(a, b) = B(a, b) = 0$, represent the steady periodic solutions already obtained. The other paths correspond to solutions of (7.4) which are non-periodic in general. The phase diagram computed for a particular case is shown in Fig. 7.3. The point $(a_3, 0)$ is a saddle and represents an unstable oscillation, and $(a_1, 0), (a_2, 0)$ are centres.

The equilibrium points may be analysed algebraically as follows. Consider a case when there are three equilibrium points, so that

$$\omega^2 < 1, \quad (7.19)$$

and let a_0 represent any one of the values $a = a_1, a_2,$ or a_3 of Fig. 7.2(a). Putting

$$a = a_0 + \xi, \quad (7.20)$$

the local linear approximation to (7.16), (7.17) is

$$\begin{aligned} \dot{\xi} &= A_1(a_0, 0)\xi + A_2(a_0, 0)b, \\ \dot{b} &= B_1(a_0, 0)\xi + B_2(a_0, 0)\xi, \end{aligned} \quad (7.21)$$

where $A_1(a, b) = \partial A(a, b)/\partial a$, and so on. It is easy to confirm that

$$A_1(a_0, 0) = B_2(a_0, 0) = 0,$$

$$A_2(a_0, 0) = -\frac{\omega^2 - 1}{2\omega} - \frac{a_0^2}{16\omega} = -\frac{z_0}{a_0},$$

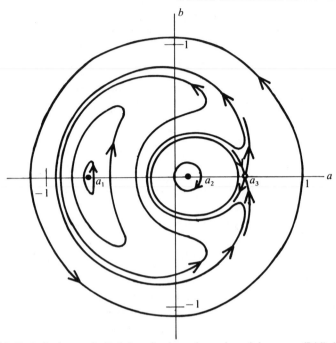

FIG. 7.3. Paths in the van der Pol plane for the undamped pendulum, eqns (7.16), (7.17); $\Gamma = 0.005$, $\omega = 0.975$, showing the equilibrium points

(where, in Fig. 7.2(a), z_0 is the ordinate at a_0), and that

$$B_1(a_0, 0) = s_0$$

(where s_0 is the slope of the curve in Fig. 7.2(a) at a_0). Therefore (7.21) can be written

$$\dot{\xi} = -\frac{z_0}{a_0}b, \quad \dot{b} = s_0\xi.$$

By considering the signs of a_0, z_0, s_0 in Fig. 7.2(a) it can be seen that $(a_1, 0)$, $(a_2, 0)$ are centres for the linear system (7.21), and that $(a_3, 0)$ is a saddle. The saddle will never be observed exactly, though if a state can be set up near enough to this point, $\dot{a}$ and $\dot{b}$ will be very small and a nearly periodic motion at the forcing frequency may linger long enough to be observed.

We may introduce a damping term into (7.4) so that it becomes

$$\ddot{x} + k\dot{x} + x - \tfrac{1}{6}x^3 = \Gamma \cos \omega t, \quad k > 0. \tag{7.22}$$

The equilibrium points (Exercise 10) are given by

$$b\{\omega^2 - 1 + \tfrac{1}{8}(a^2 + b^2)\} + k\omega a = 0,$$
$$a\{\omega^2 - 1 + \tfrac{1}{8}(a^2 + b^2)\} - k\omega b = -\Gamma;$$

so that

$$r^2\{k^2\omega^2 + (\omega^2 - 1 + \tfrac{1}{8}r^2)^2\} = \Gamma^2, \quad r = \sqrt{(a^2 + b^2)}. \qquad (7.23)$$

and the closed paths of Fig. 7.3 become spirals. In this case the theory predicts that one of two stable periodic states is approached from any initial state (we shall later show, however, that from some initial states we arrive at subharmonics). A calculated example is shown in Fig. 7.4.

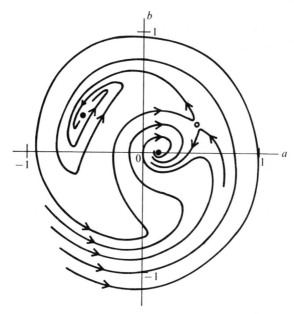

FIG. 7.4. Paths in the van der Pol plane for the damped pendulum; $\Gamma = 0{\cdot}005$, $\omega = 0{\cdot}975$, $k = 0{\cdot}005$

Typical response curves are shown in Fig. 5.3; the governing equations for the equilibrium points, and the conclusions about stability, being the same as in Chapter 5.

Referring back to Fig. 7.3, notice that the existence of a closed curve, which implies periodicity of $a(t)$ and $b(t)$, does not in general imply periodicity of $x(t)$, for since $x(t) = a(t)\cos \omega t + b(t)\sin \omega t$ the motion is periodic only if the period of $a(t)$ and $b(t)$ is a rational multiple of $2\pi/\omega$.

Though this may occur infinitely often in a family of closed paths, the theory is too rough to decide which periodic motions, possibly of very long period, are actually present. Normally the closed paths represent special cases of 'almost periodic motion'.

If P is a representative point on a given closed path with polar representation $r(t), \phi(t)$, then (7.5) can be written

$$x(t) = r(t)\cos(\omega t - \phi(t)).$$

Let $2\pi/\Omega$ be the period of $a(t)$, $b(t)$, and hence of $r(t)$, $\phi(t)$, on the path. Then since a and b are slowly varying, $\Omega \ll \omega$. When the path does not encircle the origin (Fig. 7.5(a)) the phase ϕ is restored after each

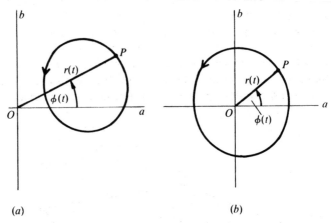

(a)   (b)

Fig. 7.5.

complete circuit. The effect is of an oscillation with frequency ω modulated by a slowly varying periodic amplitude. The phenomenon of 'beats' (Fig. 7.6) appears.

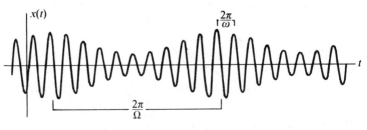

Fig. 7.6. Illustrating beats

If the path encircles the origin as in Fig. 7.5(b) the phase increases by 2π in every circuit, which effectively modulates the approximate 'frequency' of the oscillation from ω to $\omega + \Omega$.

7.3. The jump phenomenon

Equation (7.22) for damped motion of a pendulum has approximate solutions $a \cos \omega t + b \sin \omega t$, where (eqn (7.23))

$$r^2\{k^2\omega^2 + (\omega^2 - 1 + \tfrac{1}{8}r^2)^2\} = \Gamma^2, \tag{7.24}$$

and

$$r = \sqrt{(a^2 + b^2)}$$

is the amplitude. The nature of the response curves is the same as that shown in Fig. 5.3, with $\omega_0 = 1$ and F replaced by Γ.

For any fixed k, eqn (7.24) can be depicted in three dimensions with axes Γ, ω, r as shown in Fig. 7.7. The projection of the folded part of the surface is also shown, this being the Γ, ω parameter region in which three periodic solutions can occur. This is a 'catastrophe' situation (Sewell, 1976), where a continuous change in parameter values gives a discon-

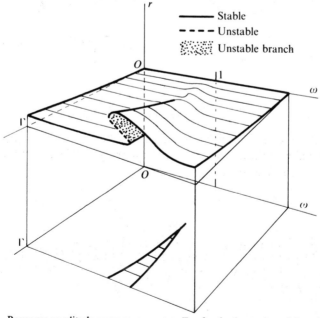

FIG. 7.7. Response amplitude r versus parameters Γ, ω for the damped pendulum (k fixed) showing the projection of the fold on the Γ, ω plane

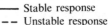

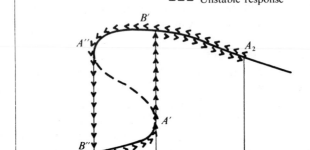

(a)

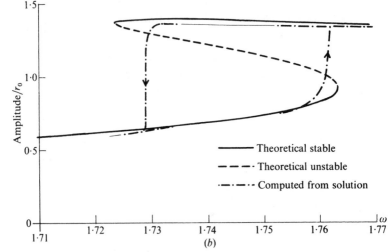

FIG. 7.8. (a) The nature of the jump phenomenon. (b) Computed amplitude–frequency diagram for the solution of $\ddot{x} + 0\cdot24\dot{x} + 4x - 0\cdot67x^3 = 0\cdot58\cos\{\omega(t)t\}$, where $\omega(t)$ is slowly-varying with

$$\omega(t) = \text{constant} \pm 0\cdot00001t,$$

showing the jump phenomenon. The precise points of departure from the theoretical curve are sensitive to the rate of change of $\omega(t)$.

tinuous change in response on crossing the boundary of the fold projection. Imagine an experiment in which, say, Γ is held constant, but the applied frequency may be varied very slowly. In Fig. 7.8 the response curve for a given value of Γ is shown.

Begin the experiment at applied frequency $\omega = \omega_1$, steadily increase to ω_2, then bring it back again to ω_1. Starting at A_1, the response moves to A' at frequency ω'. As ω increases beyond ω', the oscillation will, after some irregular motion, settle down to the condition represented by B'; that is to say, the amplitude 'jumps' at the critical frequency ω'. After this it follows the smooth curve from B' to A_2. On the way back the response point moves along the upper curve as far as A''. Here it must drop to B'' on the lower curve and then back to A_1.

Clearly a jump of a similar kind will occur whenever the parameters ω and Γ change in such a way as to pass into or out of the projection of the fold in Fig. 7.7. This so-called *jump phenomenon* has been observed in many nonlinear physical systems.

7.4. Harmonic oscillations, stability, and transients for the forced van der Pol equation

The equation considered is

$$\ddot{x} + \varepsilon(x^2 - 1)\dot{x} + x = \Gamma \cos \omega t, \tag{7.25}$$

where $\varepsilon > 0$. In the absence of a forcing term there is a single, self-excited oscillation (justified in Section 4.1 but not proved until Section 11.3). We look for responses approximately of the form

$$x(t) = a(t)\cos \omega t + b(t)\sin \omega t \tag{7.26}$$

where a, b are slowly varying functions. Then, neglecting $\ddot{a}, \ddot{b}$,

$$\dot{x}(t) = (\dot{a} + \omega b)\cos \omega t + (-\omega a + \dot{b})\sin \omega t, \tag{7.27a}$$

$$\ddot{x}(t) = (-\omega^2 a + 2\omega \dot{b})\cos \omega t + (-2\omega \dot{a} - \omega^2 b)\sin \omega t. \tag{7.27b}$$

After some algebra,

$$(x^2 - 1)\dot{x} = \{(\tfrac{3}{4}a^2 + \tfrac{1}{4}b^2 - 1)\dot{a} + \tfrac{1}{2}ab\dot{b} - \omega b(1 - \tfrac{1}{4}a^2 - \tfrac{1}{4}b^2)\} \cos \omega t$$
$$+ \{\tfrac{1}{2}ab\dot{a} + (\tfrac{1}{4}a^2 + \tfrac{3}{4}b^2 - 1)\dot{b} + \omega a(1 - \tfrac{1}{4}a^2 - \tfrac{1}{4}b^2)\} \sin \omega t$$
$$+ \text{higher harmonics.} \tag{7.28}$$

Finally substitute (7.26) to (7.28) into (7.25), ignoring the higher harmonics. As in Section 7.2, the equation is satisfied approximately if the coefficients of $\cos \omega t$, $\sin \omega t$ are equated to zero. We obtain

$$(2\omega - \tfrac{1}{2}\varepsilon ab)\dot{a} + \varepsilon(1 - \tfrac{1}{4}a^2 - \tfrac{3}{4}b^2)\dot{b} = \varepsilon\omega a(1 - \tfrac{1}{4}r^2) - (\omega^2 - 1)b, \tag{7.29a}$$

$$-\varepsilon(1 - \tfrac{3}{4}a^2 - \tfrac{1}{4}b^2)\dot{a} + (2\omega + \tfrac{1}{2}\varepsilon ab)\dot{b} = (\omega^2 - 1)a$$
$$+ \varepsilon\omega b(1 - \tfrac{1}{4}r^2) + \Gamma, \tag{7.29b}$$

where

$$r = \sqrt{(a^2 + b^2)}. \tag{7.30}$$

The equilibrium points occur when $\dot{a} = \dot{b} = 0$, that is, when the right-hand side is zero. We can reduce the parameters $(\omega, \Gamma, \varepsilon)$ to two by multiplying through by $1/\varepsilon\omega$ and putting

$$v = (\omega^2 - 1)/\varepsilon\omega, \quad \gamma = \Gamma/\varepsilon\omega; \tag{7.31}$$

(the quantity v is a measure of the 'detuning'). The equations for the equilibrium points become

$$a(1 - \tfrac{1}{4}r^2) - vb = 0, \tag{7.32a}$$

$$va + b(1 - \tfrac{1}{4}r^2) = -\gamma. \tag{7.32b}$$

By squaring and adding these equations we obtain

$$r^2\{v^2 + (1 - \tfrac{1}{4}r^2)^2\} = \gamma^2, \tag{7.33}$$

and when this is solved the roots a, b can be recovered from $(7.32a,b)$. Equation (7.33) may have either 1 or 3 real roots (since $r > 0$) depending on the parameter values v and γ (or $\omega, \Gamma, \varepsilon$). This dependence can be conveniently represented on a single figure, the 'response diagram', Fig. 7.9(a).

There are two questions which can be settled by a study of (7.29): that of the stability of the periodic solutions just found and, connected with this, the changes in the general behaviour of the system when one of the parameters ω or Γ varies. We shall simplify equations (7.29) by supposing that ε is small enough for $\varepsilon\dot{a}, \varepsilon\dot{b}$ to be of negligible effect. Then (7.29) becomes

$$\dot{a} = \tfrac{1}{2}\varepsilon(1 - \tfrac{1}{4}r^2)a - \frac{\omega^2 - 1}{2\omega}b \equiv A(a, b), \quad \text{say}, \tag{7.34a}$$

$$\dot{b} = \frac{\omega^2 - 1}{2\omega}a + \tfrac{1}{2}\varepsilon(1 - \tfrac{1}{4}r^2)b + \frac{\Gamma}{2\omega} \equiv B(a, b), \quad \text{say}. \tag{7.34b}$$

Now consider the stability of an equilibrium point at $a = a_0, b = b_0$. In the neighbourhood of the point put

$$a = a_0 + \xi, \quad b = b_0 + \eta. \tag{7.35}$$

The corresponding linearized equations are

$$\begin{aligned}
\dot{\xi} &= A_1(a_0, b_0)\xi + A_2(a_0, b_0)\eta, \\
\dot{\eta} &= B_1(a_0, b_0)\xi + B_2(a_0, b_0)\eta.
\end{aligned} \tag{7.36}$$

The point is a stable equilibrium point (Fig. 2.9) if

$$\det \begin{pmatrix} A_1 & A_2 \\ B_1 & B_2 \end{pmatrix} = A_1 B_2 - A_2 B_1 > 0, \tag{7.37}$$

and

$$\operatorname{tr} \begin{pmatrix} A_1 & A_2 \\ B_1 & B_2 \end{pmatrix} = A_1 + B_2 \leqslant 0, \tag{7.38}$$

From (7.34) and (7.35),

$$A_1 = \tfrac{1}{2}\varepsilon(1 - \tfrac{3}{4}a_0^2 - \tfrac{1}{4}b_0^2), \quad A_2 = -\tfrac{1}{4}\varepsilon a_0 b_0 - \frac{\omega^2 - 1}{2\omega},$$

$$B_1 = -\tfrac{1}{4}\varepsilon a_0 b_0 + \frac{\omega^2 - 1}{2\omega}, \quad B_2 = \tfrac{1}{2}\varepsilon(1 - \tfrac{1}{4}a_0^2 - \tfrac{3}{4}b_0^2).$$

Therefore, after simplification, we have

$$A_1 B_2 - A_2 B_1 = \tfrac{1}{4}\varepsilon^2(\tfrac{3}{16}r_0^4 - r_0^2 + 1) + \left(\frac{\omega^2 - 1}{2\omega}\right)^2, \tag{7.39}$$

$$A_1 + B_2 = \tfrac{1}{2}\varepsilon(2 - r_0^2), \tag{7.40}$$

where $r_0^2 = \sqrt{(a_0^2 + b_0^2)}$.

In terms of the parameter v, (7.31), the conditions for stability, (7.37) and (7.38), become

$$\tfrac{3}{16}r_0^4 - r_0^2 + 1 + v^2 > 0, \tag{7.41}$$

$$2 - r_0^2 \leqslant 0. \tag{7.42}$$

The response curves derived from (7.33), and the stability regions given by (7.41) and (7.42), are exhibited in Fig. 7.9, together with some further detail about the nature of the equilibrium points in various regions. It can be seen that the stable and unstable regions are correctly given by the argument in Section 5.4, (vi), that the responses are stable when $d\Gamma/dr > 0$ and unstable when $d\Gamma/dr < 0$.

If we attempt to force an oscillation corresponding to an equilibrium point lying in the unstable region of Fig. 7.9(a) the system will drift away from this into another state: into a stable equilibrium point if one is available, or into a limit cycle, which corresponds to an almost-periodic motion. A guide to the transition states is provided by the phase paths of (7.34). Although we are not directly concerned with the detail of the

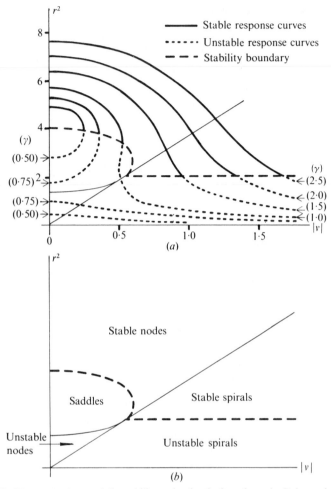

FIG. 7.9. Response curves and the stability region for the forced van der Pol equation (see eqns (7.31), (7.33), (7.41), (7.42)). _ _ _ _ _ stability boundary.

phase plane it is interesting to see the variety of patterns that can arise according to the parameter values concerned, and three typical cases are shown in Figs. 7.11, 7.12, and 7.13.

We can show more directly how the settling-down takes place from a given initial condition to a periodic oscillation by a phase-plane representation of (7.25) given by

$$\dot{x} = y, \quad \dot{y} = -\varepsilon(x^2 - 1)y - x + \Gamma \cos \omega t. \tag{7.43}$$

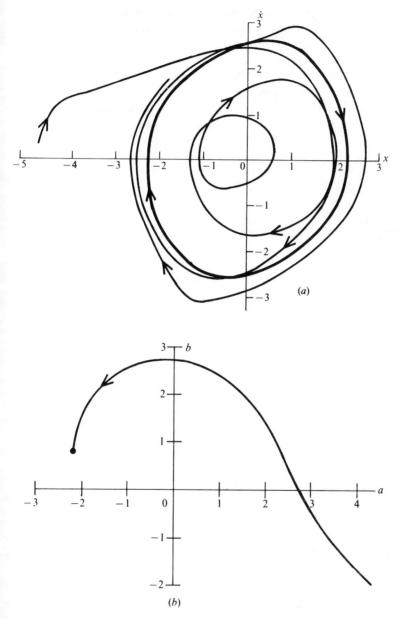

FIG. 7.10. (a) Solution of the forced van der Pol equation in the $x, \dot{x}$ plane. (b) The corresponding path in the van der Pol plane: • equilibrium point.

A periodic solution shows up as a limiting closed path (Fig. 7.10(a)). Since the system (7.43) is not autonomous, the paths may cross. The corresponding path in the van der Pol plane is shown in Fig. 7.10(b).

7.5. Frequency entrainment for the van der Pol equation

The description offered by the treatment of Sections 7.3 and 7.4 is incomplete because of the approximate representation used. For example, no subharmonics (of period $2\pi n/\omega, n > 1$) can be revealed by the representation (7.26) if a and b are slowly-varying. Since subharmonics do exist they must be arrived at from some region of initial conditions $x(0), \dot{x}(0)$; (see Section 7.7), but such regions are not identifiable on the van der Pol plane for a, b. References to 'all paths' and 'any initial condition' in the remarks below must therefore be taken as merely a broad descriptive wording.

Referring to Fig. 7.9, the phenomena to be expected for different ranges of v, γ (or $\omega, \Gamma, \varepsilon$, eqn (7.31)) fall into three main types.

(I) When $\gamma < 0.77$ about, and v is appropriately small enough, there are three equilibrium points. One is a stable node, and the other two are unstable. Starting from any initial condition, the corresponding path will in practice run into the stable node. Figures 7.11(a) and (b) illustrate this case.

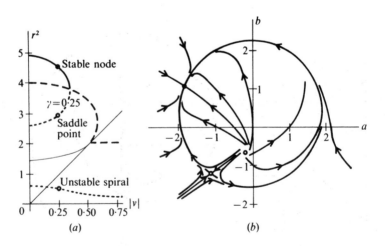

Fig. 7.11. (a) Response region for case I. (b) Corresponding paths in the van der Pol plane ($\gamma = 0.25, v = 0.25$).

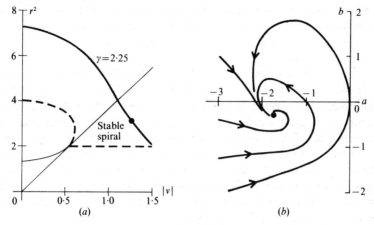

FIG. 7.12. (*a*) Response region for case II. (*b*) Corresponding paths in the van der Pol plane ($\gamma = 2 \cdot 25$, $v = 1 \cdot 25$).

(II) For any v, and γ large enough, there is a single stable point which all paths approach (Fig. 7.12).

(III) For any v, and γ small enough, the only equilibrium point is an unstable one. It can be shown (Example 11.2) that all paths approach a limit cycle (Fig. 7.13). There are no stable harmonic solutions in this case.

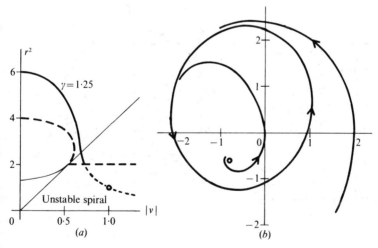

FIG. 7.13. (*a*) Response region for case III. (*b*) Corresponding paths in the van der Pol plane showing limit cycle ($\gamma = 1 \cdot 25$, $v = 1 \cdot 0$).

In cases I and II the final state is periodic with the period of the forcing function and the natural oscillation of the system appears to be completely suppressed despite its self-sustaining nature. This condition is arrived at from arbitrary initial states (at any rate, so far as this approximate theory tells us). The system is said to be *entrained* at the frequency of the forcing function. Moreover, this phenomenon does not depend critically on exact parameter values: if any of the parameters $\omega, \Gamma, \varepsilon$ of the system fluctuate a little, the system will continue to be entrained at the prevailing frequency.

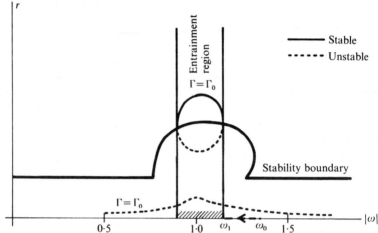

FIG. 7.14. Harmonic entrainment under varying frequency input: van der Pol equation

To illustrate this, consider an experiment in which the applied amplitude remains constant at Γ_0 but the frequency varies slowly. We treat ω close to unity and Γ_0 small as representing the physically interesting case. Figure 7.14 is a rendering of Fig. 7.9 in terms of ω and Γ instead of ν and γ for $\varepsilon = 0.1$.

Suppose the experiment starts with $\omega = \omega_0$ where there is no stable periodic oscillation, and then ω is then reduced slowly. At $\omega = \omega_1$ the stable branch is encountered, and in the space of a few cycles a periodic oscillation of frequency ω_1 is established. This effect is allied to the jump phenomenon of Section 7.3. Figure 7.15 shows the occurrence of entrainment in a particular case. The phenomenon of entrainment is related to that of *synchronization* (Minorsky, 1962), in which two coupled systems having slightly different natural frequencies may fall into a common frequency of oscillation.

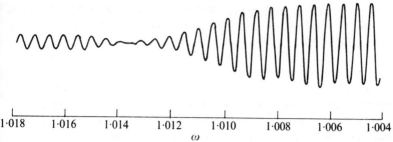

FIG. 7.15. Harmonic entrainment for $\ddot{x} + 0.1(x^2-1)\dot{x} + x = 0.025 \cos \omega t$ by varying ω slowly from 1·018 down to 1·004. (Exercise 20 predicts 1·006 as the entrainment boundary.)

7.6. Comparison of the theory with computations

The approximations discussed in the previous sections appear to be very rough. It is therefore of interest to compare, for a special case, the theoretical predictions summarized in Fig. 7.9 with exact calculations. In Fig. 7.16 the data for Fig. 7.9 is replotted for the case $\varepsilon = 1$ in terms of input frequency ω, input amplitude Γ, and output amplitude r.

Also plotted are the corresponding curves derived by direct numerical

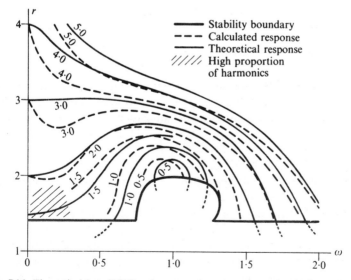

FIG. 7.16. Theoretical (eqn (7.33)) and computed response curves for the forced van der Pol equation with $\varepsilon = 1$, eqn (7.25). Values of Γ are shown on curves

solution of (7.24), giving the amplitudes of the stable periodic oscillations to which the solutions settle down after a sufficient time. These computed response curves terminate near the stability boundary, where no steady oscillations could be found.

In the hatched region the computed amplitudes cannot be compared directly with the theory since there is a heavy admixture of third, fifth, and higher harmonics, ignored by the theory. A typical steady oscillation in this region is shown in Fig. 7.17, together with its harmonic analysis. The amplitude of the first harmonic is as consistent with prediction as are the other curves of Fig. 7.16.

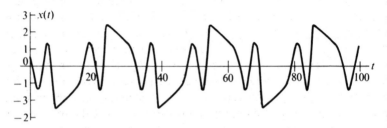

FIG. 7.17. Computed harmonic response for the forced van der Pol equation with $\varepsilon = 1$, $\Gamma = 1\cdot5$, $\omega = 0\cdot2$

Order of harmonic	1	2	3	4	5
Amplitude (from computed curve)	1·59	0·04	0·79	0·06	0·91

7.7. Subharmonics of Duffing's equation by perturbation

In considering periodic solutions of differential equations with a forcing term we have so far looked only for solutions having the period of the forcing term. But even a linear equation may have a periodic solution with a different period. For example, all solutions of

$$\ddot{x} + \frac{1}{n^2} x = \Gamma \cos t$$

are of the form

$$x(t) = a \cos \frac{1}{n} t + b \sin \frac{1}{n} t - \frac{\Gamma}{1 - n^{-2}} \cos t.$$

If n is an integer the period is $2\pi n$ instead of 2π, and the wave has the general appearance of a long-period oscillation with a forced oscillation superposed. The response is said to be a *subharmonic of order* $1/n$. Figure 7.18 shows a case when $n = 3$.

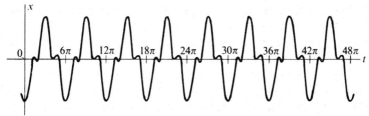

FIG. 7.18. A subharmonic of order $\frac{1}{3}$ for $\ddot{x} + \frac{1}{9}x = \cos t$

For the linear equation this periodic motion appears to be merely an anomalous case of the usual almost-periodic motion, depending on a precise relation between the forcing and natural frequencies. Also, any damping will cause its disappearance. When the equation is nonlinear, however, the generation of alien harmonics by the nonlinear terms may cause a stable subharmonic to appear for a range of the parameters, and in particular for a range of applied frequencies. Also, the forcing amplitude plays a part in generating and sustaining the subharmonic even in the presence of damping. Thus there will exist the tolerance of slightly varying conditions necessary for the consistent appearance of a subharmonic and its use in physical systems. The possibility of such a response was pointed out in Section 7.1.

Assuming that a subharmonic exists, its form can be established by using the perturbation method. Consider Duffing's equation without damping

$$\ddot{x} + \alpha x + \beta x^3 = \Gamma \cos \omega t, \qquad (7.44)$$

$\alpha, \Gamma, \omega > 0$. Take β to be the 'small parameter' of the perturbation method, so that we regard (7.44) as a member of a family of equations for which β lies in an interval including $\beta = 0$. The values of ω for which the subharmonics occur are unknown. To simplify the subsequent algebra we put

$$\omega t = \tau \qquad (7.45)$$

and look for periodic solutions of the transformed equation

$$\omega^2 x'' + \alpha x + \beta x^3 = \Gamma \cos \tau, \qquad (7.46)$$

where the derivatives are with respect to τ. There are no subharmonics of order $1/2$ (also for the damped case as in Exercise 16), and we shall look for those of order $1/3$. The solution of (7.44) will then have period $6\pi/\omega$ and the solution of (7.46) will have period 6π.

Now write, in (7.46),

$$x(\tau) = x_0(\tau) + \beta x_1(\tau) + \dots, \tag{7.47}$$

$$\omega = \omega_0 + \beta \omega_1 + \dots. \tag{7.48}$$

The condition that x has period 6π for all β implies that $x_0, x_1, \dots$ have period 6π. After substitution in (7.46) we obtain

$$\omega_0^2 x_0'' + \alpha x_0 = \Gamma \cos \tau, \tag{7.49a}$$

$$\omega_0^2 x_1'' + \alpha x_1 = -2\omega_0 \omega_1 x_0'' - x_0^3. \tag{7.49b}$$

The periodicity condition applied to (7.49a) gives $\alpha/\omega_0^2 = 1/9$, or

$$\omega_0 = 3\sqrt{\alpha}. \tag{7.50}$$

The $\frac{1}{3}$ subharmonic is therefore stimulated by applied frequencies in the neighbourhood of three times the natural frequency of the linearized equation ($\beta = 0$), as expected. Then

$$x_0(\tau) = a_{1/3} \cos \tfrac{1}{3}\tau + b_{1/3} \sin \tfrac{1}{3}\tau - \frac{\Gamma}{8\alpha} \cos \tau \tag{7.51}$$

where $a_{1/3}, b_{1/3}$ are constants, to be settled at the next stage. The solution therefore bifurcates from a solution of the linearized version of (7.46). For (7.49b), we have

$$x_0''(\tau) = -\tfrac{1}{9} a_{1/3} \cos \tfrac{1}{3}\tau - \tfrac{1}{9} b_{1/3} \sin \tfrac{1}{3}\tau + \frac{\Gamma}{8\alpha} \cos \tau \tag{7.52}$$

and

$$x^3(\tau) = \frac{3}{4} \left\{ a_{1/3} \left(a_{1/3}^2 + b_{1/3}^2 + 2\left(\frac{\Gamma}{8\alpha}\right)^2 \right) - \frac{\Gamma}{8\alpha}(a_{1/3}^2 - b_{1/3}^2) \right\} \cos \tfrac{1}{3}\tau$$

$$+ \frac{3}{4} \left\{ b_{1/3} \left(a_{1/3}^2 + b_{1/3}^2 + 2\left(\frac{\Gamma}{8\alpha}\right)^2 \right) + \frac{\Gamma}{4\alpha} a_{1/3} b_{1/3} \right\} \sin \tfrac{1}{3}\tau$$

$$+ \text{terms in } \cos \tau, \sin \tau \text{ and higher harmonics.} \tag{7.53}$$

In order that $x_1(\tau)$ should be periodic it is necessary that resonance should not occur in (7.49b). The coefficients of $\cos \tfrac{1}{3}\tau, \sin \tfrac{1}{3}\tau$ on the right must therefore be zero. Clearly this condition is insufficient to determine all three of $a_{1/3}, b_{1/3}$, and ω_1. Further stages will still give no condition to determine ω_1 so we shall consider it to be arbitrary, and determine $a_{1/3}$ and $b_{1/3}$ in terms of ω_0, ω_1 and so (to order β^2) in terms of ω. It is convenient to write

$$2\omega_0 \omega_1 \approx (\omega^2 - 9\alpha)/\beta \tag{7.54}$$

this being correct with an error of order β, so that the arbitrary status of ω is clearly displayed. (Remember, however, that ω must remain near ω_0, or $3\sqrt{\alpha}$, by (7.50).) Then the conditions that the coefficients of $\cos\frac{1}{3}\tau, \sin\frac{1}{3}\tau$ on the right of (7.49b) should be zero are

$$a_{1/3}\left(a_{1/3}^2 + b_{1/3}^2 + \frac{\Gamma}{32\alpha^2} - \frac{4}{27}\frac{(\omega^2-9\alpha)}{\beta}\right) - \frac{\Gamma}{8\alpha}(a_{1/3}^2 - b_{1/3}^2) = 0,$$

(7.55a)

$$b_{1/3}\left(a_{1/3}^2 + b_{1/3}^2 + \frac{\Gamma}{32\alpha^2} - \frac{4}{27}\frac{(\omega^2-9\alpha)}{\beta}\right) + \frac{\Gamma}{4\alpha}a_{1/3}b_{1/3} = 0.$$

(7.55b)

The process of solution can be simplified in the following way. One solution of (7.55b) is given by

$$b_{1/3} = 0,$$

(7.56)

and then (7.55a) becomes

$$a_{1/3}^2 - \frac{\Gamma}{8\alpha}a_{1/3} + \left(\frac{\Gamma^2}{32\alpha^2} - \frac{4}{27}\frac{(\omega^2-9\alpha)}{\beta}\right) = 0$$

(7.57)

(rejecting the trivial case $a_{1/3} = 0$). Thus there are either two real solutions of (7.57), or none. Now consider the case when $b_{1/3} \neq 0$. Solve (7.55b) for $b_{1/3}$ and substitute for $b_{1/3}^2$ in (7.55a). We obtain

$$b_{1/3} = \pm\sqrt{3}a_{1/3}$$

(7.58)

and (compare (7.57))

$$(-2a_{1/3})^2 - \frac{\Gamma}{8\alpha}(-2a_{1/3}) + \left(\frac{\Gamma}{32\alpha^2} - \frac{4}{27}\frac{(\omega^2-9\alpha)}{\beta}\right) = 0. \quad (7.59)$$

If the roots are displayed on the $a_{1/3}, b_{1/3}$ plane, those arising from (7.58) and (7.59) are obtained from the first pair by rotations through an angle $\pm\frac{2}{3}\pi$ (Fig. 7.19). Another interpretation is obtained by noting that the differential equation (7.46) is invariant under the change of variable $\tau \to \tau \pm 2\pi$. If $b_{1/3} = 0$ and $a_{1/3}$ represent one solution of (7.56) and (7.57), and $x^*(\tau)$ is the corresponding solution of (7.46), then

$$x^*(\tau \pm 2\pi) = a_{1/3}\cos(\tfrac{1}{3}\tau \pm \tfrac{2}{3}\pi) - \frac{\Gamma}{8\alpha}\cos\tau$$

$$= -\tfrac{1}{2}a_{1/3}\cos\tfrac{1}{3}\tau \pm \frac{\sqrt{3}}{2}a_{1/3}\sin\tfrac{1}{3}\tau - \frac{\Gamma}{8\alpha}\cos\tau$$

are also solutions. The new coefficients are clearly the solutions of (7.58),

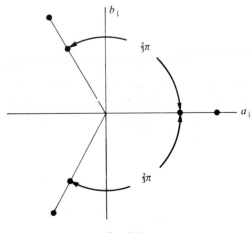

FIG. 7.19.

(7.59). The stability properties of corresponding pairs of equilibrium points are the same.

Figure 7.20 shows the subharmonic for a special case. Note that the subharmonic state will only be entered when the initial conditions are suitable.

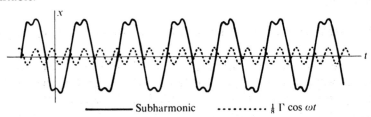

FIG. 7.20. Subharmonic order $\frac{1}{3}$, for the pendulum equation $\ddot{x} + x - \frac{1}{6}x^3 = 1\cdot5\cos 2\cdot85t$

We should expect that the perturbation method would be satisfactory only if $\alpha = O(1)$, $|\beta| \ll 1$ and $(\omega^2 - 9\alpha)/\beta \approx 2\omega_0\omega_1$ is of order 1. However, put

$$y = x/C \qquad (7.60)$$

where C is constant, into (7.44). This becomes

$$y'' + c_1 y + c_3 y^3 = \gamma \cos \tau \qquad (7.61)$$

where

$$c_1 = \alpha/\omega^2, \quad c_3 = \beta C^2/\omega^2, \quad \gamma = \Gamma/C\omega^2. \qquad (7.62)$$

Now choose

$$C^2 = (\omega^2 - 9\alpha)/\beta \qquad (7.63)$$

so that

$$9c_1 + c_3 = 1. \qquad (7.64)$$

The perturbation method applied to (7.61) in the same way as to (7.46) (and obviously leading to the same result), requires at most

$$|c_3| \ll 1, \quad (1 - 9c_1)/c_3 = O(1). \qquad (7.65)$$

But from (7.64)

$$(1 - 9c_1)/c_3 = 1. \qquad (7.66)$$

So the second condition in (7.65) is satisfied, and $|c_3| \ll 1$ requires only

$$|c_3| = \left| \beta \left(\frac{\omega^2 - 9\alpha}{\beta} \right) \frac{1}{\omega^2} \right| \ll 1,$$

or

$$\alpha \simeq \tfrac{1}{9} \omega^2. \qquad (7.67)$$

In other words, so long as ω is near enough to $3\sqrt{\alpha}$ (a near-resonant state for the subharmonic), the calculations are valid without the restriction on β. The usefulness of the approximate form (7.51) is even wider than this, and the reader is referred to Hayashi (1964) for numerical comparisons.

7.8. Stability and transients for subharmonics of Duffing's equation

Some of the subharmonics obtained in Section 7.7 may not be stable, and so will not appear in practice. When there are stable subharmonics, the question arises of how to stimulate them. We know that there are stable solutions having the period of the forcing term; there may also be subharmonics of order other than 1/3 (so far as we know at the moment). Which state of oscillation is ultimately adopted by a system depends on the initial conditions. The following method, due to Mandelstam and Papalexi, is similar to the use of the van der Pol plane in Section 7.4. It enables the question of stability to be settled and gives an idea of the domain of initial conditions leading ultimately to a subharmonic of order 1/3.

The method involves accepting the form (7.51) as a sufficiently good approximation to the subharmonic, and that the solutions for the 'transient' states of

$$\ddot{x} + \alpha x + \beta x^3 = \Gamma \cos \omega t \qquad (7.68)$$

are approximately of the form

$$x(t) = a_{1/3}(t)\cos\tfrac{1}{3}\omega t + b_{1/3}(t)\sin\tfrac{1}{3}\omega t - \frac{\Gamma}{8\alpha}\cos\omega t, \qquad (7.69)$$

where $a_{1/3}$ and $b_{1/3}$ are slowly-varying functions in the sense of Section 7.4. Further justification for the use of the form (7.69) will be found in Hayashi (1964) and McLachlan (1956).

From (7.69)

$$\dot{x}(t) = (\dot{a}_{1/3} + \tfrac{1}{3}\omega b_{1/3})\cos\tfrac{1}{3}\omega t + (-\tfrac{1}{3}\omega a_{1/3} + \dot{b}_{1/3})\sin\tfrac{1}{3}\omega t$$
$$+ (\omega\Gamma/8\alpha)\sin\omega t,$$

and

$$\ddot{x}(t) = (-\tfrac{1}{9}\omega^2 a_{1/3} + \tfrac{2}{3}\omega\dot{b}_{1/3})\cos\tfrac{1}{3}\omega t$$
$$+ (-\tfrac{2}{3}\omega\dot{a}_{1/3} - \tfrac{1}{9}\omega^2 b_{1/3})\sin\tfrac{1}{3}\omega t - (\omega^2\Gamma/8\alpha)\cos\omega t, \quad (7.70)$$

where $\ddot{a}_{1/3}, \ddot{b}_{1/3}$ have been neglected. Equations (7.69) and (7.70) are substituted into (7.68). When the terms are assembled, neglect all harmonics of higher order than $\tfrac{1}{3}$, and balance the coefficients of $\cos\tfrac{1}{3}\omega t, \sin\tfrac{1}{3}\omega t$. This leads to

$$\dot{a}_{1/3} = \frac{9\beta}{8\omega}\left\{ b_{1/3}\left(a_{1/3}^2 + b_{1/3}^2 + \frac{\Gamma^2}{32\alpha^2} - \frac{4}{27}\frac{(\omega^2 - 9\alpha)}{\beta} \right) \right.$$
$$\left. + \frac{\Gamma}{4\alpha}a_{1/3}b_{1/3} \right\} \equiv A(a_{1/3}, b_{1/3}), \quad (7.71a)$$

$$\dot{b}_{1/3} = \frac{9\beta}{8\omega}\left\{ -a_{1/3}\left(a_{1/3}^2 + b_{1/3}^2 + \frac{\Gamma^2}{32\alpha^2} - \frac{4}{27}\frac{(\omega^2 - 9\alpha)}{\beta} \right) \right.$$
$$\left. + \frac{\Gamma}{8\alpha}a_{1/3}^2 - b_{1/3}^2 \right\} \equiv B(a_{1/3}, b_{1/3}). \quad (7.71b)$$

The phase paths of these autonomous equations, representing 'transients' for $x(t)$, can be displayed on a van der Pol plane of $a_{1/3}, b_{1/3}$ (for example see Fig. 7.21). Stable solutions of the form (7.69) correspond to stable equilibrium points of the system (7.71), and unstable solutions to unstable equilibrium points. By comparing with (7.55) we see that the equilibrium points are the same as those found earlier by the perturbation method. The point

$$a_{1/3} = b_{1/3} = 0$$

is also an equilibrium point and this may be taken to acknowledge the possibility that from some range of initial conditions the harmonic, rather than a subharmonic, oscillation is approached, namely, the solution

$$x(t) = -\frac{\Gamma}{8\alpha}\cos\omega t.$$

This, as expected, is approximately equal to the forced solution of the linearized equation, for the exact solution of $\ddot{x} + \alpha x = \Gamma\cos\omega t$ is

$$\frac{\Gamma}{\alpha - \omega^2}\cos\omega t \simeq -\frac{\Gamma}{8\alpha}\cos\omega t$$

so long as $\omega^2 \approx 9\alpha$ (eqn (7.67)).

We may determine the stability of the equilibrium points as follows. It is only necessary to consider the pair for which $b_{1/3} = 0$, by the argument on p. 201. Let $a_{1/3}^*, b_{1/3}^* = 0$, be such a point, and consider the neighbourhood of this point by writing

$$a_{1/3} = a_{1/3}^* + \xi, \quad b_{1/3} = b_{1/3}^* + \eta = \eta$$

with ξ, η small. Then the system (7.71) becomes

$$\left.\begin{aligned} \dot{\xi} &= A_1(a_{1/3}^*,0)\xi + A_2(a_{1/3}^*,0)\eta, \\ \dot{\eta} &= B_1(a_{1/3}^*,0)\xi + B_2(a_{1/3}^*,0)\eta. \end{aligned}\right\} \tag{7.72}$$

Writing

$$\gamma = \frac{\Gamma}{8\alpha}, \quad v = \frac{\Gamma^2}{32\alpha^2} - \frac{4}{27}\frac{(\omega^2 - 9\alpha)}{\beta}, \tag{7.73}$$

we find that the coefficients in (7.72), with $b_{1/3}^* = 0$, are given by

$$A_1(a_{1/3}^*,0) = 0,$$

$$A_2(a_{1/3}^*,0) = \frac{9\beta}{8\omega}(a_{1/3}^{*2} + 2\gamma a_{1/3}^* + v) = \frac{9\beta}{8\omega}3\gamma a_{1/3}^*$$

(from (7.57)), $\tag{7.74}$

$$B_1(a_{1/3}^*,0) = \frac{9\beta}{8\omega}(-3a_{1/3}^{*2} + 2\gamma a_{1/3}^* - v) = \frac{9\beta}{8\omega}(-\gamma a_{1/3}^* + 2v),$$

$$B_2(a_{1/3}^*,0) = 0.$$

The equilibrium points are therefore either centres (in the linear approximation) or saddles. The origin can similarly be shown to be a centre. A typical layout for the $a_{1/3}, b_{1/3}$ plane, using the data of Fig. 7.20, is shown in Fig. 7.21.

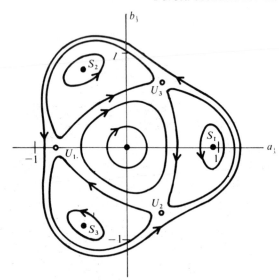

FIG. 7.21. Van der Pol plane for subharmonic order $\frac{1}{3}$; data as in Fig. 7.20. S_1, S_2, S_3, stable; U_1, U_2, U_3, unstable.

When damping is taken into account the closed paths become spirals. Figure 7.22(a) shows a typical pattern (see also Hayashi, 1964). The shaded areas are *domains of attraction* for the equilibrium points to which they relate. It can be seen that in this case the subharmonics occur over a fairly narrow range of initial conditions. Figure 7.22(b) shows a typical subharmonic.

It may be inferred that when the parameters of the system (such as the applied amplitude or frequency) are varied slowly, the system will suddenly become entrained at a subharmonic if the domain of attraction of a non-zero equilibrium point, appropriate to the instantaneous value of the changing parameter, is encountered. Figures such as Fig. 7.21 have to be treated with caution so far as the details are concerned. The nature and positions of the equilibrium points are likely to be nearly correct, but the phase paths are not necessarily very close to the correct ones.

Subharmonics provide a means for 'gearing down' from a high-frequency input by stages to a low-frequency output. In the quartz crystal clock the basic timing mechanism is the oscillating crystal. Its electrical output is passed through a succession of circuits producing submultiples of the input frequency until a frequency appropriate for a

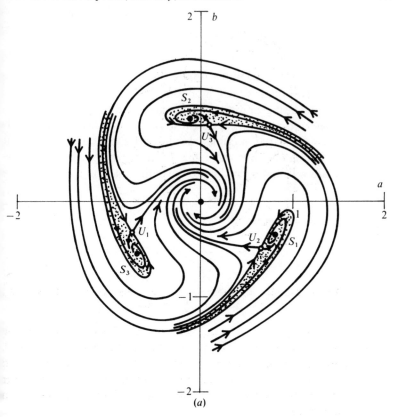

(a)

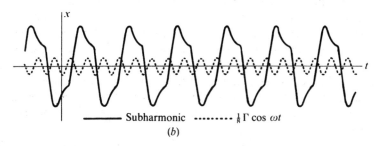

—— Subharmonic ········ $\frac{1}{8}\Gamma\cos\omega t$

(b)

FIG. 7.22. (a) van der Pol plane; subharmonics order $\frac{1}{3}$ of

$$\ddot{x} + 0{\cdot}002\dot{x} + x - \tfrac{1}{6}x^3 = 1{\cdot}5\cos 2{\cdot}85t.$$

S_1, S_2, S_3, stable; U_1, U_2, U_3, unstable.
(b) Subharmonic near S_1.

mechanical drive is reached. Subharmonic entrainment is essential for this process since small variations in the electronic system must be accommodated to maintain exact timing.

7.9. Entrainment by periodic stimuli

Consider a two-dimensional autonomous system having a stable limit-cycle $\mathscr{C}$ and a single unstable equilibrium point (Fig. 7.23). Then $\mathscr{C}$ is approached from every initial state (except the equilibrium state), and all oscillations ultimately become periodic with the period T of $\mathscr{C}$. All oscillations ultimately have the same waveform, but in general their ultimate phase will vary according to the initial conditions.

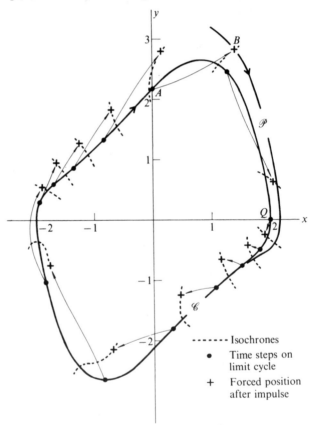

FIG. 7.23. Limit cycle $\mathscr{C}$ for van der Pol's equation $\ddot{x} + (x^2 - 1)\dot{x} + x = 0$. The disturbing impulse has the form $f(t) = 1, 0 < t < 0.5$. $\mathscr{C}$ is marked in steps of 0.5 time units, measured from the origin of phase, Q.

We shall consider how a new periodic motion can be set up by imposing a suitable periodic disturbance on the system. Consider the van der Pol oscillator, representing a self-excited system, with equation $\ddot{x} + e(x^2 - 1)\dot{x} + x = 0$; or

$$\dot{x} = y, \quad \dot{y} = -e(x^2 - 1)y - x, \quad e > 0,$$

on the usual phase plane (Fig. 7.23). Suppose that it is in a state of oscillation represented by $\mathscr{C}$, the limit cycle, and that when it is at A, at time $t = 0$, it is subjected to a disturbance of duration τ:

$$\ddot{x} + e(x^2 - 1)\dot{x} + x = \begin{cases} f(t), & 0 < t < \tau, \\ 0, & \text{elsewhere.} \end{cases} \tag{7.75}$$

$f(t)$ may temporarily be thought of as an impulse. Then at $t = \tau$, the duration of the 'impulse', the system has been moved to B, and subsequently it approaches $\mathscr{C}$ again along some phase path $\mathscr{P}$. After a time it is effectively on $\mathscr{C}$ again with the characteristic periodic motion, but not in general with the same phase as it had before disturbance.

To show the pattern of phase change resulting from disturbances, choose an arbitrary origin Q on $\mathscr{C}$. The motion starting at Q at a standard time t_0 will be said to have zero phase. Then the oscillations starting at A and Q respectively, at the same time, are identical but for a phase difference equal to the time lapse from Q to A along $\mathscr{C}$. Suppose that a time scale is marked out along $\mathscr{C}$, with origin at Q. On return to Q the time lapse is T; after going round again it is $2T$, and so on. To turn these time lapses into phases in the usual sense we define the phase α of a point on $\mathscr{C}$ by

$$\alpha = (\text{time lapse from } Q) \text{ modulo } T$$

in such a way that α is continuous on $\mathscr{C}$ except at Q, and $0 \leqslant \alpha \leqslant T$. If the solution with initial state Q at $t = t_0$ is denoted by $x_Q(t)$, then the solution $x_A(t)$ starting at A at $t = t_0$ is given by

$$x_A(t) = x_Q(t - \alpha_A).$$

Now consider motions with initial condition at a general point B at time $t = t_0$. These ultimately become periodic with period T since the phase path approaches $\mathscr{C}$, and then will have a certain phase α_B relative to Q, so that

$$x_B(t) \to x_Q(t - \alpha_B) \quad \text{as} \quad t \to \infty.$$

Thus with every point B there is associated an ultimate phase α_B measured relative to Q. This phase may be made unique by a continuity requirement based on the phases calculated for points on $\mathscr{C}$. States having the same phase may be connected by curves, called *isochrones* (Fig. 7.23).

Consider two initial states C and D lying on different isochrones and having ultimate phases α_C and α_D respectively. The phase of D relative to C is defined by

$$phase\ shift = \alpha_D - \alpha_C. \tag{7.76}$$

This is reasonable since, as $t \to \infty$,

$$x_C(t) \to x_Q(t - \alpha_C)$$

and

$$x_D(t) \to x_Q(t - \alpha_D) \equiv x_Q\{(t - \alpha_C) - (\alpha_D - \alpha_C)\}.$$

Since the system is autonomous the phase shift is independent of the (common) time at which states C and D occur, and also of the choice of origin Q. It is therefore a unique function of the positions C and D on the phase plane.

We shall now examine the phase shift resulting from the application of the standard 'impulse' to the system as in equation (7.75) at various points on the limit-cycle. Assume that the system is fairly heavily damped so that its return to $\mathscr{C}$ is effectively completed in a reasonably short time, say within a few cycles: in this way we avoid consideration of 'ultimate phase shifts' and speak simply of 'phase shifts'.

Let the system be at A (Fig. 7.23) with phase α_A. After the impulse it is at B, with phase α_B. This disturbance takes a time τ; the arrival at the reset state B is delayed by this amount and therefore the phase shift due to the impulse, P_A, is given by

$$P_A = \alpha_B - \alpha_A - \tau. \tag{7.77}$$

P_A is a function only of position A. The plot of P_A against α_A for various A and a given impulse function f is called a *phase-response curve*. Figure 7.24 shows the phase-response curve corresponding to the conditions of Fig. 7.23 for the van der Pol system.

The phase response curve has application in biology. Living organisms have certain stable, approximately-daily rhythms (circadian rhythms) of body chemistry appearing as cycles of fatigue, hunger, activity, and so on, pointing to an underlying oscillatory chemical dynamics. These oscillations may sometimes be shifted in time (Pavlidis, 1973) by a strong stimulus; a change of phase occurs, and persists, all things being equal, the original period being maintained. It is only this

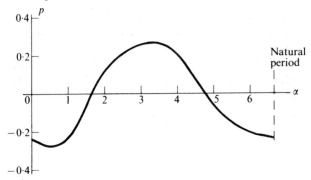

FIG. 7.24. Phase response curve for the equation of Fig. 7.23; the origin of phase is at Q in Fig. 7.23

change of phase which can be observed, but the effect suggests a dynamical structure of which the van der Pol system is a vastly simplified model (Cronin, 1977).

When a periodic stimulus is applied, of a different period from the self-sustained oscillation, the system may under some circumstances become entrained at the forcing frequency. The simplest case occurs as follows. After the system has returned to its limit-cycle following an 'impulse', the next impulse may occur at the same point on the waveform that the previous one did; in this way each cycle becomes a repetition of the last and the motion has the period of the disturbance. This state of affairs will occur when the impulse produces a phase shift of zero or a multiple of T. This is *harmonic entrainment* in the sense of the present chapter.

Superharmonic entrainment, where the period of the excited oscillation is a submultiple of the forcing period, will not occur exactly, but an approximation is obtained in the harmonic case above when several 'cycles', say n, close to $\mathscr{C}$ are executed between impulses. The resulting wave will have period T/n approximately, except for distortion near the times when the impulse is applied.

Another possibility is that the impulse may be delivered $2, 3, \ldots$ times at different points on the phase plane before an effectively zero shift is attained and the cycle repeats itself. In this case *subharmonic entrainment* of order $\frac{1}{2}, \frac{1}{3}, \ldots$ occurs.

Corresponding to this description of entrainment, it is found that the *periods* of biological rhythms can alter under the influence of periodically applied stimuli, becoming entrained, harmonically or subharmonically, with the stimuli.

Exercises

1. Show that the solutions of eqns (7.16) and (7.17), for the undamped Duffing equation in the van der Pol plane, are given exactly by

$$r^2\{(\omega^2-1)+\tfrac{1}{16}r^2\}+2\Gamma a = \text{constant}, \quad r = \sqrt{(a^2+b^2)}.$$

Show that these approximate to circles when r is large. Estimate the period on such a path of $a(t), b(t)$.

2. Express eqns (7.16) and (7.17) in polar coordinates. Deduce the approximate period of $a(t)$ and $b(t)$ for large r. Find the approximate equations for these distant paths. Show how frequency modulation occurs, by deriving an expression for $x(t)$.

3. Consider the equation

$$\ddot{x}+\text{sgn}\,(x) = \Gamma \cos \omega t.$$

Assume solutions of the form $x = a\cos \omega t + b\sin \omega t$. Show that solutions of period $2\pi/\omega$ exist when $|\Gamma| \leqslant 4/\pi^2\omega^2$, and that

$$a = 2\{1 \pm \sqrt{(4-\pi^2\omega^2\Gamma)}\}/\pi\omega^2, \quad b = 0.$$

$$\left(\text{sgn}\,\{x(t)\} = \frac{4a}{\pi\sqrt{(a^2+b^2)}}\cos \omega t + \frac{4b}{\pi\sqrt{(a^2+b^2)}}\sin \omega t + \text{higher harmonics.}\right)$$

4. Show that solutions, period 2π, of the equation $\ddot{x}+x^3 = \Gamma \cos t$ are given approximately by $x = a\cos t$, where a is a solution of $3a^3-4a = 4\Gamma$.

5. Show that solutions, period 2π, of $\ddot{x}+k\dot{x}+x+x^3 = \Gamma \cos t$ are given approximately by $x = a\cos t + b\sin t$, where

$$ka-\tfrac{3}{4}br^2 = 0, \quad kb+\tfrac{3}{4}ar^2 = \Gamma, \quad r = \sqrt{(a^2+b^2)}.$$

Deduce that the response curves are given by

$$r^2(k^2+\tfrac{9}{16}r^4) = \Gamma^2.$$

6. Obtain approximate solutions, period $2\pi/\omega$, to the equation

$$\ddot{x}+\alpha x+\beta x^2 = \Gamma \cos \omega t,$$

by assuming the form $x = c+a\cos \omega t$, and deducing equations for c and a.
   Show that if β is small, $\Gamma = O(\beta)$, and $\omega^2-\alpha = O(\beta)$, then there is a solution with $c \approx -\beta a^2/2\alpha$, $a \approx \Gamma/(\alpha-\omega^2)$.

7. Consider the equation $\ddot{x}+x^3 = \Gamma \cos t$. Substitute $x = a\cos t + b\sin t$, and obtain the solution $x = a\cos t$, where $\tfrac{3}{4}a^3-a = \Gamma$.
   Now fit x^3, by a least squares procedure, by a straight line of the form px where p is constant on $-A \leqslant x \leqslant A$, so that

$$\int_{-A}^{A} (x^3-px)^2\,dx$$

is a minimum with respect to p. Deduce that this linear approximation to the restoring force is compatible with an oscillation, period 2π, of amplitude A, provided $\frac{3}{5}A^3 - A = \Gamma$.

Compare a with A when $\Gamma = 0.1$.

8. Consider the equation

$$\ddot{x} + 0.16x^2 = 1 + 0.2 \cos t.$$

By linearizing the restoring force about the equilibrium points of the unforced system, show that there are two modes of oscillation period 2π, given by

$$x \approx 2.5 - \cos t, \quad x \approx -2.5 - 0.11 \cos t.$$

Find to what extent the predicted modes differ when a substitution of the form $x = c + a\cos t + b\sin t$ is used instead.

9. By examining the transients in the linearized equations obtained in the first part of Exercise 8, show that the two solutions, period 2π, obtained are respectively stable and unstable.

10. Show that the equations giving the equilibrium points in the van der Pol plane for solutions period $2\pi/\omega$ for the forced, damped pendulum equation

$$\ddot{x} + k\dot{x} + x - \tfrac{1}{6}x^3 = \Gamma \cos \omega t, \quad k > 0$$

are

$$k\omega a + b\{\omega^2 - 1 + \tfrac{1}{8}(a^2 + b^2)\} = 0,$$

$$-k\omega b + a\{\omega^2 - 1 + \tfrac{1}{8}(a^2 + b^2)\} = -\Gamma.$$

Deduce that

$$r^2(\omega^2 - 1 + \tfrac{1}{8}r^2)^2 + \omega^2 k^2 r^2 = \Gamma^2, \quad \omega k r^2 = \Gamma b,$$

where $r = \sqrt{(a^2 + b^2)}$. Show that there is a single equilibrium point when $\omega^2 > 1$.

11. For the equation $\ddot{x} + x - \tfrac{1}{6}x^3 = \Gamma \cos \omega t$, show that the solutions of period $2\pi/\omega$ are stable whenever $\omega^2 > 1$; and when $\omega^2 < 1$ and $|\Gamma| > \tfrac{2}{3}\sqrt{(\tfrac{8}{3})(1 - \omega^2)^{3/2}}$.

12. For the equation $\ddot{x} + \alpha x + \beta x^2 = \Gamma \cos t$, substitute $x = c(t) + a(t)\cos t + b(t)\sin t$, and show that, neglecting $\ddot{a}$ and $\ddot{b}$,

$$\dot{a} = \tfrac{1}{2}b(\alpha - 1 + 2\beta c),$$

$$\dot{b} = -\tfrac{1}{2}a(\alpha - 1 + 2\beta c) + \Gamma,$$

$$\ddot{c} = -\alpha c - \beta\{c^2 + \tfrac{1}{2}(a^2 + b^2)\}.$$

Deduce that if $|\Gamma|$ is large there are no solutions of period 2π, and that if $\alpha < 1$ and Γ is sufficiently small there are two solutions of period 2π.

13. Substitute $x = c(t) + a(t)\cos t + b(t)\sin t$ into the equation

$$\ddot{x} + \alpha x^2 = 1 + \Gamma \cos t$$

(compare Exercise 8), and show that if $\ddot{a}$ and $\ddot{b}$ are neglected, then

$$2\dot{a} = b(2\alpha c - 1), \quad 2\dot{b} = a(1 - 2\alpha c) + \Gamma, \quad \ddot{c} + \alpha(c^2 + \tfrac{1}{2}a^2 + \tfrac{1}{2}b^2) = 1.$$

Use a graphical argument to show that there are two equilibrium points when $\alpha < \tfrac{1}{4}$. Explain in general terms why, unlike $\ddot{a}$ and $\ddot{b}$, $\ddot{c}$ cannot be suppressed.

14. In the Duffing equation

$$\ddot{x} + k\dot{x} + x - \tfrac{1}{6}x^3 = \Gamma \cos \omega t,$$

substitute $x = a(t)\cos \omega t + b(t)\sin \omega t$ to investigate the solutions of period $2\pi/\omega$. Assume that a and b are slowly-varying and that $k\dot{a}, k\dot{b}$ can be neglected. Show that the paths in the van der Pol plane are given by

$$\dot{a} = -\frac{b}{2\omega}\{\omega^2 - 1 + \tfrac{1}{8}(a^2 + b^2)\} - \tfrac{1}{2}ka,$$

$$\dot{b} = \frac{a}{2\omega}\{\omega^2 - 1 + \tfrac{1}{8}(a^2 + b^2)\} - \tfrac{1}{2}kb + \frac{\Gamma}{2\omega}.$$

Find the linear approximation in the neighbourhood of the equilibrium point when $\omega^2 > 1$, and show that it is a stable node or spiral when $k > 0$.

15. For the equation $\ddot{x} + \alpha x + \beta x^3 = \Gamma \cos \omega t$, show that the restoring force $\alpha x + \beta x^3$ is represented in the linear least-square approximation on $-A \leqslant x \leqslant A$ by $(\alpha + \tfrac{3}{5}\beta A^2)x$. Obtain the general solution of the approximating equation corresponding to a solution of amplitude A. Deduce that there may be a subharmonic of order $\tfrac{1}{3}$ if $\alpha + \tfrac{3}{5}\beta A^2 = \tfrac{1}{9}\omega^2$ has a real solution A. Compare eqn (7.57) for the case when $\Gamma/8\alpha$ is small. Deduce that when $\alpha \approx \tfrac{1}{9}\omega^2$ (close to subharmonic resonance), the subharmonic has the approximate form

$$A \cos(\tfrac{1}{3}\omega t + \phi) - \frac{\Gamma}{8\alpha}\cos \omega t$$

where ϕ is a constant.

(The interpretation is that when $\Gamma/8\alpha$ is small enough for the oscillation to lie in $[-A, A]$, A can be adjusted so that the slope of the straight-line fit on $[-A, A]$ is appropriate to the generation of a natural oscillation which is a subharmonic. The phase cannot be determined by this method.)

Show that the amplitude predicted for the equation $\ddot{x} + 0.15x - 0.1x^3 = 0.1\cos t$ is $A = 0.085$. Compare this with the solution of (7.57).

16. Use the perturbation method to show that

$$\ddot{x} + k\dot{x} + \alpha x + \beta x^3 = \Gamma \cos \omega t$$

has no subharmonic of order $\frac{1}{2}$ when β is small and $k = O(\beta)$. (Assume that

$$(a\cos\tfrac{1}{2}\tau + b\sin\tfrac{1}{2}\tau + c\cos\tau)^3 = \tfrac{3}{4}c(a^2 - b^2) + \tfrac{3}{4}a(a^2 + b^2 + 2c^2)\cos\tfrac{1}{2}\tau$$
$$+ \tfrac{3}{4}b(a^2 + b^2 + 2c^2)\sin\tfrac{1}{2}\tau + \text{higher harmonics.})$$

17. Use the perturbation method to show that

$$\ddot{x} + k\dot{x} + \alpha x + \beta x^3 = \Gamma\cos\omega t$$

has no subharmonic of order other than $\frac{1}{3}$ when β is small and $k = O(\beta)$.

18. Look for subharmonics of order $\frac{1}{2}$ for the equation

$$\ddot{x} + \varepsilon(x^2 - 1)\dot{x} + x = \Gamma\cos\omega t$$

by substituting

$$x = a\cos\tfrac{1}{2}\omega t + b\sin\tfrac{1}{2}\omega t.$$

Show that there are no non-zero equilibrium points unless $|\omega| = 2$ and $\frac{2}{9}\Gamma^2 - 4 < 0$, in which case the circle $r = \sqrt{(4 - \frac{2}{9}\Gamma^2)}$ consists of equilibrium points. (The algebra can be simplified by writing $x^2\dot{x} = \frac{1}{3}\mathrm{d}(x^3)/\mathrm{d}t$, and using the hint in Exercise 16.)

19. Extend the analysis of the equation

$$\ddot{x} + \varepsilon(x^2 - 1)\dot{x} + x = \Gamma\cos\omega t$$

in Exercise 18 by assuming that

$$x = a(t)\cos\tfrac{1}{2}\omega t + b(t)\sin\tfrac{1}{2}\omega t - \tfrac{1}{3}\Gamma\cos\omega t,$$

where a and b are slowly-varying. Show that when $\ddot{a}, \ddot{b}, \varepsilon\dot{a}, \varepsilon\dot{b}$, are neglected,

$$\omega\dot{a} = (1 - \tfrac{1}{4}\omega^2)b - \tfrac{1}{8}\varepsilon\omega a(a^2 + b^2 + \tfrac{2}{9}\Gamma^2 - 4),$$
$$\omega\dot{b} = -(1 - \tfrac{1}{4}\omega^2)a - \tfrac{1}{8}\varepsilon\omega b(a^2 + b^2 + \tfrac{2}{9}\Gamma^2 - 4),$$

on the van der Pol plane for the subharmonic.

By using $\rho = a^2 + b^2$ and ϕ the polar angle on the plane show that

$$\dot{\rho} = -\tfrac{1}{4}\varepsilon\rho(\rho + K), \quad \dot{\phi} = (1 - \tfrac{1}{4}\omega^2)/\omega; \quad K = \tfrac{2}{9}\Gamma^2 - 4.$$

Deduce that
(i) When $\omega \neq 2$ and $K \geqslant 0$, all paths spiral in to the origin, which is the only equilibrium point (so no subharmonic exists).
(ii) When $\omega = 2$ and $K \geqslant 0$, all paths are radial straight lines entering the origin (so there is no subharmonic).
(iii) When $\omega \neq 2$ and $K < 0$, all paths spiral on to a limit cycle, which is a circle, radius $\sqrt{(-K)}$ and centre the origin (so x is not periodic).
(iv) When $\omega = 2$ and $K < 0$, the circle centre the origin and radius $\sqrt{(-K)}$ consists entirely of equilibrium points, and all paths are radial straight lines approaching these points (each such point represents a subharmonic).

(Since subharmonics are expected only in case (iv), and for a critical value of ω, entrainment cannot occur. For practical purposes, even if the theory were exact we could never expect to observe the subharmonic, though solutions near to it may occur.)

20. Given eqns (7.33), (7.41), and (7.42) for the response curves and the stability boundaries for van der Pol's equation (Fig. 7.9), eliminate r^2 to show that the boundary of the entrainment region in the γ, v plane is given by

$$\gamma^2 = 8\{1 + 9v^2 - (1 - 3v^2)^{3/2}\}/27.$$

Show that, for small $v, \gamma \approx \pm 2v$.

21. Consider the equation

$$\ddot{x} + \varepsilon(x^2 + \dot{x}^2 - 1)\dot{x} + x = \Gamma \cos \omega t.$$

To obtain solutions of period $2\pi/\omega$, substitute

$$x = a(t) \cos \omega t + b(t) \sin \omega t$$

and deduce that, if $\ddot{a}, \ddot{b}, \varepsilon\dot{a}, \varepsilon\dot{b}$, can be neglected, then

$$\dot{a} = \tfrac{1}{2}\varepsilon\{a - vb - \tfrac{1}{4}\mu a(a^2 + b^2)\},$$
$$\dot{b} = \tfrac{1}{2}\varepsilon\{va + b - \tfrac{1}{4}\mu b(a^2 + b^2)\} + \tfrac{1}{2}\varepsilon\gamma;$$

where

$$\mu = 1 + 3\omega^2, \quad v = (\omega^2 - 1)/\varepsilon\omega, \quad \text{and} \quad \gamma = \Gamma/\varepsilon\omega.$$

Show that the stability boundaries are given by

$$1 + v^2 - \mu r^2 + \tfrac{3}{16}\mu^2 r^4 = 0, \quad 2 - \mu r^2 = 0.$$

(See Fig. 7.25, plotted in terms of ω, Γ, and r. The response curves computed directly from the original differential equation (compare Fig. 7.16) are also shown.)

22. Show that the equation

$$\ddot{x}(1 - x\dot{x}) + (\dot{x}^2 - 1)\dot{x} + x = 0$$

has an *exact* periodic solution $x = \cos t$. Show that the corresponding forced equation:

$$\ddot{x}(1 - x\dot{x}) + (\dot{x}^2 - 1)\dot{x} + x = \Gamma \cos \omega t$$

has an *exact* solution of the form $a \cos \omega t + b \sin \omega t$, where

$$a(1 - \omega^2) - \omega b + \omega^3 b(a^2 + b^2) = \Gamma,$$
$$b(1 - \omega^2) + \omega a - \omega^3 a(a^2 + b^2) = 0.$$

Deduce that the amplitude $r = \sqrt{(a^2 + b^2)}$ satisfies

$$r^2\{(1 - \omega^2)^2 + \omega^2(1 - r^2\omega^2)^2\} = \Gamma^2.$$

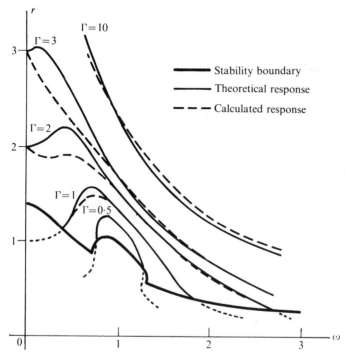

FIG. 7.25. Theoretical and computed response curves, and the stability boundaries, for
$$\ddot{x} + (x^2 + \dot{x}^2 - 1)\dot{x} + x = \Gamma \cos \omega t$$

23. The frequency–amplitude relation for the damped forced pendulum is (eqn 7.24)

$$r^2\{k^2\omega^2 + (\omega^2 - 1 + \tfrac{1}{8}r^2)^2\} = \Gamma^2.$$

Show that the vertex of the cusp bounding the fold in Fig. 7.7 occurs where

$$\omega^2 = \tfrac{1}{2}[2 + 3k^2 - \sqrt{(9k^4 - 12k^2)}].$$

Find the corresponding value for Γ^2.

24. (Combination tones.) Consider the equation

$$\ddot{x} + \alpha x + \beta x^2 = \Gamma_1 \cos \omega_1 t + \Gamma_2 \cos \omega_2 t, \quad \alpha > 0, \quad |\beta| \ll 1,$$

where the forcing term contains two distinct frequencies ω_1 and ω_2. To find an approximation to the response, construct the iterative process leading to the sequence of approximations $x^0(t), x^1(t), \ldots$, and starting with

$$\ddot{x}^0 + \alpha x^0 = \Gamma_1 \cos \omega_1 t + \Gamma_2 \cos \omega_2 t,$$
$$\ddot{x}^1 + \alpha x^1 = \Gamma_1 \cos \omega_1 t + \Gamma_2 \cos \omega_2 t - \beta(x^0)^2,$$

show that a particular solution is given approximately by

$$x(t) = -\frac{\beta}{2\alpha}(a^2 + b^2) + a\cos\omega_1 t + b\cos\omega_2 t$$

$$+ \frac{\beta a^2}{2(4\omega_1^2 - \alpha)}\cos 2\omega_1 t + \frac{\beta b^2}{2(4\omega_2^2 - \alpha)}\cos 2\omega_2 t$$

$$+ \frac{\beta ab}{(\omega_1 + \omega_2)^2 - \alpha}\cos(\omega_1 + \omega_2)t + \frac{\beta ab}{(\omega_1 - \omega_2)^2 - \alpha}\cos(\omega_1 - \omega_2)t,$$

where

$$a \approx \Gamma_1/(\alpha - \omega_1^2), \quad b \approx \Gamma_2/(\alpha - \omega_2^2).$$

(The presence of 'sum and difference tones' with frequencies $\omega_1 \pm \omega_2$ can be detected in sound resonators having suitable nonlinear characteristics, or as an auditory illusion attributed to the nonlinear detection mechanism in the ear (McLachlan, 1956). The iterative method of solution can be adapted to simpler forced oscillation problems involving a single input frequency.)

25. Apply the method of Exercise 24 to the Duffing equation

$$\ddot{x} + \alpha x + \beta x^3 = \Gamma_1 \cos\omega_1 t + \Gamma_2 \cos\omega_2 t.$$

26. Obtain the isochrones (Section 7.9) for the system

$$\dot{x} = (1 - x^2 - y^2)x - y, \quad \dot{y} = (1 - x^2 - y^2)y + x.$$

(Hint: Express the system in polar coordinates.) Show that for any given stimulus the phase-response curve represents a constant which depends only on the stimulus.

27. Investigate the resonant solutions of Duffing's equation in the form

$$\ddot{x} + x + \varepsilon^3 x^3 = \cos t, \quad |\varepsilon| \ll 1,$$

by the method of multiple scales (Section 6.4) using a slow time $\eta = \varepsilon t$ and a solution of the form

$$x(\varepsilon, t) = \frac{1}{\varepsilon}X(\varepsilon, t, \eta) = \frac{1}{\varepsilon}\sum_{n=0}^{\infty}\varepsilon^n X_n(t, \eta).$$

Show that $X_0 = a_0(\eta)\cos t + b_0(\eta)\sin t$, where

$$8a_0' + 3b_0(a_0^2 + b_0^2) = 0, \quad 8b_0' - 3a_0(a_0^2 + b_0^2) = 4.$$

(This example illustrates that even a small nonlinear term may inhibit the growth of resonant solutions.)

28. Repeat the multiple scale procedure of the previous exercise for the equation

$$\ddot{x} + x + \varepsilon^3 x^2 = \cos t, \quad |\varepsilon| \ll 1,$$

which has an unsymmetrical, quadratic departure from linearity. Use a slow time $\eta = \varepsilon^2 t$ and an expansion $x(\varepsilon, t) = \varepsilon^{-2}\sum_{n=0}^{\infty}\varepsilon^n X_n(t, \eta).$

8 Linear systems

THIS CHAPTER may be omitted by readers familiar with the theory of nth order linear systems, or merely referred to on the occasions when its results are needed in Chapters 9 and 10. The need to consider linear systems arises in connection with nonlinear systems which are in some sense close to linear. In such cases certain characteristics of their solutions, notably stability properties, may follow those of the approximating linear systems.

8.1. Structure of solutions of the general linear system

The general homogeneous, first-order linear system of n dimensions is

$$\dot{\mathbf{x}} = A(t)\mathbf{x} \qquad (8.1)$$

where $A(t)$ is an $n \times n$ matrix whose elements $a_{ij}(t)$ are functions of time, and $\mathbf{x}(t)$ is a column vector of the n dependent variables. The explicit appearance of t shows that the system is non-autonomous. In component form,

$$
\begin{pmatrix} \dot{x}_1 \\ \dot{x}_2 \\ \vdots \\ \dot{x}_n \end{pmatrix} = \begin{pmatrix} a_{11} & a_{12} & \cdots & a_{1n} \\ a_{21} & a_{22} & \cdots & a_{2n} \\ \vdots & \vdots & \vdots & \vdots \\ a_{n1} & a_{n2} & \cdots & a_{nn} \end{pmatrix} \begin{pmatrix} x_1 \\ x_2 \\ \vdots \\ x_n \end{pmatrix},
$$

or, when expanded,

$$\dot{x}_i = \sum_{j=1}^{n} a_{ij}(t)x_j; \quad i,j = 1,2,\ldots,n. \qquad (8.2)$$

We shall assume that each $a_{ij}(t)$ is continuous on $-\infty < t < \infty$, so that the system is regular (Appendix A). In that case, given any initial conditions $\mathbf{x}(t_0) = \mathbf{x}_0$, there is a unique solution satisfying this condition. Moreover, it exists for all t, $-\infty < t < \infty$ (the same is not true in general of nonlinear equations).

If $\mathbf{x}_1(t), \mathbf{x}_2(t), \ldots, \mathbf{x}_m(t)$ are real or complex solutions of (8.1), then so is $\alpha_1\mathbf{x}_1(t) + \alpha_2\mathbf{x}_2(t) + \ldots + \alpha_m\mathbf{x}_m(t)$ where $\alpha_1, \alpha_2, \ldots, \alpha_m$ are any constants, real or complex. A central question to be settled is the dimension of the solution space: the minimum number of solution functions which, with their linear combinations, generate every solution of (8.1).

DEFINITION 8.1   *Let* $\psi_1(t), \psi_2(t), \ldots, \psi_m(t)$ *be vector functions (real or complex), continuous on* $-\infty < t < \infty$, *none being identically zero. If there exist constants (real or complex)* $\alpha_1, \alpha_2, \ldots, \alpha_m$, *not all zero, such that*

$$\alpha_1 \psi_1(t) + \alpha_2 \psi_2(t) + \ldots + \alpha_m \psi_m(t) = \mathbf{0}$$

for $-\infty < t < \infty$, *the functions are linearly dependent. Otherwise they are linearly independent.*

(As a consequence of the definition, note that the functions $(1, 1)$, (t, t) are linearly independent, although the constant vectors $(1, 1)$, (t_0, t_0) are linearly dependent for every t_0.)

Example 8.1 $\cos t$ *and* $\sin t$ *are linearly independent on* $-\infty < t < \infty$.

$$\alpha_1 \cos t + \alpha_2 \sin t = \sqrt{(\alpha_1^2 + \alpha_2^2)} \sin (t + \beta)$$

where $\tan \beta = \alpha_1/\alpha_2$. There is no choice of α_1, α_2, except for $\alpha_1 = \alpha_2 = 0$, which makes this function zero for all t.

Example 8.2   *The vector functions* $(t, t), (t^2, t^2), (t^3, t^3)$ *are linearly independent on every interval.*

This is obvious, but note the difference between this case and the case of the *constant* vectors $(a, b), (c, d), (e, f)$ which are always linearly dependent.

Example 8.3   *The functions* $\cos t, \sin t, 2 \sin t$ *are linearly dependent.*

Since, for example,

$$0(\cos t) + 2(\sin t) - 1(2 \sin t) \equiv 0.$$

If functions are linearly dependent, at least one of them can be expressed as a linear combination of the others. From now on we shall be concerned with the linear dependence of solutions of (8.1). Given any set of n solution (column) vectors $\phi_1, \phi_2, \ldots, \phi_n$, real or complex, where ϕ_j has elements $\phi_{1j}(t), \phi_{2j}(t), \ldots, \phi_{nj}(t)$, we shall use the notation

$$\Phi(t) = (\phi_1, \phi_2, \ldots, \phi_n) = \begin{pmatrix} \phi_{11} & \phi_{12} & \cdots & \phi_{1n} \\ \phi_{21} & \phi_{22} & \cdots & \phi_{2n} \\ \vdots & \vdots & & \vdots \\ \phi_{n1} & \phi_{n2} & \cdots & \phi_{nn} \end{pmatrix}$$

for the matrix of these solutions. Then from (8.1),

$$\dot{\Phi}(t) = A(t)\Phi(t). \tag{8.3}$$

Remember that the constituent solutions may be real or complex; for example it is convenient to be able to choose a matrix of solutions for

$$\dot{x}_1 = x_2, \quad \dot{x}_2 = -x_1$$

of the form

$$\Phi(t) = \begin{pmatrix} e^{it} & e^{-it} \\ ie^{it} & -ie^{-it} \end{pmatrix}.$$

Correspondingly, the constants occurring in linear dependence, and the eigenvalues and eigenvectors, may all be complex. This possibility will not always be specially pointed out in the rest of this chapter.

THEOREM 8.1 *Any $n+1$ nontrivial solutions of (8.1) are linearly dependent.*

Proof Let the solutions be $\phi_1(t), \phi_2(t), \ldots, \phi_{n+1}(t)$. Let t_0 be any value of t. Then the $n+1$ *constant* vectors $\phi_1(t_0), \phi_2(t_0), \ldots, \phi_{n+1}(t_0)$ are linearly dependent (as are any set of $n+1$ constant vectors of length n); that is, there exist constants $\alpha_1, \alpha_2, \ldots, \alpha_{n+1}$, not all zero, such that

$$\sum_{j=1}^{n+1} \alpha_j \phi_j(t_0) = \mathbf{0}.$$

Let

$$\mathbf{x}(t) = \sum_{j=1}^{n+1} \alpha_j \phi_j(t).$$

Then $\mathbf{x}(t_0) = \mathbf{0}$, and $\mathbf{x}(t)$ is a solution of (1). Therefore, by the Uniqueness Theorem (Appendix A), $\mathbf{x}(t) = \mathbf{0}$ for all t: that is to say, the solutions $\phi_j(t), j = 1, 2, \ldots, n+1$ are linearly dependent.

THEOREM 8.2 *There exists a set of n linearly independent solutions of (8.1)*

Proof By the Existence Theorem (Appendix A), there is a set of n solutions $\psi_1(t), \psi_2(t), \ldots, \psi_n(t)$ whose matrix $\Psi(t)$ satisfies

$$\Psi(0) = I,$$

where I is the identity matrix. Since the columns of $\Psi(0): \psi_1(0), \ldots, \psi_n(0)$, are linearly independent, so are the solutions.

These two theorems settle the dimension of the solution space: every solution is a linear combination of the solutions $\psi_j, j = 1, 2, \ldots, n$ of Theorem 8.2; but since these solutions are themselves linearly independent, we cannot do without any of them. Instead of the special solutions $\psi_j(t)$ we may take any set of n linearly independent solutions as the basis, as is shown in Theorem 8.3.

THEOREM 8.3   *Let* $\phi_1(t), \phi_2(t), \ldots, \phi_n(t)$ *be any set of linearly independent solutions* (*real or complex*) *of* (8.1). *Then every solution of* (8.1) *is a linear combination of these solutions.*

Proof Let $\phi(t)$ be any non-trivial solution of (8.1). Then by Theorem 8.1, $\phi, \phi_1, \phi_2, \ldots, \phi_n$ are linearly dependent. The coefficient of ϕ in any linear relation of linear dependence is non-zero, or we should violate the requirement that the ϕ_j are linearly independent. Therefore ϕ is a linear combination of $\phi_1, \ldots, \phi_n$.

DEFINITION 8.2   *Let* $\phi_1(t), \phi_2(t), \ldots, \phi_n(t)$ *be* n *linearly independent solutions of* (8.1). *Then the matrix*

$$\Phi(t) = (\phi_1, \phi_2, \ldots, \phi_n) = \begin{pmatrix} \phi_{11} & \phi_{12} & \cdots & \phi_{1n} \\ \phi_{21} & \phi_{22} & \cdots & \phi_{2n} \\ \vdots & \vdots & & \vdots \\ \phi_{n1} & \phi_{n2} & \cdots & \phi_{nn} \end{pmatrix}$$

is called a fundamental matrix of the system (8.1).

Note that any two fundamental matrices Φ_1 and Φ_2 are related by

$$\Phi_2(t) = \Phi_1(t)C, \tag{8.4}$$

where C is a non-singular $n \times n$ matrix (since by Theorem 8.2, each column of Φ_2 is a linear combination of the columns of Φ_1, and vice versa).

THEOREM 8.4   *Given any* $n \times n$ *solution matrix* $\Phi(t) = (\phi_1(t), \ldots, \phi_n(t))$, *then either* (i) *for all* t, $\det\{\Phi(t)\} = 0$, *or* (ii) *for all* t, $\det\{\Phi(t)\} \neq 0$. *Case* (i) *occurs if and only if the solutions are linearly dependent, and case* (ii) *implies that* Φ *is a fundamental matrix.*

Proof Suppose $\det\{\Phi(t_0)\} = 0$ for some t_0. Then the columns of $\Phi(t_0)$ are linearly dependent: there exist constants $\alpha_1, \ldots, \alpha_n$, not all zero, such that $\alpha_1 \phi_1(t_0) + \ldots + \alpha_n \phi_n(t_0) = \mathbf{0}$. Define $\phi(t)$ by

$$\phi(t) = \alpha_1 \phi_1(t) + \ldots + \alpha_n \phi_n(t).$$

Then $\phi(t)$ is a solution satisfying the initial condition $\phi(t_0) = \mathbf{0}$. By the Uniqueness Theorem (Appendix A), $\phi(t) = \mathbf{0}$ for all t. Therefore

$$\alpha_1 \phi_1(t) + \alpha_2 \phi_2(t) + \ldots + \alpha_n \phi_n(t) = \mathbf{0}$$

for all t, and $\det\{\Phi(t)\}$ is therefore either zero everywhere, or non-zero everywhere.

Example 8.4

$$\begin{pmatrix} e^t \\ e^{2t} \end{pmatrix}, \quad \begin{pmatrix} e^{3t} \\ e^{4t} \end{pmatrix}$$

cannot be a pair of solutions of a second-order linear homogeneous system on $-\infty < t < \infty$.

For, writing

$$\Phi(t) = \begin{pmatrix} e^t & e^{2t} \\ e^{2t} & e^{4t} \end{pmatrix},$$

$\det \{\Phi(t)\} = e^{5t} - e^{4t}$, which is zero at $t = 0$ but nowhere else.

THEOREM 8.5 *The solution of* (8.1) *satisfying the initial conditions* $\mathbf{x}(t_0)$ = $\mathbf{x}_0$ *is given by* $\mathbf{x}(t) = \Phi(t)\Phi^{-1}(t_0)\mathbf{x}_0$, *where* Φ *is any fundamental matrix for* (8.1).

Proof The solution must be of the form

$$\mathbf{x}(t) = \Phi(t)\mathbf{a} \tag{8.5}$$

where $\mathbf{a}$ is a constant vector, by Theorem 8.3. The initial conditions give $\mathbf{x}_0 = \Phi(t_0)\mathbf{a}$. The columns of $\Phi(t_0)$ (regarded as columns of constants) are linearly independent by Theorem 8.4, so $\Phi(t_0)$ has an inverse $\Phi^{-1}(t_0)$. Therefore $\mathbf{a} = \Phi^{-1}(t_0)\mathbf{x}_0$ and the result follows from (8.5).

Example 8.5 *Verify that*

$$\begin{pmatrix} 2 \\ e^t \end{pmatrix}, \begin{pmatrix} e^{-t} \\ 1 \end{pmatrix}$$

are particular solutions of

$$\begin{pmatrix} \dot{x}_1 \\ \dot{x}_2 \end{pmatrix} = \begin{pmatrix} 1 & -2e^{-t} \\ e^t & -1 \end{pmatrix}\begin{pmatrix} x_1 \\ x_2 \end{pmatrix},$$

and find the solution $\mathbf{x}(t)$ *such that*

$$\mathbf{x}(0) = \begin{pmatrix} 3 \\ 1 \end{pmatrix}.$$

Direct substitution confirms that the given functions are solutions. They are linearly independent, so

$$\Phi(t) = \begin{pmatrix} 2 & e^{-t} \\ e^t & 1 \end{pmatrix}$$

is a fundamental matrix. We have

$$\Phi(0) = \begin{pmatrix} 2 & 1 \\ 1 & 1 \end{pmatrix}, \quad \Phi^{-1}(0) = \begin{pmatrix} 1 & -1 \\ -1 & 2 \end{pmatrix}.$$

Hence

$$\mathbf{x}(t) = \begin{pmatrix} 2 & e^{-t} \\ e^t & 1 \end{pmatrix} \begin{pmatrix} 1 & -1 \\ -1 & 2 \end{pmatrix} \begin{pmatrix} 3 \\ 1 \end{pmatrix} = \begin{pmatrix} 4 - e^{-t} \\ 2e^t - 1 \end{pmatrix}.$$

The general *inhomogeneous* linear system is

$$\dot{\mathbf{x}} = A(t)\mathbf{x} + \mathbf{f}(t), \tag{8.6}$$

where $\mathbf{f}(t)$ is a column vector. The *associated homogeneous system* is

$$\dot{\boldsymbol{\phi}} = A(t)\boldsymbol{\phi}. \tag{8.7}$$

The following properties are readily verified.

(I) Let $\mathbf{x} = \mathbf{x}_p(t)$ be any solution of (8.6) (called a *particular solution* of the given system) and $\boldsymbol{\phi} = \boldsymbol{\phi}_c(t)$ any solution of (8.7) (called a *complementary function* for the given system). Then $\mathbf{x}_p(t) + \boldsymbol{\phi}_c(t)$ is a solution of (8.6).

(II) Let $\mathbf{x}_{p_1}(t)$ and $\mathbf{x}_{p_2}(t)$ be any solutions of (8.6). Then $\mathbf{x}_{p_1}(t) - \mathbf{x}_{p_2}(t)$ is a solution of (8.7); that is, it is a complementary function.

Theorem 8.6 follows immediately:

THEOREM 8.6   *Let* $\mathbf{x}_p(t)$ *be any one particular solution of* $\dot{\mathbf{x}} = A(t)\mathbf{x} + \mathbf{f}(t)$. *Then every solution of this equation is of the form* $\mathbf{x}(t) = \mathbf{x}_p(t) + \boldsymbol{\phi}_c(t)$, *where* $\boldsymbol{\phi}$ *is a complementary function, and conversely.*

The strategy for finding all solutions of (8.6) is therefore to obtain, somehow, any *one* solution of (8.6), then to find *all* the solutions of the associated homogeneous system (8.7).

Example 8.6   *Find all solutions of the system*

$$\dot{x}_1 = x_2, \quad \dot{x}_2 = -x_1 + t. \tag{8.8}$$

The corresponding homogeneous system is $\dot{\phi}_1 = \phi_2, \dot{\phi}_2 = -\phi_1$, which is equivalent to $\ddot{\phi}_1 + \phi_1 = 0$. The linearly independent solutions $\phi_1 = \cos t, \sin t$ correspond respectively to $\phi_2 = -\sin t, \cos t$. Therefore all solutions of the homogeneous system are given by the linear combinations of

$$\begin{pmatrix} \cos t \\ -\sin t \end{pmatrix}, \quad \begin{pmatrix} \sin t \\ \cos t \end{pmatrix};$$

these are given in matrix form by

$$\boldsymbol{\phi}(t) = \begin{pmatrix} \cos t & \sin t \\ -\sin t & \cos t \end{pmatrix} \begin{pmatrix} a_1 \\ a_2 \end{pmatrix},$$

where a_1, a_2 are arbitrary.

It can be confirmed that $x_1 = t, x_2 = 1$ is a particular solution of the original system (8.8). Therefore all solutions of (8.8) are given by

$$\mathbf{x}(t) = \begin{pmatrix} t \\ 1 \end{pmatrix} + \begin{pmatrix} \cos t & \sin t \\ -\sin t & \cos t \end{pmatrix} \begin{pmatrix} a_1 \\ a_2 \end{pmatrix},$$

where a_1, a_2 are arbitrary.

THEOREM 8.7 *The solution of the system $\dot{\mathbf{x}} = A(t)\mathbf{x} + \mathbf{f}(t)$ with initial conditions $\mathbf{x}(t_0) = \mathbf{x}_0$ is given by*

$$\mathbf{x}(t) = \Phi(t)\Phi^{-1}(t_0)\mathbf{x}_0 + \Phi(t)\int_{t_0}^{t} \Phi^{-1}(s)\mathbf{f}(s)\,\mathrm{d}s, \qquad (8.9)$$

where $\Phi(t)$ is any fundamental solution matrix of $\dot{\phi} = A(t)\phi$.

Proof Let $\mathbf{x}(t)$ be the required solution, for which the following form is postulated

$$\mathbf{x}(t) = \Phi(t)\Phi^{-1}(t_0)\{\mathbf{x}_0 + \phi(t)\}. \qquad (8.10)$$

The inverses of $\Phi(t)$ and $\Phi^{-1}(t_0)$ exist since, by Theorem 8.4, they are non-singular. Then by the initial condition, $\mathbf{x}(t_0) = \mathbf{x}_0$, or $\mathbf{x}_0 + \phi(t_0)$ by (8.10) and so

$$\phi(t_0) = \mathbf{0}. \qquad (8.11)$$

To find the equation satisfied by $\phi(t)$, substitute (8.10) into the equation, which becomes

$$\dot{\Phi}(t)\Phi^{-1}(t_0)\{\mathbf{x}_0 + \phi(t)\} + \Phi(t)\Phi^{-1}(t_0)\dot{\phi}(t)$$
$$= A(t)\Phi(t)\Phi^{-1}(t_0)\{\mathbf{x}_0 + \phi(t)\} + \mathbf{f}(t). \qquad (8.12)$$

Since $\Phi(t)$ is a solution matrix of the homogeneous equation, $\dot{\Phi}(t) = A(t)\Phi(t)$. Equation (8.12) then becomes

$$\Phi(t)\Phi^{-1}(t_0)\dot{\phi}(t) = \mathbf{f}(t).$$

Therefore

$$\dot{\phi}(t) = \Phi(t_0)\Phi^{-1}(t)\mathbf{f}(t),$$

whose solution satisfying (8.11) is

$$\phi(t) = \Phi(t_0)\int_{t_0}^{t} \Phi^{-1}(s)\mathbf{f}(s)\,\mathrm{d}s.$$

Therefore, by (8.10),

$$\mathbf{x}(t) = \Phi(t)\Phi^{-1}(t_0)\mathbf{x}_0 + \Phi(t)\int_{t_0}^{t} \Phi^{-1}(s)\mathbf{f}(s)\,\mathrm{d}s.$$

8.2 The system with constant coefficients

The following alternative representation of the solutions of the inhomogeneous equation is available when A is constant.

LEMMA   *Let $\Phi(t)$ be any fundamental matrix of the system $\dot{\phi} = A\phi$, A constant. Then for any two parameters s, t_0,*

$$\Phi(t)\Phi^{-1}(s) = \Phi(t-s+t_0)\Phi^{-1}(t_0). \tag{8.13}$$

In particular

$$\Phi(t)\Phi^{-1}(s) = \Phi(t-s)\Phi^{-1}(0). \tag{8.14}$$

Proof   Since $\dot{\Phi} = A\Phi$, if we define $U(t)$ by $U(t) = \Phi(t)\Phi^{-1}(s)$, then $\dot{U}(t) = AU(t)$, and $U(s) = I$.

Now consider $V(t) = \Phi(t-s+t_0)\Phi^{-1}(t_0)$. Then $\dot{V}(t) = AV(t)$ (for since A is constant $\Phi(t)$ and $\Phi(t-s+t_0)$ satisfy the same equation), and $V(s) = I$.

Therefore the corresponding columns of U and V satisfy the same equation with the same initial conditions, and are therefore identical by the Uniqueness Theorem (Appendix A).

THEOREM 8.8   *Let A be a constant matrix. The solution of the system $\dot{\mathbf{x}} = A\mathbf{x} + \mathbf{f}(t)$, with initial conditions $\mathbf{x}(t_0) = \mathbf{x}_0$, is given by*

$$\mathbf{x}(t) = \Phi(t)\Phi^{-1}(t_0)\mathbf{x}_0 + \int_{t_0}^{t} \Phi(t-s+t_0)\Phi^{-1}(t_0)\mathbf{f}(s)\,ds, \tag{8.15}$$

where $\Phi(t)$ is any fundamental matrix for $\dot{\phi} = A\phi$. In particular, if $\Psi(t)$ is the fundamental matrix satisfying $\Psi(0) = I$, then

$$\mathbf{x}(t) = \Psi(t)\mathbf{x}_0 + \int_{t_0}^{t} \Psi(t-s)\mathbf{f}(s)\,ds. \tag{8.16}$$

Proof   (8.15) follows from (8.9) and (8.13), and (8.16) from (8.9) and (8.14).

Example 8.7   *Express the solution of the second-order equation $\ddot{x} - x = h(t)$, with $x(0) = 0$, $\dot{x}(0) = 1$ as an integral.*

An equivalent first-order pair is

$$\dot{x} = y, \quad \dot{y} = x + h(t),$$

and

$$\begin{pmatrix} \dot{x} \\ \dot{y} \end{pmatrix} = \begin{pmatrix} 0 & 1 \\ 1 & 0 \end{pmatrix}\begin{pmatrix} x \\ y \end{pmatrix} + \begin{pmatrix} 0 \\ h(t) \end{pmatrix} = A\begin{pmatrix} x \\ y \end{pmatrix} + \mathbf{f}(t).$$

A fundamental matrix for the homogeneous system can be verified to be

$$\Phi(t) = \begin{pmatrix} e^t & e^{-t} \\ e^t & -e^{-t} \end{pmatrix}.$$

Then, following Theorem 8.8,

$$\Phi(0) = \begin{pmatrix} 1 & 1 \\ 1 & -1 \end{pmatrix}, \quad \Phi^{-1}(0) = \begin{pmatrix} \frac{1}{2} & \frac{1}{2} \\ \frac{1}{2} & -\frac{1}{2} \end{pmatrix},$$

and

$$\begin{pmatrix} x \\ y \end{pmatrix} = \begin{pmatrix} e^t & e^{-t} \\ e^t & -e^{-t} \end{pmatrix} \begin{pmatrix} \frac{1}{2} & \frac{1}{2} \\ \frac{1}{2} & -\frac{1}{2} \end{pmatrix} \begin{pmatrix} 0 \\ 1 \end{pmatrix} + \int_0^t \begin{pmatrix} e^{t-s} & e^{-t+s} \\ e^{t-s} & -e^{-t+s} \end{pmatrix} \begin{pmatrix} \frac{1}{2} & \frac{1}{2} \\ \frac{1}{2} & -\frac{1}{2} \end{pmatrix} \begin{pmatrix} 0 \\ h(s) \end{pmatrix} ds$$

$$= \begin{pmatrix} \sinh t \\ \cosh t \end{pmatrix} + \int_0^t \begin{pmatrix} h(s)\sinh (t-s) \\ h(s)\cosh (t-s) \end{pmatrix} ds.$$

The most difficult problem is to construct an explicit fundamental matrix for the homogeneous system. When A is a function of t this is impossible in general, since the solutions of equations with time-dependent coefficients are not generally expressible in terms of known functions. When A is constant, however, the solutions are of elementary type.

Consider the system

$$\dot{\mathbf{x}} = A\mathbf{x} \tag{8.17}$$

where A is constant with real elements. We look for solutions of the form

$$\mathbf{x} = \mathbf{r}e^{\lambda t} \tag{8.18}$$

where λ is a constant. In order to satisfy (8.17) we must have

$$A\mathbf{r}e^{\lambda t} - \lambda\mathbf{r}e^{\lambda t} = (A - \lambda I)\mathbf{r}e^{\lambda t} = \mathbf{0}$$

for all t, or

$$(A - \lambda I)\mathbf{r} = \mathbf{0} \tag{8.19}$$

Given a value for λ, this is equivalent to n linear equations for the components of $\mathbf{r}$, and has non-trivial solutions if and only if

$$\det (A - \lambda I) = 0. \tag{8.20a}$$

In component form (8.20a) becomes

$$\begin{vmatrix} a_{11} - \lambda & a_{12} & \cdots & a_{1n} \\ a_{21} & a_{22} - \lambda & \cdots & a_{2n} \\ \cdots & \cdots & \cdots & \cdots \\ a_{n1} & a_{n2} & \cdots & a_{nn} - \lambda \end{vmatrix} = 0. \tag{8.20b}$$

Equation (8.20) is a polynomial equation of degree n for λ, called the *characteristic equation*. It therefore has n roots, real or complex, some of which may be repeated roots. If λ is a complex root, so is $\bar{\lambda}$ since A is a real matrix. The values of λ given by (8.20) are the *eigenvalues* of A. Equation (8.17) *therefore has a solution of the form* (8.18) *if and only if λ is an eigenvalue of A*.

Now suppose that *the eigenvalues are all different*, so that there are exactly n distinct eigenvalues, real, complex or zero, $\lambda_1, \lambda_2, \ldots, \lambda_n$. For each λ_i there exist non-zero solutions $\mathbf{r} = \mathbf{r}_i$ of equation (8.19). These are the *eigenvectors* corresponding to λ_i. It is known that in the present case of n distinct eigenvalues, all the eigenvectors of a particular eigenvalue are simply multiples of each other: therefore we have essentially n solutions of (8.17):

$$(\mathbf{r}_1\, e^{\lambda_1 t}, \mathbf{r}_2\, e^{\lambda_2 t}, \ldots, \mathbf{r}_n\, e^{\lambda_n t}),$$

where $\mathbf{r}_i$ is any one of the eigenvectors of λ_i. The solutions are linearly independent. Therefore we have

THEOREM 8.9   *For the system* $\dot{\mathbf{x}} = A\mathbf{x}$, *A a real, constant matrix, whose eigenvalues* $\lambda_1, \lambda_2, \ldots, \lambda_n$ *are all different*,

$$\Phi(t) = (\mathbf{r}_1\, e^{\lambda_1 t}, \mathbf{r}_2\, e^{\lambda_2 t}, \ldots, \mathbf{r}_n\, e^{\lambda_n t}) \tag{8.21}$$

is a fundamental matrix (*complex in general*), *where* $\mathbf{r}_i$ *is any eigenvector corresponding to* λ_i.

Example 8.8   *Find a fundamental matrix for the system*

$$\dot{x}_1 = x_2 - x_3, \quad \dot{x}_2 = x_3, \quad \dot{x}_3 = x_2.$$

We have $\dot{\mathbf{x}} = A\mathbf{x}$ with

$$A = \begin{pmatrix} 0 & 1 & -1 \\ 0 & 0 & 1 \\ 0 & 1 & 0 \end{pmatrix}.$$

To find the eigenvalues, we require

$$\det(A - \lambda I) = \begin{vmatrix} -\lambda & 1 & -1 \\ 0 & -\lambda & 1 \\ 0 & 1 & -\lambda \end{vmatrix} = -\lambda(\lambda - 1)(\lambda + 1),$$

so the eigenvalues are $\lambda = 0, 1, -1$. The equations for the eigenvectors, $\mathbf{r} = (r_1, r_2, r_3)^t$ are as follows.

For $\lambda = 0$, $r_2 - r_3 = 0$, $r_3 = 0$, $r_2 = 0$. One solution is $\mathbf{r} = (1, 0, 0)^t$.

For $\lambda = 1, -r_1 + r_2 - r_3 = 0, -r_2 + r_3 = 0, \quad r_2 - r_3 = 0.$ One solution is $\mathbf{r} = (0, 1, 1)^t$.

For $\lambda = -1, r_1 + r_2 - r_3 = 0, r_2 + r_3 = 0, r_2 + r_3 = 0.$ One solution is $\mathbf{r} = (2, -1, 1)^t$.

A fundamental matrix is therefore given by

$$\Phi(t) = \begin{pmatrix} 1 & 0 & 2e^{-t} \\ 0 & e^t & -e^{-t} \\ 0 & e^t & e^{-t} \end{pmatrix}.$$

When eigenvalues are not all distinct the formal situation is more complicated, and for the theory the reader is referred, for example, to Wilson (1971). We illustrate some possibilities in the following examples.

Example 8.9 *Find a fundamental matrix for the system*

$$\dot{x}_1 = x_1, \quad \dot{x}_2 = x_2, \quad \dot{x}_3 = x_3.$$

The matrix for the system is

$$A = \begin{pmatrix} 1 & 0 & 0 \\ 0 & 1 & 0 \\ 0 & 0 & 1 \end{pmatrix}.$$

The characteristic equation is $\det(A - \lambda I) = 0$, which becomes $(1 - \lambda)^3 = 0$. Thus $\lambda = 1$ is a threefold repeated root. The equation for the eigenvectors, $(A - \lambda I)\mathbf{r} = \mathbf{0}$, gives no restriction on $\mathbf{r}$, so the linearly independent vectors $(1, 0, 0)^t, (0, 1, 0)^t, (0, 0, 1)^t$ as eigenvectors. This leads to the fundamental matrix Φ given by

$$\Phi(t) = \begin{pmatrix} e^t & 0 & 0 \\ 0 & e^t & 0 \\ 0 & 0 & e^t \end{pmatrix}.$$

Example 8.10 *Find a fundamental matrix for the system*

$$\dot{x}_1 = x_1 + x_2, \quad \dot{x}_2 = x_2, \quad \dot{x}_3 = x_3.$$

We have

$$A = \begin{pmatrix} 1 & 1 & 0 \\ 0 & 1 & 0 \\ 0 & 0 & 1 \end{pmatrix}.$$

The characteristic equation becomes $(1 - \lambda)^3 = 0$, so $\lambda = 1$ is a threefold root. The equation for the eigenvectors gives simply $r_2 = 0$. We choose the two linearly independent eigenvectors, $(1, 0, 0)^t$ and $(0, 0, 1)^t$, satisfying this condition. The corresponding solutions are

$$(e^t, 0, 0)^t, \quad (0, 0, e^t)^t.$$

It is easy to confirm that a third solution, independent of the other two, is $(te^t, e^t, 0)^t$. A fundamental matrix Φ is therefore given by

$$\Phi(t) = \begin{pmatrix} e^t & 0 & te^t \\ 0 & 0 & e^t \\ 0 & e^t & 0 \end{pmatrix}.$$

Example 8.11   *Find a fundamental matrix for the system*

$$\dot{x}_1 = x_1 + x_2, \quad \dot{x}_2 = x_2 + x_3, \quad \dot{x}_3 = x_3.$$

We have

$$A = \begin{pmatrix} 1 & 1 & 0 \\ 0 & 1 & 1 \\ 0 & 0 & 1 \end{pmatrix}.$$

Once again, $\lambda = 1$ is a threefold root of the characteristic equation. The equation for the eigenvectors gives $r_2 = r_3 = 0$. One solution is $\mathbf{r} = (1, 0, 0)^t$ and there is no other which is linearly independent of this one. The corresponding solution is $\mathbf{x} = (e^t, 0, 0)^t$.

It can be confirmed that two more linearly independent solutions are

$$\mathbf{x} = (te^t, e^t, 0)^t, \quad \mathbf{x} = (\tfrac{1}{2}t^2 e^t, te^t, e^t)^t$$

leading to the fundamental matrix

$$\Phi(t) = \begin{pmatrix} e^t & te^t & \tfrac{1}{2}t^2 e^t \\ 0 & e^t & te^t \\ 0 & 0 & e^t \end{pmatrix}.$$

Inspection of these examples suggests the following theorem on the form of linearly independent solutions associated with multiple eigenvalues. We state the theorem without proof.

THEOREM 8.10   *Corresponding to an eigenvalue of $A, \lambda = \lambda_i$, of multiplicity $m \leqslant n$ there are m linearly independent solutions of the system (8.17). These are of the form*

$$\mathbf{p}_1(t)\, e^{\lambda_i t}, \dots, \mathbf{p}_m(t)\, e^{\lambda_i t}, \qquad (8.22)$$

where the $\mathbf{p}_j(t)$ are vector polynomials of degree less than m.

Note that when an eigenvalue is complex, the eigenvectors and the polynomials in (8.22) will be complex, and the arrays consist of complex-valued solutions. Since the elements of A are real, the characteristic equation (8.20) has real coefficients. Therefore, if λ_i is one eigenvalue, $\bar{\lambda}_i$ is another. The corresponding polynomial coefficients in (8.22) are similarly, complex conjugates. In place, therefore, of the pair of complex solutions $\phi_i(t), \bar{\phi}_i(t)$ corresponding to λ_i and $\bar{\lambda}_i$ respectively, we could take the real solutions, $\text{Re}\{\phi_i(t)\}, \text{Im}\{\phi_i(t)\}$.

8.3 Equations with periodic coefficients (Floquet theory)

A second special case where more may be said about the nature of the fundamental solutions is that of the system

$$\dot{\mathbf{x}} = P(t)\mathbf{x}, \tag{8.23}$$

where $P(t)$ is periodic with principal period T; that is, T is the smallest positive number for which

$$P(t+T) = P(t), \quad -\infty < t < \infty. \tag{8.24}$$

($P(t)$, of course, also has periods $2T, 3T, \ldots$) The *solutions* are not necessarily periodic, as can be seen from the one-dimensional example

$$\dot{x} = P(t)x = (1 + \sin t)x;$$

the coefficient $P(t)$ has period 2π, but all solutions are given by

$$x = c\,e^{t - \cos t},$$

where c is any constant, so only the solution $x = 0$ is periodic.

In the following discussions remember that the displayed solution-vectors may consist of complex solutions.

THEOREM 8.11 (*Floquet's theorem.*) *The regular system* $\dot{\mathbf{x}} = P(t)\mathbf{x}$, *where* P *is an* $n \times n$ *matrix function with principal period* T, *has at least one non-trivial solution* $\mathbf{x} = \boldsymbol{\chi}(t)$ *such that*

$$\boldsymbol{\chi}(t+T) = \mu\boldsymbol{\chi}(t), \quad -\infty < t < \infty \tag{8.25}$$

where μ *is a constant.*

Proof Let $\Phi(t) = (\phi_{ij}(t))$ be a fundamental matrix for the system. Then $\dot{\Phi}(t) = P(t)\Phi(t)$. Since $P(t+T) = P(t)$, $\Phi(t+T)$ satisfies the same equation, and by Theorem 8.4, $\det \Phi(t+T) \neq 0$, so $\Phi(t+T)$ is another fundamental matrix. The columns (solutions) in $\Phi(t+T)$ are linear combinations of those in $\Phi(t)$ by Theorem 8.3:

$$\phi_{ij}(t+T) = \sum_{k=1}^{n} \phi_{ik}(t)e_{kj}$$

for some constants e_{kj}, or

$$\Phi(t+T) = \Phi(t)E \tag{8.26}$$

where $E = (e_{kj})$. E is nonsingular, since $\det \Phi(t + T) = \det \Phi(t) \det (E)$, and $\det (E) \neq 0$.

Let μ be an eigenvalue of E:

$$\det (E - \mu I) = 0, \tag{8.27}$$

and let **s** be an eigenvector corresponding to μ:

$$(E - \mu I)\mathbf{s} = \mathbf{0}. \tag{8.28}$$

Consider the solution $\mathbf{x} = \Phi(t)\mathbf{s} = \boldsymbol{\chi}(t)$ (being a linear combination of the columns of Φ, $\boldsymbol{\chi}$ is a solution of (8.23)). Then

$$\begin{aligned}
\boldsymbol{\chi}(t + T) &= \Phi(t + T)\mathbf{s} \\
&= \Phi(t)E\mathbf{s} = \Phi(t)\mu\mathbf{s}, \qquad \text{(by (8.28))} \\
&= \mu\boldsymbol{\chi}(t).
\end{aligned}$$

The eigenvalues of E are called *characteristic numbers* of equation (8.23). The importance to us of this theorem is the possibility of a characteristic number with a special value implying the existence of a periodic solution (though not necessarily of period T).

Example 8.12 *Find a fundamental matrix for the periodic differential equation*

$$\begin{pmatrix} \dot{x}_1 \\ \dot{x}_2 \end{pmatrix} = \begin{pmatrix} 1 & 1 \\ 0 & h(t) \end{pmatrix} \begin{pmatrix} x_1 \\ x_2 \end{pmatrix} \tag{8.29}$$

where $h(t) = (\cos t + \sin t)/(2 + \sin t - \cos t)$, and determine the characteristic numbers.

From (8.29),

$$(2 + \sin t - \cos t)\dot{x}_2 = (\cos t + \sin t)x_2;$$

which has solutions

$$x_2 = b(2 + \sin t - \cos t) \tag{8.30a}$$

where b is any constant. Then x_1 satisfies

$$\dot{x}_1 - x_1 = x_2 = b(2 + \sin t - \cos t)$$

and therefore

$$x_1 = a\mathrm{e}^t - b(2 + \sin t). \tag{8.30b}$$

A fundamental matrix $\Phi(t)$ can be obtained by putting, say, $a = 0$, $b = 1$, and $a = 1, b = 0$:

$$\Phi(t) = \begin{pmatrix} -2 - \sin t & \mathrm{e}^t \\ 2 + \sin t - \cos t & 0 \end{pmatrix}.$$

$P(t)$ for (8.29) has period $T = 2\pi$, and E in (8.26) must satisfy $\Phi(t+2\pi) = \Phi(t)E$ for all t. Therefore $\Phi(2\pi) = \Phi(0)E$ and

$$E = \Phi^{-1}(0)\Phi(2\pi) = \begin{pmatrix} 1 & 0 \\ 0 & e^{2\pi} \end{pmatrix}.$$

The eigenvalues μ of E satisfy

$$\begin{vmatrix} 1-\mu & 0 \\ 0 & e^{2\pi}-\mu \end{vmatrix} = 0,$$

so $\mu = 1$ and $e^{2\pi}$.

From (8.25), since one eigenvalue is unity there exist solutions such that $\chi(t+2\pi) = \chi(t)$; that is, solutions with period 2π. We have already found these: they correspond to $a = 0$.

THEOREM 8.12 *The constants μ in Theorem 8.11 are independent of the choice of Φ.*

Proof Let $\Phi(t)$, $\Phi^*(t)$ be two fundamental matrices; then

$$\Phi^*(t) = \Phi(t)C, \tag{8.31}$$

where C is some constant, nonsingular matrix (nonsingular since $\Phi(t)$ and $\Phi^*(t)$ are nonsingular by Theorem 8.4). Let T be the principal period of $P(t)$. Then

$$\begin{aligned} \Phi^*(t+T) &= \Phi(t+T)C, & \text{(by (8.31))} \\ &= \Phi(t)EC, & \text{(by (8.26))} \\ &= \Phi^*(t)C^{-1}EC, & \text{(by (8.31))} \\ &= \Phi^*(t)D, \end{aligned}$$

say. But E and D are *similar* matrices and so have the same eigenvalues.

We can therefore properly refer to '*the* characteristic numbers of the system'. Note that when Φ is chosen as real, E is real, and the characteristic equation for the numbers μ has real coefficients. Therefore if μ (complex) is a characteristic number, then so is $\bar{\mu}$.

DEFINITION 8.3 *A solution of (8.23) satisfying (8.25) is called a **normal solution.***

DEFINITION 8.4 (*Characteristic exponent*) *Let μ be a characteristic number of (8.23), corresponding to the principal period T of $P(t)$. Then ρ, defined by*

$$e^{\rho T} = \mu, \tag{8.32}$$

is called a characteristic exponent of the system. Note that ρ is defined only to an additive multiple of $2\pi i/T$. It will be fixed by requiring $-\pi < \mathrm{Im}\,(\rho T) \leqslant \pi$, or by $\rho = 1/T\,\mathrm{Log}\,(\mu)$, where the principal value of the logarithm is taken.

THEOREM 8.13 *Suppose that E of Theorem 8.10 has n distinct eigenvalues, $\mu_i, i = 1, 2, \ldots, n$. Then (8.23) has n linearly independent normal solutions of the form*

$$\mathbf{x}_i = \mathbf{p}_i(t)\,e^{\rho_i t} \tag{8.33}$$

(ρ_i are the characteristic exponents corresponding to μ_i), where the $\mathbf{p}_i(t)$ are functions with period T.

Proof To each μ_i corresponds a solution $\mathbf{x}_i(t)$ satisfying (8.25): $\mathbf{x}_i(t + T) = \mu_i \mathbf{x}_i(t) = e^{\rho_i T} x_i(t)$. Therefore, for every t,

$$\mathbf{x}_i(t + T)\,e^{-\rho_i(t+T)} = \mathbf{x}_i(t)\,e^{-\rho_i t}. \tag{8.34}$$

Writing

$$\mathbf{p}_i(t) = e^{-\rho_i t} \mathbf{x}_i(t),$$

(8.34) implies that $\mathbf{p}_i(t)$ has period T.

The linear independence of the $\mathbf{x}_i(t)$ is implied by their method of construction in Theorem 8.11: from (8.28), they are given by $\mathbf{x}_i(t) = \Phi(t)\mathbf{s}_i$; $\mathbf{s}_i$ are the eigenvectors corresponding to the *different* eigenvalues μ_i, and are therefore linearly independent. Since $\Phi(t)$ is nonsingular it follows that the $\mathbf{x}_i(t)$ are also linearly independent.

When the eigenvalues of E are not all distinct, the coefficients corresponding to the $\mathbf{p}_i(t)$ are more complicated. Under the conditions of Theorem 8.13, periodic solutions of period T exist when E has an eigenvalue

$$\mu = 1.$$

The corresponding normal solutions have period T, the principal period of $P(t)$. This can be seen from (8.25) or from the fact that the corresponding ρ is zero.

There are solutions with other periods whenever E has an eigenvalue μ which is one of the mth roots of unity:

$$\mu = 1^{1/m}, \quad m \text{ a positive integer.} \tag{8.35a}$$

In this case, from (8.25),

$$\chi(t + mT) = \mu\chi\{t + (m-1)T\} = \ldots = \mu^m \chi(t) = \chi(t), \tag{8.35b}$$

and $\chi(t)$ has period mT.

8.4 The Wronskian

In the preceding theory, $\det \Phi(t)$ appeared repeatedly, where Φ is a fundamental matrix of the regular system $\dot{\mathbf{x}} = A(t)\mathbf{x}$. This has a simple representation, as follows.

DEFINITION 8.4 *Let $(\boldsymbol{\phi}_1(t), \boldsymbol{\phi}_2(t), \ldots, \boldsymbol{\phi}_n(t))$ be a matrix whose columns are any solutions of the system $\dot{\mathbf{x}} = A(t)\mathbf{x}$. Then*

$$W(t) = \det(\boldsymbol{\phi}_1(t), \boldsymbol{\phi}_2(t), \ldots, \boldsymbol{\phi}_n(t)) \tag{8.36}$$

is called the Wronskian of this set of solutions, taken in order.

THEOREM 8.14 *Let $\Phi(t)$ be any matrix of solutions for $\dot{\mathbf{x}} = A(t)\mathbf{x}$; then, for any t_0,*

$$W(t) = W(t_0)\exp\left(\int_{t_0}^{t} \mathrm{tr}\{A(s)\}\,\mathrm{d}s\right), \tag{8.37}$$

where $\mathrm{tr}\{A(s)\}$ is the trace of $A(t)$: the sum of the elements of its principal diagonal.

Proof If the solutions are linearly dependent, $W(t) \equiv 0$ by Theorem 8.4, and the result is true trivially.

If not, let $\Phi(t)$ be any fundamental matrix of solutions, with $\Phi(t) = (\phi_{ij}(t))$. Then $\det(\phi_{ij}(t))$ is equal to the sum of n determinants $\Delta_k, k = 1, 2, \ldots, n$, where Δ_k is the same as $\det(\phi_{ij}(t))$, except for having $\dot{\phi}_{kj}(t), j = 1, 2, \ldots, n$ in place of $\phi_{kj}(t)$ in its kth row. Consider one of the Δ_k, say Δ_1:

$$\Delta_1 = \begin{vmatrix} \dot{\phi}_{11} & \dot{\phi}_{12} & \cdots & \dot{\phi}_{1n} \\ \phi_{21} & \phi_{22} & \cdots & \phi_{2n} \\ \cdots & \cdots & \cdots & \cdots \\ \phi_{n1} & \phi_{n2} & & \phi_{nn} \end{vmatrix} =$$

$$\begin{vmatrix} \sum_{m=1}^{n} a_{1m}\phi_{m1} & \sum_{m=1}^{n} a_{1m}\phi_{m2} & \cdots & \sum_{m=1}^{n} a_{1n}\phi_{mn} \\ \phi_{21} & \phi_{22} & \cdots & \phi_{2n} \\ \cdots & \cdots & \cdots & \cdots \\ \phi_{n1} & \phi_{n2} & \cdots & \phi_{nn} \end{vmatrix}$$

(from eqn (8.2))

$$= \sum_{m=1}^{n} a_{1m} \begin{vmatrix} \phi_{m1} & \phi_{m2} & \cdots & \phi_{mn} \\ \phi_{21} & \phi_{22} & \cdots & \phi_{2n} \\ \cdots & \cdots & \cdots & \cdots \\ \phi_{n1} & \phi_{n2} & & \phi_{nn} \end{vmatrix}$$

$$= a_{11} \begin{vmatrix} \phi_{11} & \phi_{12} & \cdots & \phi_{1n} \\ \phi_{21} & \phi_{22} & \cdots & \phi_{2n} \\ \cdots & \cdots & \cdots & \cdots \\ \phi_{n1} & \phi_{n2} & & \phi_{nn} \end{vmatrix} = a_{11} W(t),$$

since all the other determinants have repeated rows, and therefore vanish. In general $\Delta_k = a_{kk} W(t)$. Therefore

$$\frac{dW(t)}{dt} = \text{tr}\{A(t)\} W(t),$$

which is a differential equation for W having solution (8.37).

Another theorem involving the trace, of use in Section 8.5, is the following.

THEOREM 8.15   *For the system* $\dot{\mathbf{x}} = P(t)\mathbf{x}$, *where* $P(t)$ *has principal period* T, *let the characteristic numbers of the system be* $\mu_1, \mu_2, \ldots, \mu_n$. *Then*

$$\mu_1 \mu_2 \ldots \mu_n = \exp\left(\int_0^T \text{tr}\{P(s) \, ds \right),$$

a repeated characteristic number being counted according to its multiplicity.

Proof   Let $\Psi(t)$ be the fundamental matrix of the system for which

$$\Psi(0) = I. \tag{8.38}$$

Then, (eqn (8.26)),

$$\Psi(T) = \Psi(0)E = E, \tag{8.39}$$

in the notation of Theorem 8.11. The characteristic numbers μ are the eigenvalues of E:

$$\det(E - \mu I) = 0.$$

This is an nth degree polynomial in μ, and the product of the roots is equal to the constant term: that is, equal to the value taken when $\mu = 0$. Thus, by (8.39),

$$\mu_1 \mu_2 \ldots \mu_n = \det(E) = \det \Psi(T) = W(T),$$

but by Theorem 8.14 with $t_0 = 0$ and $t = T$,

$$W(t) = W(0) \int_0^T \text{tr}\,\{P(s)\}\,\mathrm{d}s$$

and $W(0) = 1$ by (8.38).

8.5. Hill's and Mathieu's equations

The following problem introduces an important class of linear equations with periodic coefficients. A pendulum of length a with a bob of mass m is suspended from a support which is constrained to move vertically with displacement $\zeta(t)$ (Fig. 8.1).

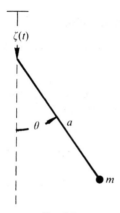

Fig. 8.1.

The kinetic energy $\mathscr{T}$ and potential energy $\mathscr{V}$ are given by

$$\mathscr{T} = \tfrac{1}{2}m[(\dot{\zeta} - a\sin\theta\,\dot{\theta})^2 + a^2\cos^2\theta\,\dot{\theta}^2],$$
$$\mathscr{V} = -mg(\zeta + a\cos\theta).$$

Lagrange's equation

$$\frac{\mathrm{d}}{\mathrm{d}t}\left(\frac{\partial \mathscr{T}}{\partial \dot{\theta}}\right) - \frac{\partial \mathscr{T}}{\partial \theta} = -\frac{\partial \mathscr{V}}{\partial \theta}$$

becomes

$$a\ddot{\theta} + (g - \ddot{\zeta})\sin\theta = 0$$

which, for oscillations of small amplitude, simplifies to

$$a\ddot{\theta} + (g - \ddot{\zeta})\theta = 0.$$

As a standard form for this equation we may write

$$\ddot{x} + (\alpha + p(t))x = 0. \tag{8.40}$$

When $p(t)$ is periodic it is known as Hill's equation. For the special case $p(t) = \beta \cos t,$

$$\ddot{x} + (\alpha + \beta \cos t)x = 0 \tag{8.41}$$

is called Mathieu's equation. This type of forced motion, in which $p(t)$ acts as an energy source, is an instance of *parametric excitation.*

Consider the system equivalent to Mathieu's equation (8.41):

$$\begin{pmatrix} \dot{x} \\ \dot{y} \end{pmatrix} = \begin{pmatrix} 0 & 1 \\ -\alpha - \beta \cos t & 0 \end{pmatrix} \begin{pmatrix} x \\ y \end{pmatrix}. \tag{8.42}$$

In the notation of the previous sections,

$$P(t) = \begin{pmatrix} 0 & 1 \\ -\alpha - \beta \cos t & 0 \end{pmatrix} \tag{8.43}$$

is a periodic coefficient matrix with least period 2π and Theorems 8.11 to 8.13, and 8.15, apply; in particular, the general structure of the solution is shown in Theorem 8.13. It is important for the study of stability (Chapter 9) to be able to decide *whether, for given values of α and β, there are any unbounded solutions of the system* (8.42). We are not very interested in periodic solutions as such, though we shall need them in order to decide this point.

From eqn (8.43),

$$\mathrm{tr}\,\{P(t)\} = 0. \tag{8.44}$$

Therefore, by Theorem 8.15,

$$\mu_1 \mu_2 = \mathrm{e}^0 = 1, \tag{8.45}$$

where μ_1, μ_2 are the characteristic numbers of $P(t)$. Therefore they are solutions of a quadratic characteristic equation (8.27), with real coefficients (see p. 233), which by (8.45) has the form

$$\mu^2 - \phi(\alpha, \beta)\mu + 1 = 0.$$

The solutions μ are given by

$$\mu_1, \mu_2 = \tfrac{1}{2}[\phi \pm \sqrt{(\phi^2 - 4)}]. \tag{8.46}$$

In (8.46), $\phi(\alpha, \beta)$ is not known explicitly (see Exercise 18), but we can make the following deductions by reference to Theorem 8.13.

(i) $\phi > 2$. The characteristic numbers are real, different, and positive, and by (8.45), one of them, say μ_1, exceeds unity. The corresponding characteristic exponents (8.32) are real and have the form $\rho_1 = \sigma > 0$, $\rho_2 = -\sigma < 0$. The general solution is therefore of the form (Theorem 8.13)

$$x(t) = c_1 e^{\sigma t} p_1(t) + c_2 e^{-\sigma t} p_2(t),$$

where c_1, c_2 are constants and p_1, p_2 have principal period 2π. The parameter region $\phi(\alpha, \beta) > 2$ therefore contains unbounded solutions, and is called an *unstable parameter region*.

(ii) $\phi = 2$. Then $\mu_1 = \mu_2 = 1$, $\rho_1 = \rho_2 = 0$. By (8.33), there is *one solution of period 2π on the curves $\phi(\alpha, \beta) = 2$*. (The other is unbounded.)

(iii) $-2 < \phi < 2$. The characteristic numbers are complex, and $\mu_2 = \bar{\mu}_1$. Since also $|\mu_1| = |\mu_2| = 1$, we must have $\rho_1 = i\nu, \rho_2 = -i\nu, \nu$ real. The general solution is of the form

$$x(t) = c_1 e^{i\nu t} p_1(t) + c_2 e^{-i\nu t} p_2(t)$$

and *all solutions in the parameter region $-2 < \phi(\alpha, \beta) < 2$ are bounded*. This is called the *stable parameter region*. The solutions are oscillatory, but not in general periodic, since the two frequencies ν and 2π are present.

(iv) $\phi = -2$. Then $\mu_1 = \mu_2 = -1$ ($\rho_1 = \rho_2 = \frac{1}{2}i$), so by Theorem 8.11, eqn (8.25), there *is one solution with period 4π at every point on $\phi(\alpha, \beta) = -2$*. (The other solution is in fact unbounded.)

(v) $\phi < -2$. Then μ_1 and μ_2 are real and negative. Since, also, $\mu_1 \mu_2 = 1$, the general solution is of the form

$$x(t) = c_1 e^{(\sigma + \frac{1}{2}i)t} p_1(t) + c_2 e^{(-\sigma + \frac{1}{2}i)t} p_2(t)$$

where $\sigma > 0$ and p_1, p_2 have period 2π. For later purposes it is important to notice that *the solutions have the alternative form*

$$c_1 e^{\sigma t} q_1(t) + c_2 e^{-\sigma t} q_2(t), \tag{8.47}$$

where q_1, q_2 have period 4π.

From (i) to (v) it can be seen that certain curves, of the form

$$\phi(\alpha, \beta) = \pm 2,$$

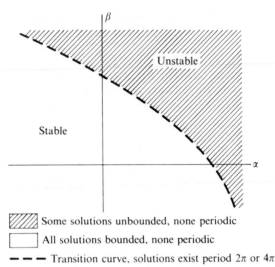

Some solutions unbounded, none periodic

All solutions bounded, none periodic

— — — Transition curve, solutions exist period 2π or 4π

Fig. 8.2.

separate parameter regions where unbounded solutions exist ($|\phi(\alpha, \beta)|$ > 2) from regions where all solutions are bounded ($|\phi(\alpha, \beta)| < 2$) (Fig. 8.2). We do not know the function $\phi(\alpha, \beta)$ explicitly, but we do know that these are also the curves on which periodic solutions, period 2π or 4π, occur. Therefore, *if we can establish, by any method, the parameter values for which such periodic solutions can occur, then we have also found the boundaries between the stable and unstable region.* These boundaries are called *transition curves.*

First, we find what parameter values α, β give periodic solutions of period 2π. Represent such a solution by the complex Fourier series

$$x(t) = \sum_{n=-\infty}^{\infty} c_n e^{int}.$$

After substituting into Mathieu's equation, (8.41) (with $\cos t = \frac{1}{2}(e^{it} + e^{-it})$), we obtain after rearrangement

$$\sum_{n=-\infty}^{\infty} \{\tfrac{1}{2}\beta c_{n+1} + (\alpha - n^2)c_n + \tfrac{1}{2}\beta c_{n-1}\} e^{int} = 0$$

for all t. This can be satisfied only if the coefficients are all zero:

$$\tfrac{1}{2}\beta c_{n+1} + (\alpha - n^2)c_n + \tfrac{1}{2}\beta c_{n-1} = 0, \quad n = 0, \pm 1, \pm 2, \dots. \quad (8.48)$$

This infinite set of homogeneous equations for $\{c_n\}$ has non-zero solutions if the infinite determinant (Whittaker and Watson, 1962) formed by the coefficients is zero; when $\alpha \neq n^2$ for any n the following form is convenient:

$$\begin{vmatrix} \cdot & \cdot & \cdot & \cdot & \cdot & \cdot & \cdot \\ \cdot & \gamma_1 & 1 & \gamma_1 & 0 & 0 & \cdot \\ \cdot & 0 & \gamma_0 & 1 & \gamma_0 & 0 & \cdot \\ \cdot & 0 & 0 & \gamma_1 & 1 & \gamma_1 & \cdot \\ \cdot & \cdot & \cdot & \cdot & \cdot & \cdot & \cdot \end{vmatrix} = 0, \qquad (8.49)$$

where

$$\gamma_n = \beta/2(\alpha - n^2), \quad n = 0, 1, 2, \dots.$$

This is not explicitly the relation $\phi(\alpha, \beta) = 2$, but it must be equivalent to it, since it specifies transition lines on the α, β plane.

To find the equations for transition lines corresponding to solutions of (8.41) with period 4π, start with the series

$$x(t) = \sum_{n=-\infty}^{\infty} d_n \, e^{\frac{1}{2}int}.$$

The equations for the coefficients are

$$\tfrac{1}{2}\beta d_{n+2} + (\alpha - \tfrac{1}{4}n^2)d_n + \tfrac{1}{2}\beta d_{n-2} = 0, \qquad (8.50)$$

$n = 0, \pm 1, \pm 2, \dots$. This set of equations splits into two independent sets for $\{d_n\}$ with n even and odd respectively. The set for n even gives a Hill determinant (as these are called) of the form in (8.49). We may therefore specify $d_{2m+1} = 0$ for all m and solve (8.49); this produces the functions of period 2π already obtained. Otherwise, specify $d_{2m} = 0$, all m; for the remaining equations in (8.50) we require

$$\begin{vmatrix} \cdot & \cdot & \cdot & \cdot & \cdot & \cdot & \cdot \\ \cdot & \gamma_2 & 1 & \gamma_2 & 0 & 0 & 0 & \cdot \\ \cdot & 0 & \gamma_1 & 1 & \gamma_1 & 0 & 0 & \cdot \\ \cdot & 0 & 0 & \gamma_1 & 1 & \gamma_1 & 0 & \cdot \\ \cdot & 0 & 0 & 0 & \gamma_2 & 1 & \gamma_2 & \cdot \\ \cdot & \cdot & \cdot & \cdot & \cdot & \cdot & \cdot \end{vmatrix} = 0 \qquad (8.51)$$

where

$$\gamma_1 = \beta/2\{\alpha - \tfrac{1}{4}1^2\}, \quad \gamma_2 = \beta/2\{\alpha - \tfrac{1}{4}3^2\}, \dots,$$

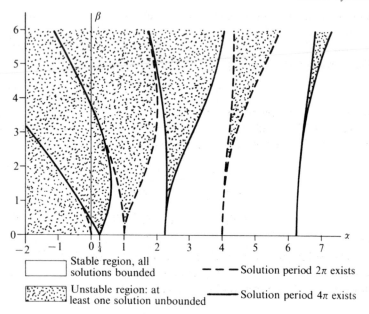

FIG. 8.3. Stability diagram for Mathieu's equation

(provided $\alpha \neq \frac{1}{4}(2m+1)^2$ for any m), which specifies the transition curves corresponding to solutions of period 4π.

Given α, β, the determinants in (8.49), (8.51) can rapidly be computed from recurrence relations between the determinants of orders differing by 2 at each step, starting with the mid-determinant (see Exercises 19 and 20). Zeros can be found by specifying α (or β) and searching for the corresponding β (or α) which gives a result sufficiently close to zero. Convergence is rapid enough to produce the data for Fig. 8.3 to two or three decimals in a very short time. The transition curves corresponding to eqn (8.49) pass through the points $\beta = 0$, $\alpha = n^2$, $n = 1, 2, \ldots$, and on these curves solutions of period 2π exist. The curves associated with eqn (8.51) pass through $\beta = 0$, $\alpha = \frac{1}{4}m^2$, $m = 1, 3, 5, \ldots$, and on these solutions of period 4π exist.

8.6 Transition curves for Mathieu's equation by perturbation

For small values of $|\beta|$ the perturbation method can be used to establish the transition curves. In the equation

$$\ddot{x} + (\alpha + \beta \cos t)x = 0, \tag{8.52}$$

suppose that the transition curves are given by

$$\alpha = \alpha(\beta) = \alpha_0 + \beta\alpha_1 + \dots, \tag{8.53}$$

and that the corresponding solutions have the form

$$x(t) = x_0(t) + \beta x_1(t) + \dots, \tag{8.54}$$

where $x_0, x_1, \dots$ have period 2π or 4π.

When (8.53) and (8.54) are substituted into (8.52) and the coefficients of powers of β equated to zero in the usual way, we have

$$\ddot{x} + \alpha_0 x_0 = 0, \tag{8.55a}$$

$$\ddot{x}_1 + \alpha_0 x_1 = -(\alpha_1 + \cos t)x_0, \tag{8.55b}$$

$$\ddot{x}_2 + \alpha_0 x_2 = -x_0\alpha_2 - (\alpha_1 + \cos t)x_1, \tag{8.55c}$$

and so on.

The solutions of (8.55a) have minimum period 2π if $\alpha_0 = n^2$, $n = 0, 1, \dots$; and minimum period 4π if $\alpha_0 = (n + \frac{1}{2})^2$, $n = 0, 1, \dots$. Both classes of solution can be obtained together by considering

$$\alpha_0 = \tfrac{1}{4}n^2, \quad n = 0, 1, \dots. \tag{8.56}$$

We consider only $n = 0$ and 1 (see also Exercise 21).

(i) $n = 0$. Equation (8.55a) has the periodic solution $x_0(t) = a_0$. Then, from (8.55b),

$$\ddot{x}_1 = -(\alpha_1 + \cos t)a_0,$$

which has periodic solutions only if $\alpha_1 = 0$, the solutions being of the form

$$x_1(t) = a_0 \cos t + a_1.$$

Equation (8.55c) becomes

$$\ddot{x}_2 = -a_0\alpha_2 - \tfrac{1}{2}a_0 - \tfrac{1}{2}a_0 \cos 2t - a_1 \cos t,$$

which has a periodic solution (of period 2π) only if $a_0\alpha_2 + \frac{1}{2}a_0 = 0$ or if $\alpha_2 = -\frac{1}{2}$. Therefore, for small β

$$\alpha = -\tfrac{1}{2}\beta^2 + O(\beta^3). \tag{8.57}$$

(ii) $n = 1$. Here $\alpha_0 = \frac{1}{4}$, and $x_0(t) = a_0 \cos \frac{1}{2}t + b_0 \sin \frac{1}{2}t$. (8.55b) becomes

$$\ddot{x}_1 + \tfrac{1}{4}x_1 = -a_0(\alpha_1 - \tfrac{1}{2})\cos \tfrac{1}{2}t - b_0(\alpha_1 - \tfrac{1}{2})\sin \tfrac{1}{2}t - \tfrac{1}{2}a_0 \cos \tfrac{3}{2}t - \tfrac{1}{2}b_0 \sin \tfrac{3}{2}t.$$

There are periodic solutions (of period 4π) only if $b_0 = 0, \alpha_1 = -\frac{1}{2}$, or if $a_0 = 0, \alpha_1 = \frac{1}{2}$. Thus, to a first approximation, for small β

$$\alpha = \tfrac{1}{4} \pm \tfrac{1}{2}\beta \qquad (8.58)$$

describes the transition curves near $\alpha = \frac{1}{4}, \beta = 0$. Figure 8.3 may be compared with these results.

8.7. Mathieu's equation with damping

Consider Mathieu's equation with a damping term:

$$\ddot{x} + \kappa\dot{x} + (v + \beta\cos t)x = 0, \qquad (8.59)$$

where $v > 0, \kappa > 0$. If we write

$$x(t) = e^{-\frac{1}{2}\kappa t}\eta(t), \qquad (8.60)$$

(8.59) transforms into

$$\ddot{\eta} + (v - \tfrac{1}{4}\kappa^2 + \beta\cos t)\eta = 0, \qquad (8.61)$$

which is a form of Mathieu's equation, (8.52), with

$$\alpha = v - \tfrac{1}{4}\kappa^2. \qquad (8.62)$$

We shall try to construct part of a diagram of the stable regions for (8.59) similar to Fig. 8.3. Certainly, whenever α and β have values such that only bounded solutions of (8.61) exist, then only bounded solutions of (8.59) exist, by (8.60). Also by (8.60), even some exponential growth in η is allowed, x still remaining bounded. Therefore the 'regions of stability' for (8.59) will be extensions of those for Mathieu's equation.

We shall examine particularly the first unstable region for (8.61) which (by (8.62) and Fig. 8.3) occurs near $\alpha = \frac{1}{4}$ or near $v = \frac{1}{4}(1 + \kappa^2)$. On the boundaries of this unstable region we know from the previous section that there are periodic solutions of (8.61) of period 4π. Therefore this corresponds to case (iv), p. 239, in which $\phi(\alpha, \beta) = -2$. Inside the corresponding unstable region we must therefore have case (v), $\phi(\alpha, \beta) < -2$ and the solutions have the form (8.47). Therefore, by (8.60), the corresponding $x(t)$ have the form

$$x(t) = c_1 e^{(\sigma - \frac{1}{2}\kappa)t}q_1(t) + c_2 e^{(-\sigma - \frac{1}{2}\kappa)t}q_2(t), \qquad (8.63)$$

where $\sigma > 0$ and q_1, q_2 have period 4π. The boundary of the unstable region for $x(t)$ occurs where $\sigma - \frac{1}{2}\kappa = 0$, and so by (8.63), the boundary can be detected by the existence of solutions, $x(t)$, of period 4π at points on it.

We shall suppose that

$$|\beta| \ll 1, \tag{8.64}$$

and that κ is also small:

$$\kappa = \beta\kappa_1. \tag{8.65}$$

As in Section 8.6, let the unknown boundary be given by

$$v = v(\beta) = v_0 + \beta v_1 + \ldots, \tag{8.66}$$

and the solutions, of period 4π, by

$$x(t) = x_0(t) + \beta x_1(t) + \ldots. \tag{8.67}$$

The perturbation method requires

$$\ddot{x}_0 + v_0 x_0 = 0, \tag{8.68a}$$

$$\ddot{x}_1 + v_0 x_1 = -\kappa_1 \dot{x}_0 - (v_1 + \cos t)x_0, \tag{8.68b}$$

and so on. Equation (8.68a), with the periodicity condition, requires $v_0 = \frac{1}{4}$; then $x_0(t) = a_0 \cos \frac{1}{2}t + b_0 \sin \frac{1}{2}t$. By substituting these results into (8.68b) and equating the coefficients of $\cos \frac{1}{2}t$, $\sin \frac{1}{2}t$ to zero on the right to avoid secular terms we find

$$(v_1 + \tfrac{1}{2})a_0 + \tfrac{1}{2}\kappa_1 b_0 = 0, \quad \tfrac{1}{2}\kappa_1 a_0 - (v_1 - \tfrac{1}{2})b_0 = 0.$$

The existence of a non-zero solution requires $(v_1 - \frac{1}{2})(v_1 + \frac{1}{2}) + (\frac{1}{2}\kappa_1)^2 = 0$, or $v_1 = \pm\frac{1}{2}\sqrt{(1 - \kappa_1^2)}$. Thus the stability boundary near $v = \frac{1}{4}$ is given by

$$v = \tfrac{1}{4} \pm \tfrac{1}{2}\beta\sqrt{(1 - \kappa_1^2)} \tag{8.69}$$

(so long as $\kappa_1^2 < 1$; otherwise the whole region becomes stable for $|\beta| \ll 1$). The stable region is given by

$$v < \tfrac{1}{4} - \tfrac{1}{2}\beta\sqrt{(1 - \kappa_1^2)}, \quad v > \tfrac{1}{4} + \tfrac{1}{2}\beta\sqrt{(1 - \kappa_1^2)},$$

or by

$$(v - \tfrac{1}{4})^2 - \tfrac{1}{4}\beta^2(1 - \kappa_1^2) > 0. \tag{8.70}$$

Note that, by virtue of eqn (8.60), there are regions in which the solutions of the damped Mathieu equation tend to zero; and from the analysis of the various regions in Section 8.5 and the form of the corresponding solutions these regions are exactly the stable regions of (8.59).

Exercises

1. Determine the linear dependence or independence of the following:
   (i) $(1, 1, -1), (2, 1, 1), (0, 1, -3)$.
   (ii) $(t, 2t), (3t, 4t), (5t, 6t)$.
   (iii) $(e^t, e^{-t}), (e^{-t}, e^t)$. Could these both be solutions of a 2×2 homogeneous linear system?
   (iv) $(\sin(t-\alpha), \cos(t-\alpha)), (\sin(t-\beta), \cos(t-\beta)), (\sin(t-\gamma), \cos(t-\gamma))$.

2. Construct a complex and a real fundamental matrix Φ for the system $\dot{x} = y$, $\dot{y} = -x - 2y$. Deduce a fundamental matrix Ψ satisfying $\Psi(0) = I$.

3. Construct a fundamental matrix for the system $\dot{x}_1 = -x_1, \dot{x}_2 = x_1 + x_2 + x_3$, $\dot{x}_3 = -x_2$.

4. Construct a fundamental matrix for the system $\dot{x}_1 = x_2, \dot{x}_2 = x_1$, and deduce the solution satisfying $x_1 = 1, x_2 = 0$, at $t = 0$.

5. Construct a fundamental matrix for the system $\dot{x}_1 = x_2, \dot{x}_2 = x_3, \dot{x}_3 = -2x_1 + x_2 + 2x_3$, and deduce the solution of $\dot{x}_1 = x_2 + e^t, \dot{x}_2 = x_3, \dot{x}_3 = -2x_1 + x_2 + 2x_3$, with $\mathbf{x}(0) = (1, 0, 0)^t$.

6. Show that the differential equation $x^{(n)} + a_1 x^{(n-1)} + \ldots + a_n x = 0$ is equivalent to the system

$$\dot{x}_1 = x_2, \quad \dot{x}_2 = x_3, \ldots, \quad \dot{x}_{n-1} = x_n, \quad \dot{x}_n = -a_n x_1 - \ldots - a_1 x_n,$$

with $x = x_1$. Show that the equation for the eigenvalues is

$$\lambda^n + a_1 \lambda^{n-1} + \ldots + a_n = 0.$$

7. Find all the solutions of the system $\dot{x}_1 = x_1 - x_2 + e^t, \dot{x}_2 = x_1 + x_2 - e^t$.

8. Find the solution of the system

$$\begin{pmatrix} \dot{x}_1 \\ \dot{x}_2 \end{pmatrix} = \begin{pmatrix} 0 & 1 \\ 1 & 0 \end{pmatrix} \begin{pmatrix} x_1 \\ x_2 \end{pmatrix} + \begin{pmatrix} h_1(t) \\ h_2(t) \end{pmatrix}$$

subject to $\mathbf{x}(0) = \mathbf{x}_0$, using the fundamental matrix for the homogeneous problem.

9. Express the solution of the system $\dot{x}_1 = x_1 + x_2 + h_1(t), \dot{x}_2 = -x_2 + h_2(t), \dot{x}_3 = x_2 + h_3(t)$, with $x_1(0) = a_1, x_2(0) = a_2, x_3(0) = a_3$, in the form of an integral.

10. A bird population, $p(t)$, is governed by the differential equation $\dot{p} = \mu(t)p - kp$, where k is the death-rate and $\mu(t)$ represents a variable birth-rate with period 1 year. Derive a condition which ensures that the mean annual population remains constant. Assuming that this condition is fulfilled, does it seem likely that, *in practice*, the average population will remain constant? (This is asking a question about a particular kind of stability.)

11. The system

$$\dot{x}_1 = (-\sin 2t)x_1 + (\cos 2t - 1)x_2, \quad \dot{x}_2 = (\cos 2t + 1)x_1 + (\sin 2t)x_2$$

has a fundamental matrix of normal solutions:

$$\begin{pmatrix} e^t(\cos t - \sin t) & e^{-t}(\cos t + \sin t) \\ e^t(\cos t + \sin t) & e^{-t}(-\cos t + \sin t) \end{pmatrix}$$

Obtain the corresponding E-matrix (eqn (8.26)), the characteristic numbers, and the characteristic exponents.

12. Let the system $\dot{\mathbf{x}} = P(t)\mathbf{x}$ have a matrix of coefficients P with minimum period T (and therefore also with periods $2T, 3T, \ldots$). Follow the argument of Theorem 8.11, using period mT, $m > 1$, to show that $\Phi(t + mT) = \Phi(t)E^m$. Assuming that if the eigenvalues of E are μ_i, then those of E^m are μ_i^m, deduce that no new periodic solutions are disclosed by this process.

13. Obtain Wronskians for the following linear systems:
(i) $\dot{x}_1 = x_1 \sin t + x_2 \cos t$, $\dot{x}_2 = -x_1 \cos t + x_2 \sin t$,
(ii) $\dot{x}_1 = f(t)x_2$, $\dot{x}_2 = g(t)x_1$.

14. The equations of motion for two dissimilar pendulums coupled through a certain type of support become, after linearizing,

$$\ddot{\theta} + \frac{\alpha}{1+\beta}\theta + \frac{\gamma}{1+\beta}\ddot{\phi} = 0, \quad \ddot{\phi} + \frac{\alpha}{1+\gamma}\phi + \frac{\beta}{1+\gamma}\ddot{\theta} = 0$$

$\alpha, \beta, \gamma > 0$, where θ and ϕ are the inclinations and β and γ are certain parameters associated respectively with the pendulums. Obtain the eigenvalues and the eigenvectors for the system, and construct the normal modes.

15. By substituting $x = c + a\cos t + b\sin t$ into Mathieu's equation
$$\ddot{x} + (\alpha + \beta\cos t)x = 0,$$

obtain by harmonic balance an approximation to the transition curve near $\alpha = 0$, $\beta = 0$ (compare with Section 8.6).

By substituting $x = c + a\cos\frac{1}{2}t + b\sin\frac{1}{2}t$, find the transition curves near $\alpha = \frac{1}{4}$, $\beta = 0$.

16. The figure represents a mass m attached to two identical linear elastic strings of stiffness λ and natural length l. The ends of the strings pass through frictionless guides A and B at a distance $2L$, $l < L$, apart. The particle is set into lateral

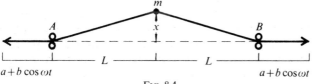

$$a + b\cos\omega t \qquad\qquad a + b\cos\omega t$$

FIG. 8.4.

motion at the mid-point, and symmetrical displacements $a + b \cos \omega t$, $a > b$, are imposed on the ends of the string. Show that, for $x \ll L$,

$$\ddot{x} + \left(\frac{2\lambda(L - l + a)}{mL} + \frac{2\lambda b}{mL} \cos \omega t \right) x = 0.$$

Analyse the motion in terms of suitable parameters, using the information of Sections 8.5 and 8.6 on the growth or decay, periodicity and near-periodicity of the solutions of Mathieu's equation in the regions of its parameter plane.

17. A pendulum with a light, rigid suspension is placed upside-down on end, and the point of suspension is caused to oscillate vertically with displacement y given by $y = \varepsilon \cos \omega t$, $\varepsilon \ll 1$. Show that the equation of motion is

$$\ddot{\theta} + \left(-\frac{g}{a} + \frac{1}{a} \ddot{y} \right) \sin \theta = 0,$$

where a is the length of the suspension, g is gravitational acceleration, and θ the inclination to the vertical. Linearize the equation for small amplitudes and show that the vertical position is stable (that is, the motion of the pendulum restricts itself to the neighbourhood of the vertical: it does not topple over) provided $\varepsilon^2 C^2 / 2ag > 1$.

18. Let $\Phi(t) = (\phi_{ij}(t))$, $i,j = 1,2$, be the fundamental matrix for the system $\dot{x}_1 = x_2, \dot{x}_2 = -(\alpha + \beta \cos t)x_1$, satisfying $\Phi(0) = I$ (Mathieu's equation). Show that the characteristic numbers μ satisfy the equation

$$\mu^2 - \mu\{\phi_{11}(2\pi) + \phi_{22}(2\pi)\} + 1 = 0.$$

19. In eqn (8.49), for the transition curves of Mathieu's equation for solutions period 2π, let

$$D_{m,n} = \begin{vmatrix} 1 & \gamma_m & & & & \\ \gamma_{m-1} & 1 & \gamma_{m-1} & & & \\ & & \cdots & & & \\ & & & \gamma_0 & 1 & \gamma_0 & \\ & & & & \cdots & \\ & & & & \gamma_{n-1} & 1 & \gamma_{n-1} \\ & & & & & \gamma_n & 1 \end{vmatrix}$$

for $m \geqslant 0, n \geqslant 0$. Show that

$$D_{m,n} = D_{m-1,n} - \gamma_m \gamma_{m-1} D_{m-2,n}.$$

Let $E_n = D_{n,n}$ and verify that

$$E_0 = 1, \quad E_1 = 1 - 2\gamma_0\gamma_1, \quad E_2 = (1 - \gamma_1\gamma_2)^2 - 2\gamma_0\gamma_1(1 - \gamma_1\gamma_2).$$

Prove that, for $n \geqslant 1$,

$$E_{n+2} = (1 - \gamma_{n+1}\gamma_{n+2})E_{n+1} + \gamma_{n+1}\gamma_{n+2}(1 - \gamma_{n+1}\gamma_{n+2})E_n + \gamma_n^2\gamma_{n+1}^3\gamma_{n+2}E_{n-1}.$$

This recurrence relation can be used to evaluate the determinant (8.49).

20. In equation (8.51), for the transition curves for Mathieu's equation for solutions of period 4π, let

$$F_{m,n} = \begin{vmatrix} 1 & \gamma_m & & & & \\ \gamma_{m-1} & 1 & \gamma_{m-1} & & & \\ & & \cdots & & & \\ & & & \gamma_{n-1} & 1 & \gamma_{n-1} \\ & & & & \gamma_n & 1 \end{vmatrix}$$

Show as in the last exercise that $G_n = F_{n,n}$ satisfies the same recurrence relation as E_n for $n \geqslant 2$. Verify that

$$G_1 = 1 - \gamma_1^2,$$
$$G_2 = (1 - \gamma_1\gamma_2)^2 - \gamma_1^2,$$
$$G_3 = (1 - \gamma_1\gamma_2 - \gamma_2\gamma_3)^2 - \gamma_1^2\gamma_2\gamma_3(1 - \gamma_2\gamma_3).$$

21. Show, by the perturbation method, that the transition curves for Mathieu's equation

$$\ddot{x} + (\alpha + \beta \cos t)x = 0,$$

$\alpha \approx 1, \beta \approx 0$, are given by $\alpha = 1 + \frac{1}{12}\beta^2$, $\alpha = 1 - \frac{5}{12}\beta^2$.

22. Consider Hill's equation $\ddot{x} + f(t)x = 0$, where f has period 2π, and

$$f(t) = \alpha + \sum_{r=1}^{\infty} \beta_r \cos rt$$

is its Fourier expansion, with $\alpha \approx \frac{1}{4}$ and $|\beta_r| \ll 1$, $r = 1, 2, \ldots$. Assume an approximate solution $e^{\sigma t}q(t)$, where σ is real and q has period 4π as in (8.47). Show that

$$\ddot{q} + 2\sigma\dot{q} + \left(\sigma^2 + \alpha + \sum_{r=1}^{\infty} \beta_r \cos rt\right)q = 0.$$

Take $q \approx \sin(\frac{1}{2}t + \gamma)$ as the approximate form for q and match terms in $\sin \frac{1}{2}t$, $\cos \frac{1}{2}t$, on the assumption that these terms dominate. Deduce that

$$\sigma^2 = -(\alpha + \frac{1}{4}) + \frac{1}{2}\sqrt{(4\alpha + \beta_1^2)}$$

and that the transition curves near $\alpha = \frac{1}{4}$ are given by $\alpha = \frac{1}{4} \pm \frac{1}{2}\beta_1$. ($\beta_n$ is similarly the dominant coefficient for transition curves near $\alpha = \frac{1}{4}n^2$, $n \geqslant 1$.)

23. Obtain, as in Section 8.5, the boundary of the stable region in the neighbourhood of $v = 1$, $\beta = 0$ for Mathieu's equation with damping,

$$\ddot{x} + \kappa\dot{x} + (v + \beta\cos t)x = 0$$

where $\kappa = O(\beta^2)$.

24. Solve Meissner's equation

$$\ddot{x} + (\alpha + \beta f(t))x = 0$$

where $f(t) = 1$, $0 \leqslant t < \pi$; $f(t) = -1$, $\pi \leqslant t < 2\pi$; and $f(t+2\pi) = f(t)$ for all t. Find the conditions on α, β, for periodic solutions by putting $x(0) = x(2\pi)$, $\dot{x}(0) = \dot{x}(2\pi)$ and by making x and $\dot{x}$ continuous at $t = \pi$. Find a determinant equation for α and β.

25. By using the harmonic balance method of Chapter 4, show that the van der Pol equation with parametric excitation,

$$\ddot{x} + \varepsilon(x^2 - 1)\dot{x} + (1 + \beta\cos t)x = 0$$

has a 2π-periodic solution with approximately the same amplitude as the unforced van der Pol equation.

26. Consider the linear 3×3 system $\dot{\mathbf{x}} = A\mathbf{x}$. Investigate the types of phase diagram (in three-dimensional space) associated with the main classifications of the eigenvalues of A.

27. The male population M and female population F of a bird community have a constant death-rate k and a variable birth-rate $\mu(t)$ which has period T, so that

$$\dot{M} = -kM + \mu(t)F, \quad \dot{F} = -kF + \mu(t)F.$$

If the births are seasonal, so that

$$\mu(t) = \begin{cases} \delta, & 0 < t \leqslant \varepsilon; \\ 0, & \varepsilon < t \leqslant T, \end{cases}$$

show that periodic solutions of period T exist for M and F if $kT = \delta\varepsilon$.

9 Stability

THE WORD 'stability', as a descriptive term, is used freely in earlier chapters. In Chapter 2 *equilibrium points* are classified as stable or unstable according to their appearance in the phase diagram. Roughly speaking, if every initial state close enough to equilibrium leads to states which continue permanently to be close, then the equilibrium is stable. Some equilibrium states may be thought of as being more stable than others; for example, all initial states near a stable node lead ultimately into the node, but in the case of a centre there always remains a residual oscillation after disturbance of the equilibrium. *Limit cycles* are classified as stable or unstable according to whether or not 'nearby' paths 'spiral' into the limit cycle. In Chapter 7 we considered the stability of *forced periodic oscillations*, reducing the question to that of the stability of the corresponding equilibrium points on a van der Pol plane.

These cases are formally different, as well as being imprecisely defined. Moreover, they are not exhaustive: the definition of stability chosen must be that appropriate to the types of phenomena we want to distinguish between. The question to be answered is usually of this kind. If a system is in some way disturbed, will its subsequent behaviour differ from its undisturbed behaviour by an acceptably small amount? In practice, physical and other systems are always subject to small, unpredictable influences: to variable switch-on conditions, to maladjustment, to variation in physical properties and the like. If such variations produce large changes in the operating conditions the system is probably unusable, and its normal operating condition would be described as unstable. Even if the effect of small disturbances does not grow, the system may not be 'stable enough' for its intended purpose: for proper working it might be necessary for the system to re-approach the normal operating condition, rather than to sustain a permanent superimposed deviation, however small. We might even require it to approach its normal state without ultimately suffering any time delay.

Some of these possibilities are treated in the following chapter. We are concerned with *regular systems* throughout (see Appendix A). The treatment is not restricted to second-order systems.

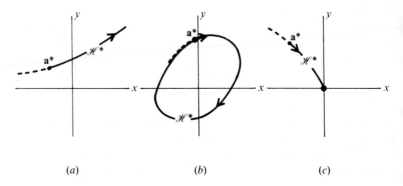

Fig. 9.1. Typical half-paths $\mathcal{H}^*$. In (b), the half-path is a closed curve, repeated indefinitely.

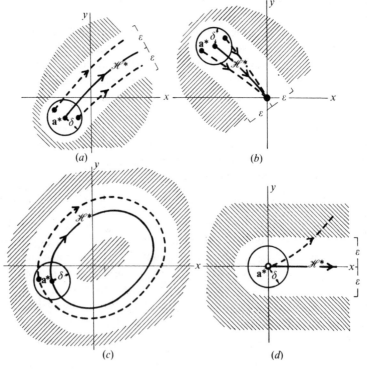

Fig. 9.2. ———— $\mathcal{H}^*$ ———— standard path, — — — neighbouring paths. (d) Represents an unstable case.

9.1. Poincaré stability

This is a relatively undemanding criterion, which applies to autonomous systems. We shall treat it intuitively, in the two-dimensional case only. It agrees with the tentative classification of equilibrium points as stable or unstable in Chapter 2, and of limit cycles in Section 3.4.

We identify a phase path (or an equilibrium point) of the system $\dot{\mathbf{x}} = \mathbf{X}(\mathbf{x})$ representing a solution $\mathbf{x}^*(t)$ whose stability is in question, called the *standard path*. We are usually interested in this path only from a particular point $\mathbf{a}^*$ onwards: then we have a *positive half-path* $\mathscr{H}^*$ (or semi-path) *with initial point* $\mathbf{a}^*$ (Fig. 9.1). The solution which $\mathscr{H}^*$ represents is

$$\mathbf{x}^*(t), \quad t \geqslant t_0,$$

where

$$\mathbf{x}^*(t_0) = \mathbf{a}^*,$$

and since the system is autonomous, all choices of t_0 lead to the same $\mathscr{H}^*$.

The solution $\mathbf{x}^*(t)$, $t \geqslant t_0$ is *Poincaré stable* (orbitally stable), if all sufficiently small disturbances of the initial value $\mathbf{a}^*$ lead to half-paths remaining uniformly at a small distance from $\mathscr{H}^*$. To form an analytical definition it is necessary to reverse this statement. Firstly, choose $\varepsilon > 0$ arbitrarily and construct a strip (or a tube in three dimensions) of 'radius' ε about $\mathscr{H}^*$ (Fig. 9.2). Then for stability it must be shown that *all* paths starting within *some* distance δ of $\mathbf{a}^*$ (where $\delta \leqslant \varepsilon$ necessarily), remain *permanently* within the strip. Such a condition must hold for every ε: in general, the smaller ε, the smaller δ must be.

The formal definition is as follows.

DEFINITION 9.1 (*Poincaré, or orbital stability*). *Let $\mathscr{H}^*$ be the half-path for the solution $\mathbf{x}^*(t)$ of $\dot{\mathbf{x}} = \mathbf{X}(\mathbf{x})$ which starts at $\mathbf{a}^*$ at $t = t_0$. Suppose that for every $\varepsilon > 0$ there exists $\delta(\varepsilon) > 0$ such that if $\mathscr{H}$ is the half-path starting at $\mathbf{a}$, then*

$$|\mathbf{a} - \mathbf{a}^*| < \delta \Rightarrow \sup_{\mathbf{x} \in \mathscr{H}} \operatorname{dist}(\mathbf{x}, \mathscr{H}^*) < \varepsilon.$$

Then $\mathscr{H}^*$ (or the corresponding solution) is said to be Poincaré stable. Otherwise $\mathscr{H}^*$ is unstable.

Here, the distance between a point $\mathbf{x}$ and a curve $\mathscr{C}$ is defined by

$$\operatorname{dist}(\mathbf{x}, \mathscr{C}) = \inf_{\mathbf{y} \in \mathscr{C}} |\mathbf{x} - \mathbf{y}|;$$

that is, it is the least distance in the ordinary sense (which is also a perpendicular distance). The maximum permitted separation of $\mathcal{H}$ from $\mathcal{H}^*$ is limited, in two dimensions, by the following construction: two discs of diameter ε are rolled along either side of $\mathcal{H}^*$ to sweep out a strip; for stability, there must exist a $\delta > 0$ such that every $\mathcal{H}$ with starting point $\mathbf{a}$, $|\mathbf{a} - \mathbf{a}^*| < \delta$, lies permanently in the strip (Fig. 9.3).

Figure 9.2 illustrates schematically some cases of stability and instability. (a), (b), and (c) show stable cases. Case (d) represents the system

$$\dot{x} = x, \quad \dot{y} = 2y$$

with solutions

$$x = A\,e^t, \quad y = B\,e^{2t}$$

where A and B are constants. The paths are given by $y = Cx^2$ for various C. Consider, for example, the Poincaré stability of the half-path $y = 0$, $x > 0$. Choose $\varepsilon > 0$ and consider any neighbouring path starting at $\mathbf{a} = (x_0, y_0)$. Parametrically, the corresponding half-path is

$$x(t) = x_0\,e^{t-t_0}, \quad y(t) = y_0\,e^{t-t_0}$$

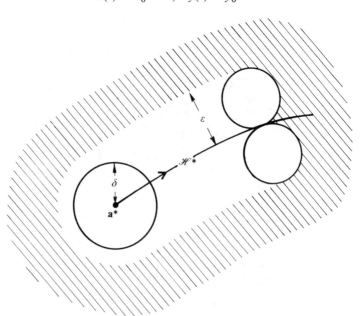

FIG. 9.3.

when the initial time is t_0. $|y(t)|$ becomes equal to, and then exceeds, ε (at $t = t_0 + \log_e(\varepsilon/|y_0|)$), no matter how close to the origin (x_0, y_0) is chosen. The solution is therefore unstable.

An example of a Poincaré stable solution is any one of the periodic solutions of the system

$$\dot{x} = y, \quad \dot{y} = -x,$$

whose phase diagram consists of circles, centre the origin. The positive half-paths consist of these circles, traversed endlessly. It is clear from Fig. 9.4 that every circular path is Poincaré stable; for if the path considered is on the circle of radius $(x_0^2 + y_0^2)^{1/2}$, starting at (x_0, y_0), the strip of arbitrary given width 2ε is that bounded by two more circles, and every half-path starting within a distance $\delta = \varepsilon$ of (x_0, y_0) lies permanently between these circles.

The same will be true of paths representing periodic solutions, and forming part of a centre, when these are not circular. However, closed

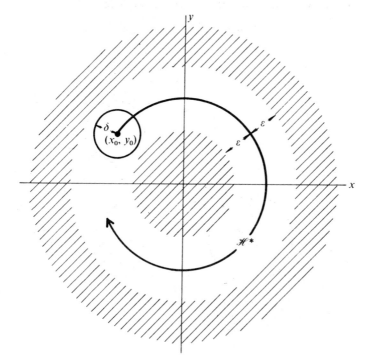

Fig. 9.4. Stability of the system $\dot{x} = y, \dot{y} = -x$

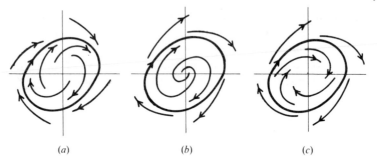

Fig. 9.5. Limit cycles: (*a*) is Poincaré stable; (*b*) and (*c*) are not

paths are not in general Poincaré stable. In Fig. 9.5 idealized limit cycles
are shown; the first is Poincaré stable and the other two are not.

Consider the system

$$\dot{x} = y, \quad \dot{y} = x,$$

the paths being given by $x^2 - y^2 = C$ where C is the parameter of the
family. The phase diagram shows a saddle point (Fig. 9.6).

On the basis of the previous discussion it can be seen that the half-
paths starting typically at A and C, and all the curved half-paths, are
Poincaré stable. Those starting typically at B and D are unstable
despite the fact that they approach the origin, since every nearby path
eventually goes to infinity.

Consider now the special half-paths which are *equilibrium points*,
representing constant solutions. That corresponding to a saddle (for
example, Fig. 9.6) is unstable, since there is no circle with centre at the
equilibrium point, of however small radius, such that every half-path
starting in it remains within an arbitrary preassigned distance ε of the
equilibrium point.

Now consider a previously-designated 'stable spiral' (Fig. 9.7). We
wish to show that the equilibrium point (not the arms of the spiral) is
stable.

Given $\varepsilon > 0$ (arbitrarily small), draw a circle, radius ε, with centre the
equilibrium point. We want to show that there is a certain disc, $|\mathbf{x}| < \delta$,
such that every half-path starting in it remains in the disc $|\mathbf{x}| = \varepsilon$ for all
time. Obviously $\delta \leqslant \varepsilon$, and it may be seen from the figure that the relative
size of δ is determined by the eccentricity of the spiral. We take it as
geometrically obvious that, for each ε, such a δ exists, and that the
equilibrium point is therefore stable. (A strict proof is difficult: a spiral

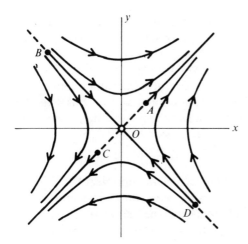

Fig. 9.6. Stability of the paths near a saddle point

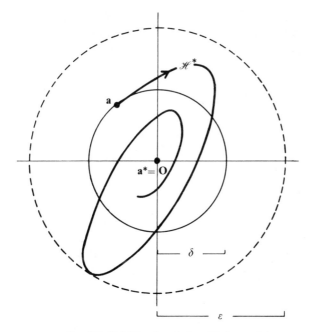

Fig. 9.7. Stability of a spiral equilibrium point

could be conceived whose 'spiral' paths looked more like a maze than a spiral.) Similarly it can be shown that a *centre* is Poincaré stable.

9.2. Solutions, paths, and norms

Poincaré stability is concerned with autonomous systems; another concept is required for non-autonomous systems. Even when a system is autonomous, however, a more sensitive criterion may be needed.

Consider, for example, the system

$$\dot{x} = y, \quad \dot{y} = -\sin x,$$

representing the motion of a pendulum. Near the origin the phase paths have the form shown in Fig. 9.8. Consider a solution $\mathbf{x}^*(t)$, $t \geqslant t_0$, with $\mathbf{x}^*(t_0) = \mathbf{a}^*$, its half-path being the heavy line starting at P; and a second solution $\mathbf{x}(t)$, $t \geqslant t_0$, $\mathbf{x}(t_0) = \mathbf{a}$, its half-path starting at Q.

$\mathbf{x}^*$ is Poincaré stable: if Q is close to P the corresponding paths remain close. However, *the representative points do not remain close*. It is known (Exercise 33, Chapter 1) that the period of $\mathbf{x}(t)$ is greater than that of $\mathbf{x}^*(t)$. Therefore the representative point Q' lags increasingly behind P': no matter how close Q and P are at time t_0, after a sufficient time Q' will be as much as half a cycle behind P' and the difference between the *solutions* at this time will therefore be large, a condition which recurs at regular intervals. If the pendulum (or other nonlinear oscillating system) were simply used to take a system through a sequence of states at more or less predetermined intervals this would be unimportant, but for

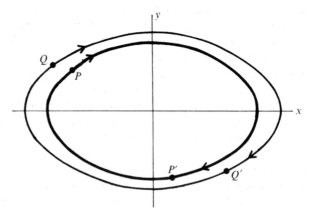

FIG. 9.8. Paths for the pendulum are Poincaré stable, but the representative points do not stay in step

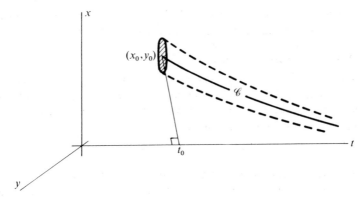

(x_0, y_0)

$\mathscr{C}$

t_0

FIG. 9.9.

timekeeping the progressively increasing error resulting from a disturbance would make the uncontrolled pendulum useless.

In two dimensions another representation is desirable which displays
solutions rather than *phase paths*, and therefore allows us to keep track of
time variation, as in Fig. 9.9. For the system $\dot{x} = X(x, y, t)$, $\dot{y} = Y(x, y, t)$,
the axes are x, y, t. A solution $x(t), y(t)$ appears as a curve in x, y, t space.
The heavy line $\mathscr{C}$ represents a solution on $t \geqslant t_0$ satisfying the initial
conditions $x(t_0) = x_0, y(t_0) = y_0$. The bundle of curves surrounding it
represents solutions with initial conditions close to those satisfied by
$x(t), y(t)$. The solution curves do not intersect at ordinary points of the
system since this would violate the uniqueness theorem (Appendix A).

When the system is autonomous, the *phase paths are the projections of
the solution curves on the x, y plane*. We know that these meet only at
equilibrium points, which represent constant solutions which are
straight lines parallel to the t axis in Fig. 9.9.

The figure suggests that $\mathscr{C}$ depicts a stable solution in an extended
sense. *Curves which start* (at t_0) *close together are close together at every
subsequent time*. If the equation generating these solutions were autonomous, then the projections of the curves on the plane would give part
of a phase diagram whose paths are close: this suggests that Poincaré
stability is also implied.

Contrast the case of the type illustrated in Fig. 9.10. The solution
curves corresponding to neighbouring paths are helices having different
rate of advance: these become widely separated although the paths do
not.

We shall formulate the definition of this new sort of stability in the

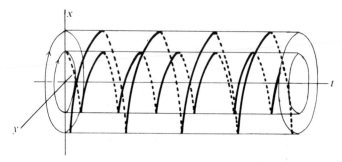

Fig. 9.10. Solution curves corresponding to paths about a centre

next section. We require a measure of the distance between solutions at particular times; that is, of the distance between the vectors of solution values. In n dimensions, let

$$\mathbf{x}(t) = (x_1(t), x_2(t), \ldots, x_n(t))^t$$

be a column vector of n real or complex functions, normally a solution vector of some system of first-order equations of dimension n. Several measures of the 'size' of $\mathbf{x}$ (for fixed t), called *norms* of $\mathbf{x}$, are used. We use the following.

The Euclidean length $|\mathbf{x}|$:

$$|\mathbf{x}| = \left(\sum_{i=1}^{n} |x_i|^2 \right)^{1/2}; \tag{9.1}$$

and

$$\|\mathbf{x}\| = \sum_{i=1}^{n} |x_i|. \tag{9.2}$$

The inequalities

$$|\mathbf{x}| < \varepsilon_1, \quad \text{and} \quad \|\mathbf{x}\| < \varepsilon_2$$

define neighbourhoods of the origin: as ε_1 or ε_2 are reduced to zero these neighbourhoods shrink steadily on to the origin. Their shapes, illustrated in two dimensions by $|\mathbf{x}| = 1, \|\mathbf{x}\| = 1$, are shown in Fig. 9.11.

The norm $\|\ldots\|$ has the following properties (shared by all acceptable norms, including $|\ldots|$):

(i) $\|\mathbf{x}\| \geqslant 0$ for all $\mathbf{x}$;

(ii) $\|\mathbf{x}\| = 0 \Leftrightarrow \mathbf{x} = \mathbf{0}$;

(iii) $\|\alpha\mathbf{x}\| = |\alpha| \|\mathbf{x}\|$, α real or complex;

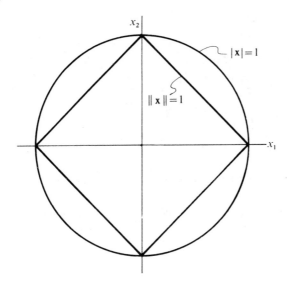

Fig. 9.11. $|\mathbf{x}| = 1$ and $\|\mathbf{x}\| = 1$, in two dimensions

(iv) $\|\mathbf{x} + \mathbf{y}\| \leqslant \|\mathbf{x}\| + \|\mathbf{y}\|$ (triangle inequality).

(v) When $\mathbf{x}$ is a function of t, then

$$\left\| \int_{t_1}^{t_2} \mathbf{x}(t)\,dt \right\| \leqslant \int_{t_1}^{t_2} \|\mathbf{x}(t)\|\,dt.$$

We also require the following inequalities:

(vi) $|\mathbf{x}| \leqslant \|\mathbf{x}\|$;　　　　　　　　　　　　　　　　　(9.3)

(vii) $\|\mathbf{x}\| \leqslant \sqrt{n}|\mathbf{x}|$　　　　　　　　　　　　　　　　　(9.4)

where n is the dimension.

We require a similar measure of size for matrices. Let A be an $n \times n$ matrix (a_{ij}) with real or complex elements. Then define

$$\|A\| = \sum_{i=1}^{n} \sum_{j=1}^{n} |a_{ij}|. \qquad (9.5)$$

This norm satisfies similar conditions to (i) to (v) above and additionally

$$\|AB\| \leqslant \|A\|\,\|B\| \qquad (9.6)$$

and

$$\|A\mathbf{x}\| \leqslant \|A\|\,\|\mathbf{x}\|, \qquad (9.7)$$

where $\|\mathbf{x}\|$ is defined as in (9.2). Relation (9.7) makes the norm (9.2) advantageous for our purposes.

9.3. Liapunov stability (stability of solutions)

Consider the regular system

$$\dot{\mathbf{x}} = \mathbf{X}(\mathbf{x}, t), \qquad (9.8)$$

or, in component form,

$$\dot{x}_i = X_i(x_1, \dots, x_n, t), \quad i = 1, 2, \dots, n.$$

The formal definition of the type of stability suggested in Section 9.2 is as follows.

DEFINITION 9.2 (*Liapunov Stability*)   *Let* $\mathbf{x}^*(t)$ *be a given real or complex solution of* (9.8). *Then*
(i) $\mathbf{x}^*(t)$ *is Liapunov stable on* $t \geq t_0$ *if, for any* $\varepsilon > 0$ *there exists* $\delta(\varepsilon, t_0)$ *such that*

$$\|\mathbf{x}(t_0) - \mathbf{x}^*(t_0)\| < \delta \Rightarrow \|\mathbf{x}(t) - \mathbf{x}^*(t)\| < \varepsilon, \qquad (9.9)$$

for all $t \geq t_0$, *where* $\mathbf{x}(t)$ *is any other solution.*
(ii) *It can be shown* (Cesari, 1971) *that if* (i) *is satisfied for initial conditions at* t_0, *then a similar condition is satisfied when any* $t_1 > t_0$ *is substituted for* t_0; *that is, if* $\mathbf{x}^*(t)$ *is stable for* $t \geq t_0$, *it is stable for* $t \geq t_1 > t_0$.
(iii) *Otherwise* $\mathbf{x}^*(t)$ *is said to be unstable (in the Liapunov sense).*

It can be shown (see Exercise 12) that when (9.8) is autonomous, Liapunov stability implies Poincaré stability.

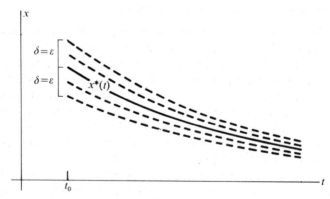

FIG. 9.12.

Example 9.1 *Show that all solutions of $\dot{x} = -x$ (a one-dimensional system) are stable.*

We will consider real solutions only, with initial condition at t_0. All real solutions are given by $x(t) = x(t_0)e^{-(t-t_0)}$. Consider the stability of $x^*(t)$ starting at $(t_0, x^*(t_0))$ (Fig. 9.12). Choose $\varepsilon > 0$ arbitrarily and consider the strip $x^*(t) - \varepsilon < x < x^*(t) + \varepsilon$. For stability, we must show that all solutions starting at t_0 and sufficiently close to $x^*(t_0)$ lie inside this strip. Since all solutions 'close up' on $x^*(t)$ as t increases it is clear that the conditions of the definition are satisfied by the choice $\delta = \varepsilon$.

The following example shows that the corresponding n-dimensional problem leads to the same conclusion.

Example 9.2 *Show that, all solutions of the n-dimensional system $\dot{\mathbf{x}} = -\mathbf{x}$ are stable for $t \geqslant 0$.*

The general solution is given by

$$\mathbf{x}(t) = \mathbf{x}(0)e^{-t}.$$

Consider the stability of $\mathbf{x}^*(t)$ for $t > 0$, where

$$\mathbf{x}^*(t) = \mathbf{x}^*(0)e^{-t}.$$

We have

$$\|\mathbf{x}(t) - \mathbf{x}^*(t)\| \leqslant \|\mathbf{x}(0) - \mathbf{x}^*(0)\|e^{-t}$$
$$\leqslant \|\mathbf{x}(0) - \mathbf{x}^*(0)\|, \quad t \geqslant 0.$$

Therefore, given any $\varepsilon > 0$,

$$\|\mathbf{x}(0) - \mathbf{x}^*(0)\| < \varepsilon \Rightarrow \|\mathbf{x}(t) - \mathbf{x}^*(t)\| < \varepsilon, \quad t > 0.$$

Thus $\delta = \varepsilon$ in Definition 9.2. (The process can be modified to hold for any t_0. However, the system is autonomous, so this is automatic.)

The implication of (ii) in Definition 9.2 is as follows. (i) taken alone implies that arbitrarily small enough changes in the initial condition at $t = t_0$ lead to uniformly small departures from the tested solution on $t_0 \leqslant t < \infty$. Such behaviour could be described as stability with respect only to variations in initial conditions at t_0. Armed with (ii), however, it can be said that the solution is stable with respect to variation in initial conditions at all times t_1 where $t_1 \geqslant t_0$, and that the stability is therefore a property of the solution as a whole. (ii) holds by virtue of (i) under very general smoothness conditions that we do not discuss here.

If (i) is satisfied, and therefore (ii), we may still want to be assured that a particular solution does not become 'less stable' as time goes on. It is possible that a system's sensitivity to disturbance might increase indefinitely with time although it remains technically stable, the symptom being that $\delta(\varepsilon, t_0)$ decreases to zero as t_0 increases.

To see that this is possible, consider the family of curves with parameter c:

$$x = f(c, t) = c\,e^{(c^2 - 1)t}/t, \quad t > 0. \tag{9.10}$$

These are certainly solutions of *some* first order equation. The equation is of no particular interest: the criterion for stability can be applied to the members of (9.10) without reference to the equation.

In Fig. 9.13, the curve separating those which tend to infinity ($c > 1$) from those tending to zero ($c < 1$) is the curve $f(1, t) = t^{-1}$, which also tends to zero. Now consider the stability of the 'solution' having $c = 0$: $f(0, t) \equiv 0$. It is apparent from the diagram that this function is stable for all $t_0 > 0$. However, it can afford no greater disturbance in 'initial condition' at time t_0 than $|f(1, t_0)| = 1/t_0$: a larger disturbance might encounter a curve which tends to infinity with t. In order that the disturbed and undisturbed curves should, for $t \geqslant t_0$, be permanently within a preassigned strip $-\varepsilon < x < \varepsilon$, starting values $x(t_0)$ must lie in the interval

$$|x(t_0)| \leqslant \min(\varepsilon, |f(1, t_0)|) = \min(\varepsilon, 1/t_0).$$

Therefore

$$\delta \leqslant \min(\varepsilon, 1/t_0)$$

in Definition 9.2, and δ tends to zero as $t_0 \to \infty$.

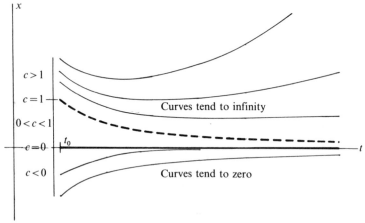

FIG. 9.13. $x(t) = c\,e^{(c^2 - 1)t}/t$

The stability is said to be *non-uniform* in this case. The definition of uniform stability is as follows.

DEFINITION 9.3 (*Uniform stability*)   *If a solution is stable for $t \geqslant t_0$ and the δ of Definition 9.2 is independent of t_0, the solution is uniformly stable on $t \geqslant t_0$.*

It is clear that any stable solutions of an autonomous system must be uniformly stable, since the system is invariant with respect to time translation.

A third desirable property is asymptotic stability. The system $\dot{\mathbf{x}} = \mathbf{0}$ has the constant solutions $\mathbf{x}(t) = \mathbf{x}(t_0)$ for given t_0. These are all stable on $t \geqslant t_0$ for any t_0 (and also uniformly stable); however, a disturbed solution shows no tendency to return to the original solution: it remains a constant distance away. On the other hand, solutions of the type examined in Figs. 9.12 and 9.13 do re-approach the undisturbed solution after being disturbed; thus the system tends to return to its original operating curve. Such solutions are said to be *asymptotically stable* (but uniformly and nonuniformly respectively in the cases of Figs. 9.12 and 9.13).

DEFINITION 9.4 (*Asymptotic stability*)   *Let $\mathbf{x}^*$ be a stable (or uniformly stable) solution for $t \geqslant t_0$. If additionally there exists $\eta(t_0) > 0$ such that*

$$\|\mathbf{x}(t_0) - \mathbf{x}^*(t_0)\| < \eta \Rightarrow \lim_{t \to \infty} \|\mathbf{x}(t) - \mathbf{x}^*(t)\| = 0, \tag{9.11}$$

then the solution is said to be asymptotically stable (or uniformly and asymptotically stable).

In the rest of this chapter we prove theorems enabling general statements to be made about the stability of solutions of certain classes of equations. Such theorems are necessary since in general it is not possible to find solutions of nonlinear systems explicitly.

9.4. Stability of linear systems

We consider first the stability (in the Liapunov sense) of the constant solutions (equilibrium points) of the two-dimensional, constant-coefficient systems classified in Chapter 2. Without loss of generality we can place the equilibrium point at the origin, since this only involves modifying the solutions by additive constants. We have then

$$\dot{x}_1 = ax_1 + bx_2, \quad \dot{x}_2 = cx_1 + dx_2 \tag{9.12}$$

in the present notation. The equilibrium points are classified as centres, nodes, spirals, and saddle points according to the relations between a, b, c, d (together with one or two degenerate cases), and their general nature is shown in Figs. 2.4 to 2.8.

The stability properties of the constant solutions $\mathbf{x}^*(t) = \mathbf{0}, t \geqslant t_0$ can be 'read off' from the phase diagram. Consider, for example, the centre. Figure 9.14(a) shows the phase plane and 9.14(b) the solution space: the solution whose stability is being considered lies along the t axis, and another typical solution, corresponding to a closed path, is shown.

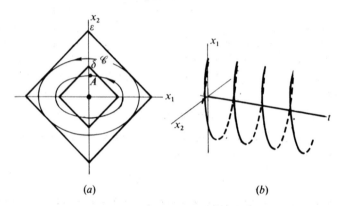

(a)   (b)

FIG. 9.14. Phase diagram and solution space near a centre

Choose a t_0, and an arbitrary $\varepsilon > 0$, and define a domain by

$$\|\mathbf{x}\| < \varepsilon \tag{9.13}$$

as indicated in Fig. 9.14. This diamond contains at least one closed path $\mathscr{C}$. Now construct a domain of initial values, near $\mathbf{x}^*(t_0) \equiv \mathbf{0}$:

$$\|\mathbf{x}(t_0) - \mathbf{x}^*(t_0)\| = \|\mathbf{x}(t_0)\| < \delta, \tag{9.14}$$

by taking δ small enough for this domain to lie entirely within $\mathscr{C}$. Then the solution with any initial condition $\mathbf{x}(t_0)$, subject to (9.14), lies permanently inside $\mathscr{C}$ and therefore satisfies (9.13) for all $t \geqslant t_0$. Since ε is arbitrary the condition (9.9) for stability is satisfied. The zero solution is uniformly stable, since given ε, δ is independent of t_0; this is also a consequence of the autonomous nature of (9.12). It is not asymptotically stable since the solutions do not approach $\mathbf{x} = \mathbf{0}$ as $t \to \infty$.

The other types of equilibrium point can be treated similarly. We have

(i) *centre*: uniformly stable,
(ii) *stable node*: uniformly and asymptotically stable,
(iii) *stable spiral*: uniformly and asymptotically stable,
(iv) *saddle point*: unstable.

(9.15)

It must be understood that the stability of the *equilibrium points only* (the constant solutions), and not that of the solutions represented by neighbouring paths, is being considered here. For example, although a centre is stable, in the nonlinear case the paths surrounding it will generally be unstable due to amplitude–frequency dependence.

The stability of a non-constant solution of a linear system (and, in fact, of a general system: see Section 9.8) can be reduced to the question of the stability of the zero solution of a related system, as follows.

Consider the nth order regular linear system, not in general autonomous,

$$\dot{\mathbf{x}} = A(t)\mathbf{x} + \mathbf{f}(t). \tag{9.16}$$

We wish to investigate the stability of a solution $\mathbf{x}^*(t)$. Let $\mathbf{x}(t)$ represent any other solution, and define $\boldsymbol{\xi}(t)$ by

$$\boldsymbol{\xi}(t) = \mathbf{x}(t) - \mathbf{x}^*(t). \tag{9.17}$$

Then $\boldsymbol{\xi}(t)$ tracks the difference between the 'test' solution and a solution having a different initial value. The initial condition for $\boldsymbol{\xi}$ is

$$\boldsymbol{\xi}(t_0) = \mathbf{x}(t_0) - \mathbf{x}^*(t_0). \tag{9.18}$$

Also, $\boldsymbol{\xi}$ satisfies the homogeneous equation derived from (9.16):

$$\dot{\boldsymbol{\xi}} = A(t)\boldsymbol{\xi}. \tag{9.19}$$

By comparison of (9.17), (9.18), and (9.19). it can be seen that the stability of $\mathbf{x}^*(t)$ is equivalent to the stability of the zero solution of (9.19). $\boldsymbol{\xi}(t)$ is called a *perturbation* of the solution $\mathbf{x}^*(t)$.

Since this new formulation of the problem is independent of the solution of (9.16) initially chosen, we can make the following statement:

THEOREM 9.1 *All solutions of the regular linear system* $\dot{\mathbf{x}} = A(t)\mathbf{x} + \mathbf{f}(t)$ *have the same stability property (unstable, stable, uniformly stable, asymptotically stable, uniformly and asymptotically stable), this being the same as that of the zero (or any other) solution of the homogeneous equation* $\dot{\boldsymbol{\xi}} = A(t)\boldsymbol{\xi}$.

Thus, in the *linear case* (only), the classifications (9.15) apply equally to the surrounding paths.

Example 9.3   *All the solutions of the equation* $\dot{x}_1 = x_2$, $\dot{x}_2 = -\omega^2 x_1 + f(t)$ *are uniformly stable, but not asymptotically stable (forced linear oscillator).*

Equation (9.19) becomes $\dot{\xi}_1 = \xi_2$, $\dot{\xi}_2 = -\omega^2 \xi_1$. The zero solution is a centre, which has the specified properties.

Example 9.4   *All the solutions of the equations* $\dot{x}_1 = x_2$, $\dot{x}_2 = -kx_2 - \omega^2 x_1 + f(t)$ *are uniformly and asymptotically stable (damped, forced linear oscillator).*

Equation (9.19) becomes

$$\dot{\xi}_1 = \xi_2, \quad \dot{\xi}_2 = -\omega^2 \xi_1 - k\xi_2.$$

The zero solution corresponds to the equilibrium point $\xi_1 = \xi_2 = 0$ which (Chapter 2) is a 'stable' node or spiral (as named in that chapter). By (9.15), these are uniformly and asymptotically stable. This result corresponds to the damping out of free oscillations set up by the initial conditions: the amplitude and phase of the forced oscillation are independent of these.

THEOREM 9.2   *For the regular system* $\dot{\mathbf{x}} = A(t)\mathbf{x}$, *the zero solution (and hence all solutions) are stable on* $t \geqslant t_0$, t_0 *arbitrary, if and only if every solution is bounded as* $t \to \infty$. *If A is constant and every solution is bounded, the solutions are uniformly stable.*

Proof   By Theorem 9.1 we need consider only the stability of the zero solution.

First, suppose that the zero solution, $\mathbf{x}^*(t) \equiv \mathbf{0}$ is stable. Choose any $\varepsilon > 0$ and let δ be that of equation (9.9). Let

$$\Psi(t) = (\boldsymbol{\psi}_1(t), \boldsymbol{\psi}_2(t), \ldots, \boldsymbol{\psi}_n(t))^t$$

be the fundamental matrix satisfying

$$\Psi(t_0) = \tfrac{1}{2}\delta I$$

where I is the unit matrix. Then each $\boldsymbol{\psi}_i(t)$ is bounded, since by (9.9)

$$\|\boldsymbol{\psi}_i(t_0)\| = \tfrac{1}{2}\delta < \delta \Rightarrow \|\boldsymbol{\psi}_i(t)\| < \varepsilon, \quad t \geqslant t_0.$$

Therefore every solution is bounded since any other solution is a linear combination of the $\boldsymbol{\psi}_i(t)$.

Suppose, conversely, that every solution is bounded. Let $\Phi(t)$ be any fundamental matrix; then there exists, by hypothesis, $M > 0$ such that $\|\Phi(t)\| < M$, $t \geqslant t_0$. Given any $\varepsilon > 0$ let $\delta = \varepsilon/M\|\Phi^{-1}(t_0)\|$. Let $\mathbf{x}(t)$ be any solution; we will test the stability of the zero solution. We have

$\mathbf{x}(t) = \Phi(t)\Phi^{-1}(t_0)\mathbf{x}(t_0)$, (Theorem 8.5), and if

$$\|\mathbf{x}(t_0)\| < \delta,$$

then

$$\|\mathbf{x}(t)\| \leqslant \|\Phi(t)\| \, \|\Phi^{-1}(t_0)\| \, \|\mathbf{x}(t_0)\| \tag{9.20}$$

$$\leqslant M \frac{\varepsilon}{M\delta} \delta = \varepsilon.$$

Thus the Definition 9.2 of stability for the zero solution is satisfied.

When A is constant, the autonomous nature of the system ensures that stability is uniform.

Linearity is essential for this theorem: it predicts that the system $\dot{x}_1 = x_2, \dot{x}_2 = -x_1$ has only stable solutions, but the solutions of $\dot{x}_1 = x_2$, $\dot{x}_2 = -\sin x_1$ are not all stable. Note that the stability of *forced* solutions (Theorem 9.1) is not determined by whether they are bounded or not; it is the boundedness of the *unforced* solutions which determines this.

According to Theorems 8.9 and 8.10 it is possible to make a simple statement about the stability of the solutions of systems with constant coefficients, $\dot{\mathbf{x}} = A\mathbf{x}$.

THEOREM 9.3 *Let A be constant in the system $\dot{\mathbf{x}} = A\mathbf{x}$, with eigenvalues $\lambda_i, i = 1, 2, \ldots, n$.*
(i) *If the system is stable, then $\mathrm{Re}\{\lambda_i\} \leqslant 0, i = 1, 2, \ldots, n$.*
(ii) *If either $\mathrm{Re}\{\lambda_i\} < 0, i = 1, 2, \ldots, n$; or if $\mathrm{Re}\{\lambda_i\} \leqslant 0, i = 1, 2, \ldots, n$ and there is no zero repeated eigenvalue; then the system is uniformly stable.*
(iii) *The system is asymptotically stable if and only if $\mathrm{Re}\{\lambda_i\} < 0$, $i = 1, 2, \ldots, n$ (and then it is also uniformly stable, by (ii)).*

In connection with (ii), note that if there is a zero repeated eigenvalue the system may be stable or unstable. For example, the system

$$\dot{x}_1 = 0, \quad \dot{x}_2 = 0$$

has a fundamental matrix

$$\begin{pmatrix} 1 & 0 \\ 0 & 1 \end{pmatrix}$$

which implies boundedness and therefore stability, but

$$\dot{x}_1 = x_2, \quad \dot{x}_2 = 0$$

has a fundamental matrix

$$\begin{pmatrix} 1 & t \\ 0 & 1 \end{pmatrix}$$

showing that the system is unstable.

COROLLARY 9.3 *If the system* $\dot{\mathbf{x}} = A\mathbf{x}$ *is asymptotically stable, there exist* $M > 0, m > 0$ *such that*

$$\|\Phi(t)\| < M\,\mathrm{e}^{-mt}, \tag{9.21}$$

where $\Phi(t)$ *is any fundamental matrix for the system.*

9.5. Stability of a class of linear systems

The system considered is

$$\dot{\mathbf{x}} = \{A + C(t)\}\mathbf{x} \tag{9.22}$$

where A is constant. Under quite general conditions the stability of the solutions of (9.22) is determined by the stability of the solutions of the system

$$\dot{\mathbf{x}} = A\mathbf{x} \tag{9.23}$$

We require the following Theorem.

THEOREM 9.4 (*Gronwall's lemma*) *If, for* $t \geqslant t_0$,
(i) $u(t)$ *and* $v(t)$ *are continuous and*

$$u(t) \geqslant 0, \quad v(t) \geqslant 0;$$

(ii) $$u(t) \leqslant K + \int_{t_0}^{t} u(s)v(s)\,\mathrm{d}s, \quad K > 0; \tag{9.24}$$

then

$$u(t) \leqslant K \exp\left(\int_{t_0}^{t} v(s)\,\mathrm{d}s \right), \quad t \geqslant t_0. \tag{9.25}$$

Proof The right-hand side of (9.24) is positive, since $K > 0$ and $u(t), v(t) \geqslant 0$. Therefore (9.24) gives

$$\frac{u(t)v(t)}{K + \displaystyle\int_{t_0}^{t} u(s)v(s)\,\mathrm{d}s} \leqslant v(t).$$

Integrate both sides from t_0 to t:

$$\log\left\{ K + \int_{t_0}^{t} u(s)v(s)\,\mathrm{d}s \right\} - \log K \leqslant \int_{t_0}^{t} v(s)\,\mathrm{d}s.$$

Therefore

$$K + \int_{t_0}^{t} u(s)v(s)\,\mathrm{d}s \leqslant K \exp\left(\int_{t_0}^{t} v(s)\,\mathrm{d}s \right).$$

The application of (9.24) again gives the required result.

THEOREM 9.5   *If*
(i) *A is constant and the eigenvalues of A have negative real parts*;
(ii) *C(t) is continuous for $t \geq t_0$ and*

$$\int_{t_0}^{t} \|C(t)\| \, dt \quad \text{is bounded for} \quad t > t_0, \tag{9.26}$$

then all solutions of $\dot{\mathbf{x}} = \{A + C(t)\}\mathbf{x}$ are asymptotically stable.

Proof   Write the system in the form

$$\dot{\mathbf{x}} = A\mathbf{x} + C(t)\mathbf{x} \tag{9.27}$$

If $\mathbf{x}(t)$ is a solution, then $C(t)\mathbf{x}(t)$ is a function of t which may play the part of $\mathbf{f}(t)$ in Theorem 8.8. Therefore (9.27) implies

$$\mathbf{x}(t) = \Phi(t)\Phi^{-1}(t_0)\mathbf{x}_0 + \int_{t_0}^{t} \Phi(t-s+t_0)\Phi^{-1}(t_0)C(s)\mathbf{x}(s)\,ds, \tag{9.28}$$

where Φ is any fundamental matrix for the system $\dot{\mathbf{x}} = A\mathbf{x}$, and $\mathbf{x}(t_0) = \mathbf{x}_0$. Using the properties of norms in Section 9.3, (9.28) gives

$$\|\mathbf{x}(t)\| \leq \|\Phi(t)\| \, \|\Phi^{-1}(t_0)\| \, \|\mathbf{x}_0\|$$
$$+ \|\Phi^{-1}(t_0)\| \int_{t_0}^{t} \|\Phi(t-s+t_0)\| \, \|C(s)\| \, \|\mathbf{x}(s)\| \, ds. \tag{9.29}$$

Since A has eigenvalues with negative real part, Theorem (9.3) and its corollary show that for some positive M and m,

$$\|\Phi(t)\| \leq M e^{-mt}, \quad t \geq t_0.$$

Therefore, putting

$$\|\Phi^{-1}(t_0)\| = \beta,$$

(9.29) implies, after some regrouping, that for $t \geq t_0$

$$\|\mathbf{x}(t)\| e^{mt} \leq M\beta\|\mathbf{x}_0\| + \int_{t_0}^{t} \{\|\mathbf{x}(s)\| e^{ms}\}\{C(s)M\beta e^{-mt_0}\}\,ds. \tag{9.30}$$

In Theorem 9.4, let

$$u(t) = \|\mathbf{x}(t)\| e^{mt}, \quad v(t) = \|C(t)\|\beta M e^{-mt_0},$$

and

$$K = M\beta\|\mathbf{x}_0\|.$$

Then from (9.30) and Theorem 9.4,

$$\|\mathbf{x}(t)\| \, e^{mt} \leqslant M\beta \|\mathbf{x}_0\| \exp\left(\beta M \, e^{-mt_0} \int_{t_0}^{t} \|C(s)\| \, ds \right)$$

or

$$\|\mathbf{x}(t)\| \leqslant M\beta \|\mathbf{x}_0\| \exp\left(bM \, e^{-mt_0} \int_{t_0}^{t} \|C(s)\| \, ds - mt \right). \qquad (9.31)$$

Therefore, by (9.26), every solution is bounded for $t \geqslant t_0$ and is therefore stable by Theorem 9.2. Also every solution tends to zero as $t \to \infty$ and is therefore asymptotically stable.

COROLLARY 9.5 *If $C(t)$ satisfies the conditions of the theorem but all solutions of $\dot{\mathbf{x}} = A\mathbf{x}$ are merely bounded, then all solutions of $\dot{\mathbf{x}} = \{A + C(t)\}\mathbf{x}$ are bounded and therefore stable.*

This follows from (9.31) by writing $m = 0$ in (9.30). Note that $\mathrm{Re}\{\lambda_i\} \leqslant 0$ for all i is not in itself sufficient to establish the boundedness of all solutions of $\dot{\mathbf{x}} = A\mathbf{x}$ (see Example 8.8).

The stability of nth order *equations* can be discussed in the same terms as that of suitable nth order *systems*. If we replace the equation

$$x_1^{(n)} + a_1(t)x_1^{(n-1)} + \ldots + a_n(t)x_1 = f(t)$$

by the equivalent system

$$\dot{x}_1 = x_2, \quad \dot{x}_2 = x_3, \ldots, \dot{x}_{n-1} = x_n, \quad \dot{x}_n = -a_1(t)x_n - \ldots - a_n x_1 + f(t),$$
$$\qquad (9.32)$$

the set of initial conditions for the *system* $(x_1(t_0), \ldots, x_n(t_0))$ correspond with the usual initial conditions appropriate to the *equation*: the set

$$(x_1(t_0), \, x_1^{(1)}(t_0), \ldots, x_1^{(n-1)}(t_0)).$$

Therefore, to discuss the stability of the solutions of equations in terms of the definitions for systems we use the presentation (9.32) rather than any other. It can, in fact, be shown (Exercise 11) that a system obtained by a general transformation of the variables need not retain the stability properties of the original system.

Example 9.5 *Show that when $a > 0$ and $b > 0$ all solutions of $\ddot{x} + a\dot{x} + (b + c\,e^{-t}\cos t)x = 0$ are asymptotically stable for $t \geqslant t_0$, for any t_0.*

The appropriate equivalent system (with $\dot{x} = y$) is

$$\begin{pmatrix} \dot{x} \\ \dot{y} \end{pmatrix} = \begin{pmatrix} 0 & 1 \\ -b & -a \end{pmatrix} \begin{pmatrix} x \\ y \end{pmatrix} + \begin{pmatrix} 0 & 0 \\ -c\,e^{-t}\cos t & 0 \end{pmatrix} \begin{pmatrix} x \\ y \end{pmatrix}.$$

In the notation of the above theorem

$$A = \begin{pmatrix} 0 & 1 \\ -b & -a \end{pmatrix}, \quad C(t) = \begin{pmatrix} 0 & 0 \\ -c\,e^{-t}\cos t & 0 \end{pmatrix}.$$

The eigenvalues of A are negative if $a > 0$ and $b > 0$. Also

$$\int_{t_0}^{\infty} \|C(t)\|\, dt = |c| \int_{t_0}^{\infty} e^{-t}|\cos t|\, dt < \infty.$$

The conditions of the theorem are satisfied, and all solutions are asymptotically stable.

Example 9.6 *Show that all solutions of $\ddot{x} + \{a + c(1 + t^2)^{-1}\}x = f(t)$ are stable if $a > 0$.*

Let $\dot{x} = y$. The equation is equivalent to

$$\begin{pmatrix} \dot{x} \\ \dot{y} \end{pmatrix} = \begin{pmatrix} 0 & 1 \\ -a & 0 \end{pmatrix}\begin{pmatrix} x \\ y \end{pmatrix} + \begin{pmatrix} 0 & 0 \\ -c(1+t^2)^{-1} & 0 \end{pmatrix}\begin{pmatrix} x \\ y \end{pmatrix} + \begin{pmatrix} 0 \\ f(t) \end{pmatrix}.$$

By Theorem 9.1, all solutions of the given system have the same stability property as the zero solution (or any other) of the corresponding homogeneous system $\dot{\xi} = \{A + C(t)\}\xi$, where

$$A = \begin{pmatrix} 0 & 1 \\ -a & 0 \end{pmatrix}, \quad C(t) = \begin{pmatrix} 0 & 0 \\ -c(1+t^2)^{-1} & 0 \end{pmatrix}.$$

The solutions of $\dot{\xi} = A\xi$ are bounded when $a > 0$ (the zero solution is a centre on the phase plane). Also

$$\int_{t_0}^{\infty} \|C(t)\|\, dt = |c| \int_{t_0}^{\infty} \frac{dt}{1+t^2} < \infty.$$

By Corollary 9.5 all solutions of $\dot{\xi} = \{A + C(t)\}\xi$ are bounded and are therefore stable. (Note that the *inhomogeneous* equation may, depending on $f(t)$, have *unbounded* solutions which are stable.)

9.6. A comparison theorem for the zero solutions of nearly-linear systems

Certain nonlinear systems can be regarded as perturbed linear systems in respect of their stability properties, the stability (or instability) of the linearized system being preserved. The following theorem refers to the stability of the *zero* solutions.

THEOREM 9.6 *If $\mathbf{h}(\mathbf{0}, t) = \mathbf{0}$, A is constant, and*
(i) *the solutions of $\dot{\mathbf{x}} = A\mathbf{x}$ are asymptotically stable;*
(ii) $\lim\limits_{\|\mathbf{x}\| \to 0} \{\|\mathbf{h}(\mathbf{x}, t)\|/\|\mathbf{x}\|\} = 0$ *uniformly in t, $0 \leqslant t < \infty$; then the zero solution, $t \geqslant t_0 \geqslant 0$, is an asymptotically stable solution of the regular*

system

$$\dot{\mathbf{x}} = A\mathbf{x} + \mathbf{h}(\mathbf{x}, t). \tag{9.33}$$

Proof Regularity implies that $\mathbf{h}$ is such that the conditions of the existence theorem (Appendix A) hold for all $\mathbf{x}$ and for $t \geqslant t_0$. However, it does not of itself imply that any particular solution actually exists for all t: for the given system this forms part of the present proof.

Let $\Psi(t)$ be the fundamental matrix for the system $\dot{\mathbf{x}} = A\mathbf{x}$ for which $\Psi(0) = I$. Then by Theorem 8.8, with $\mathbf{h}$ in place of $\mathbf{f}$, every solution of (9.33) with initial values at $t = 0$ satisfies

$$\mathbf{x}(t) = \Psi(t)\mathbf{x}(0) + \int_0^t \Psi(t-s)\mathbf{h}(\mathbf{x}(s), s)\,\mathrm{d}s. \tag{9.34}$$

By (i) and the Corollary to Theorem 9.3, there exist $M > 0, m > 0$ such that

$$\|\Psi(t)\| \leqslant M\,\mathrm{e}^{-mt}, \quad t \geqslant 0. \tag{9.35}$$

Also, by (ii), there exists δ_0 such that

$$\|\mathbf{x}\| < \delta_0 \Rightarrow \|\mathbf{h}(x, t)\| < \frac{m}{2M}\|\mathbf{x}\|, \quad t \geqslant 0. \tag{9.36}$$

Now let δ be chosen arbitrarily subject to

$$0 < \delta < \min(\delta_0, \delta_0/M), \tag{9.37}$$

and consider solutions for which

$$\|\mathbf{x}(0)\| < \delta. \tag{9.38}$$

Then by (9.37) $\|\mathbf{x}(t)\| < \delta$ and $\mathbf{x}(t)$ is continuous on some interval $0 \leqslant t < \tau$:

$$\|\mathbf{x}(t)\| < \delta, \quad 0 \leqslant t < \tau. \tag{9.39}$$

From (9.36) to (9.39), (9.34) gives

$$\|\mathbf{x}(t)\| \leqslant M\,\mathrm{e}^{-mt}\|\mathbf{x}(0)\| + \int_0^t M\,\mathrm{e}^{-mt}\mathrm{e}^{ms}\frac{m}{2M}\|\mathbf{x}(s)\|\,\mathrm{d}s, \quad 0 \leqslant t < \tau,$$

or

$$\|\mathbf{x}(t)\|\,\mathrm{e}^{mt} \leqslant M\|\mathbf{x}(0)\| + \int_0^t \tfrac{1}{2}m\|\mathbf{x}(s)\|\mathrm{e}^{ms}\,\mathrm{d}s, \quad 0 \leqslant t < \tau.$$

By Gronwall's Lemma, Theorem 9.4, with $u(t) = \|\mathbf{x}(t)\|\,\mathrm{e}^{mt}, v(t) = \tfrac{1}{2}m$, $K = M\|\mathbf{x}(0)\|$, applied to this inequality we obtain

$$\|\mathbf{x}(t)\| \leqslant M\|\mathbf{x}(0)\| e^{-\frac{1}{2}mt}$$
$$\leqslant \delta e^{-\frac{1}{2}mt}, \quad 0 \leqslant t < \tau. \tag{9.40}$$

by (9.39). Thus, ,the possibility that $\|\mathbf{x}(\tau)\| = \delta$, admitted by (9.39), is excluded by (9.40), and $\|\mathbf{x}(\tau)\| < \delta$ for *every* τ for which (9.39) holds.

We wish to show that when (9.38) holds, then $\|\mathbf{x}(t)\| < \delta$ for all $t \geqslant 0$; in other words, that τ may be infinite in (9.39). The only question to be settled is whether the *interval of existence* of $\mathbf{x}(t)$ is in fact infinite. We know (Theorem A2, Appendix A) that either the interval of existence is infinite or that $\mathbf{x}(t)$ becomes unbounded as some finite value of t is approached. The second possibility is excluded by (9.40) since unboundedness would imply that for some $\tau = \bar{\tau}$, $\|\mathbf{x}(t)\| < \delta, 0 \leqslant t < \bar{\tau}$, and also that $\|\mathbf{x}(\bar{\tau})\| = \delta$, which contradicts (9.40). Therefore the interval of existence of solutions subject to (9.38) is infinite and (9.40) holds for all τ.

Now δ is arbitrarily small, so we have proved that

$$\|\mathbf{x}(0)\| < \delta \Rightarrow \|\mathbf{x}(t)\| < \delta e^{-\frac{1}{2}mt}, \quad t \geqslant 0, \tag{9.41}$$

which implies asymptotic stability of the zero solution.

The asymptotic stability of the linearized system ensures that $\|\mathbf{x}(t)\|$ tends to decrease, reducing the relative effect of the nonlinear term. This process is self-reinforcing.

Example 9.7 *Show that van der Pol's equation $\ddot{x} + e(x^2 - 1)\dot{x} + x = 0$ has an asymptotically stable zero solution when $e < 0$.*

Replace the equation by the system

$$\begin{pmatrix} \dot{x} \\ \dot{y} \end{pmatrix} = \begin{pmatrix} 0 & 1 \\ -1 & e \end{pmatrix} \begin{pmatrix} x \\ y \end{pmatrix} + \begin{pmatrix} 0 \\ -ex^2y \end{pmatrix} = A\mathbf{x} + \mathbf{h}(x, y).$$

The eigenvalues of A are negative when $e < 0$; therefore, by Theorem 3, all the solutions of $\dot{\mathbf{x}} = A\mathbf{x}$ are asymptotically stable. Condition (ii) of Theorem 9.6 is satisfied since

$$\|\mathbf{h}(x, y)\| = |e|x^2|y| \leqslant |e|(|x| + |y|)^2 (|x| + |y|)$$
$$= |e| \|\mathbf{x}\|^3.$$

Example 9.8 *Show that the zero solution of the equation $\ddot{x} + k\dot{x} + \sin x = 0$ is asymptotically stable for $k > 0$.*

The equivalent system is

$$\begin{pmatrix} \dot{x} \\ \dot{y} \end{pmatrix} = \begin{pmatrix} 0 & 1 \\ -1 & -k \end{pmatrix} \begin{pmatrix} x \\ y \end{pmatrix} + \begin{pmatrix} 0 \\ x - \sin x \end{pmatrix}$$

which satisfies the condition of Theorem 9.6.

9.7. The stability of forced oscillations by solution-perturbation

Consider the general forced system

$$\dot{\mathbf{x}} = f(\mathbf{x}, t). \tag{9.42}$$

The stability of a solution $\mathbf{x}^*(t)$ can be reduced to consideration of the zero solution of a related system. Let $\mathbf{x}(t)$ be any other solution, and write

$$\mathbf{x}(t) = \mathbf{x}^*(t) + \boldsymbol{\xi}(t). \tag{9.43}$$

Then $\boldsymbol{\xi}(t)$ represents a perturbation, or disturbance, of the original solution: it seems reasonable to see what happens to $\boldsymbol{\xi}(t)$, since the question of stability is whether such (small) disturbances grow or not. Equation (9.42) can be written in the form

$$\dot{\mathbf{x}}^* + \dot{\boldsymbol{\xi}} = \mathbf{f}(\mathbf{x}^*, t) + \{\mathbf{f}(\mathbf{x}^* + \boldsymbol{\xi}, t) - \mathbf{f}(\mathbf{x}^*, t)\}.$$

Since $\mathbf{x}^*$ satisfies (9.42), this becomes

$$\begin{aligned}
\dot{\boldsymbol{\xi}} &= \mathbf{f}(\mathbf{x}^* + \boldsymbol{\xi}, t) - \mathbf{f}(\mathbf{x}^*, t) \\
&= \mathbf{h}(\boldsymbol{\xi}, t),
\end{aligned} \tag{9.44}$$

say, since $\mathbf{x}^*(t)$ is assumed known. By (9.43), the stability properties of $\mathbf{x}^*(t)$ are the same as those of the zero solution of (9.44), $\boldsymbol{\xi}(t) \equiv \mathbf{0}$.

The right-hand side of (9.44) may have a linear approximation for small $\boldsymbol{\xi}$, in which case

$$\dot{\boldsymbol{\xi}} = \mathbf{h}(\boldsymbol{\xi}, t) \approx A(t)\boldsymbol{\xi}, \tag{9.45}$$

and the properties of this linear system may correctly indicate that of the zero solution of the exact system (9.44). The approximation (9.45) is called the *first variational equation*. This process is not rigorous: it is generally necessary to invoke an approximation not only at the stage (9.45), but also in representing $\mathbf{x}^*(t)$, which, of course, will not generally be known exactly.

We will carry this process out directly on an equation rather than on a system. As an example we take the Duffing equation in the standardized form

$$\ddot{x} + k\dot{x} + x + \varepsilon x^3 = \Gamma \cos \omega t. \tag{9.46}$$

From Chapter 7 we know that periodic forced oscillations exist which are approximately of the form $a \cos \omega t + b \sin \omega t$, or $r \cos(\omega t + \alpha)$, where $r > 0$ is given by eqn (7.23):

$$\{(\omega^2 - 1 - \tfrac{3}{4}\varepsilon r^2)^2 + \omega^2 k^2\} r^2 = \Gamma^2. \tag{9.47}$$

Therefore take as the solution to be examined

$$x^*(t) \approx r \cos{(\omega t + \alpha)} \qquad (9.48)$$

and consider the equation satisfied by the variation, $\xi(t)$, defined by

$$\xi(t) = x(t) - x^*(t). \qquad (9.49)$$

By substituting (9.49) into (9.46), taking account of the fact that $x^*(t)$ satisfies (9.46), and neglecting all but first order terms in ξ, we obtain the *linear* equation of first variation

$$\ddot{\xi} + k\dot{\xi} + \{1 + 3\varepsilon x^{*2}(t)\}\xi = 0. \qquad (9.50)$$

The stability of $x^*(t)$ is expected to be that of the solutions of (9.50) (subject to the approximation made: we do not have the theorems to support this). Now replace $x^*(t)$ by the form (9.48):

$$\ddot{\xi} + k\dot{\xi} + \{1 + \tfrac{3}{2}\varepsilon r^2 + \tfrac{3}{2}\varepsilon r^2 \cos{(2\omega t + 2\alpha)}\}\xi = 0;$$

and write

$$2\omega t + 2\alpha = \tau. \qquad (9.51)$$

We now have

$$\xi'' + \left(\frac{k}{2\omega}\right)\xi' + \left\{\frac{1 + \tfrac{3}{2}\varepsilon r^2}{4\omega^2} + \frac{\tfrac{3}{2}\varepsilon r^2}{4\omega^2}\cos{\tau}\right\}\xi = 0, \qquad (9.52)$$

where $'$ indicates the derivative with respect to τ.

Equation (9.52) is the damped Mathieu equation (8.59), with

$$\kappa = k/2\omega, \quad \nu = (1 + \tfrac{3}{2}\varepsilon r^2)/4\omega^2, \quad \beta = \tfrac{3}{2}\varepsilon r^2/4\omega^2. \qquad (9.53)$$

Suppose that k is small, of order ε (or of order β since $\beta = O(\varepsilon)$) and that $\omega \approx 1$ (near to resonance). Then $\nu \approx \tfrac{1}{4}$ and the results of Section 8.5 are available to discuss the stability of the solutions of (9.52) in this region. Equation (8.70) shows that the solutions are stable when

$$\left(\frac{1 + \tfrac{3}{2}\varepsilon r^2}{4\omega^2} - \frac{1}{4}\right)^2 - \frac{1}{4}\left(\frac{\tfrac{3}{2}\varepsilon r^2}{4\omega^2}\right)^2\left[1 - \left(\frac{k}{2\omega}\frac{4\omega^2}{\tfrac{3}{2}\varepsilon r^2}\right)^2\right] > 0,$$

or when

$$(1 - \omega^2)^2 + 3\varepsilon(1 - \omega^2)r^2 + \tfrac{27}{16}\varepsilon^2 r^4 + k^2\omega^2 > 0. \qquad (9.54)$$

Since the solutions of the *damped* Mathieu equation (8.50) tend to zero in the stable regions (see Section 8.7), asymptotic stability is predicted, confirming the analysis of Chapter 7.

We can also confirm the remark made in Section 5.4, (vi), that stability is to be expected when

$$\frac{d(\Gamma^2)}{d(r^2)} > 0, \tag{9.55}$$

or when an increase or decrease in magnitude of Γ results in an increase or decrease respectively in the amplitude. From (9.47) it is easy to verify that $d(\Gamma^2)/d(r^2)$ is equal to the expression on the left of (9.54), and the speculation is therefore confirmed.

In general, when periodic solutions of the original equation are expected, the reduced equation (9.45) is of the more complicated Hill type. The stability regions for this equation and examples of the corresponding stability estimates may be found in Hayashi (1964).

Exercises

1. Use the phase diagram for the pendulum equation, $\ddot{x} + \sin x = 0$, to say which paths are not Poincaré stable. (See Fig. 1.2.)

2. Show that all the paths of $\dot{x} = x, \dot{y} = y$ are Poincaré unstable.

3. Find the limit cycles of the system

$$\dot{x} = -y + x \sin r, \quad \dot{y} = x + y \sin r, \quad r = \sqrt{(x^2 + y^2)}.$$

Which cycles are Poincaré stable?

4. Find the phase paths for $\dot{x} = x, \dot{y} = y \log y$, in the half-plane $y > 0$. Which paths are Poincaré stable?

5. Prove the following:
   (i) $\|\mathbf{x} + \mathbf{y}\| \leqslant \|\mathbf{x}\| + \|\mathbf{y}\|$;
   (ii) $\left\| \int_{t_1}^{t_2} \mathbf{x}(t) \, dt \right\| \leqslant \int_{t_1}^{t_2} \|\mathbf{x}(t)\| \, dt$;
   (iii) $\|AB\| \leqslant \|A\| \, \|B\|$, where A and B are $n \times n$ matrices.

6. Prove that (i) $|\mathbf{x}| \leqslant \|\mathbf{x}\|$; (ii) $\|\mathbf{x}\| \leqslant \sqrt{n}|\mathbf{x}|$, where n is the dimension.

7. Use the results in Section 9.2 to prove that the series

$$I + \sum_{n=1}^{\infty} \frac{1}{m!} A^m$$

where A is an arbitrary $n \times n$ matrix, converges.

8. Prove that $\|A\mathbf{x}\| \leqslant \|A\| \, \|\mathbf{x}\|$.

9. Show that every non-trivial solution of $\dot{x} = x$ is unbounded and unstable, but that every solution of $\dot{x} = 1$ is unbounded and stable.

10. Show that the solutions of the system $\dot{x} = 1$, $\dot{y} = 0$, are Poincaré and Liapunov stable, but that the system $\dot{x} = y$, $\dot{y} = 0$, is Poincaré but not Liapunov stable.

11. Solve the equations

$$\dot{x} = -y(x^2 + y^2), \quad \dot{y} = x(x^2 + y^2),$$

and show that the zero solution is Liapunov stable and that all other solutions are unstable.

Replace the coordinates x, y by r, ϕ where

$$x = r \cos (r^2 t + \phi), \quad y = r \sin (r^2 t + \phi)$$

and deduce that

$$\dot{r} = 0, \quad \dot{\phi} = 0.$$

Show that in this coordinate system the solutions are stable. (Change of coordinates can affect the stability of a system.)

12. Prove that Liapunov stability implies Poincaré stability of a solution (but not conversely: see Exercise 10).

13. Determine the stability of the solutions of
(i) $\dot{x}_1 = x_2 \sin t$, $\dot{x}_2 = 0$.
(ii) $\dot{x}_1 = 0$, $\dot{x}_2 = x_1 + x_2$.

14. Determine the stability of the solutions of

(i) $\begin{pmatrix} \dot{x}_1 \\ \dot{x}_2 \end{pmatrix} = \begin{pmatrix} -2 & 1 \\ 1 & -2 \end{pmatrix} \begin{pmatrix} x_1 \\ x_2 \end{pmatrix} + \begin{pmatrix} 1 \\ -2 \end{pmatrix} e^t$;

(ii) $\ddot{x} + e^{-t} \dot{x} + x = 0$.

15. Show that every solution of the system

$$\dot{x} = -t^2 x, \quad \dot{y} = -ty$$

is asymptotically stable.

16. The motion of a heavy particle on a smooth surface of revolution with vertical axis z and shape $z = f(r)$ in cylindrical polar coordinates is

$$\frac{1}{r^4} \{1 + f'^2(r)\} \frac{d^2 r}{d\theta^2} + \left[\frac{1}{r^4} f'(r) f''(r) - \frac{2}{r^5} \{1 + f'^2(r)\} \right] \left(\frac{dr}{d\theta} \right)^2 - \frac{1}{r^3} = -\frac{g}{h^2} f'(r),$$

where h is the angular momentum ($h = r^2 \dot{\theta}$), a constant. Show that plane, horizontal motion $r = a$, $z = f(a)$, is stable for perturbations leaving h unaltered provided $3 + af''(a)/f'(a) > 0$.

17. In a predator–prey model, the prey x and the predator y satisfy

$$\dot{x} = -\alpha x + \beta x^2 - \gamma xy, \quad \dot{y} = -\delta y + \varepsilon xy$$

where α, β, γ, δ, and ε are positive constants. How does the system behave when there is exactly one equilibrium point, at $(0,0)$? Prove that the solution $x = 0$, $y = 0$ is asymptotically stable. Discuss the implications of this result for the prey and the predator.

18. Is Theorem 9.5 applicable to the damped Mathieu equation?

19. A pendulum, with bob of mass m and suspension of length a, hangs from a support which is constrained to move with vertical and horizontal displacements $\zeta(t)$ and $\eta(t)$ respectively. Show that the inclination θ of the pendulum satisfies the equation

$$a\ddot{\theta} + (g - \ddot{\zeta})\sin\theta + \ddot{\eta}\cos\theta = 0.$$

   Let $\zeta = \alpha\sin\omega t$ and $\eta = \beta\sin 2\omega t$, where $\omega = \sqrt{(g/a)}$. Show that after linearizing this equation for small amplitudes, the resulting equation has a solution

$$\theta = (8\beta/\alpha)\cos\omega t.$$

Determine the stability of this solution.

20. The equation

$$\ddot{x} + (\tfrac{1}{4} - 2\varepsilon b\cos^2\tfrac{1}{2}t)x + \varepsilon x^3 = 0$$

has the exact solution $x^*(t) = \sqrt{(2b)}\cos\tfrac{1}{2}t$. Show that the solution is stable by constructing the variational equation (Section 9.8).

21. Consider the equation $\ddot{x} + (\alpha + \beta\cos t)x = 0$, where $|\beta| \ll 1$ and $\alpha = \tfrac{1}{4} + \beta c$. In the unstable region near $\alpha = \tfrac{1}{4}$ (Section 8.5) this equation has solutions of the form $c_1 e^{\sigma t} q_1(t) + c_2 e^{-\sigma t} q_2(t)$, where σ is real, $\sigma > 0$ and q_1, q_2 have period 4π. Construct the equation for q_1, q_2, and show that $\sigma = \pm\beta\sqrt{(\tfrac{1}{4} - c^2)}$.

22. By using the method of Section 9.7 show that a solution of the equation

$$\ddot{x} + \varepsilon(x^2 - 1)\dot{x} + x = \Gamma\cos\omega t$$

where $|\varepsilon| \ll 1$, $\omega = 1 + \varepsilon\omega_1$, of the form $x^* = r_0\cos(\omega t + \alpha)$ (α constant) is asymptotically stable when

$$4\omega_1^2 + \tfrac{3}{16}r_0^4 - r_0^2 + 1 < 0.$$

(Use the result of Exercise 21.)

23. The equation

$$\ddot{x} + \alpha x + \varepsilon x^3 = \varepsilon\gamma\cos\omega t$$

has the exact subharmonic solution

$$x = (4\gamma)^{1/3} \cos \tfrac{1}{3}\omega t$$

when

$$\omega^2 = 9\left(\alpha + \frac{3}{4^{1/3}} \varepsilon \gamma^{2/3}\right).$$

Show that the solution is stable.

24. Analyse the stability of the equation

$$\ddot{x} + \varepsilon x \dot{x}^2 + x = \Gamma \cos \omega t$$

for small ε. (First find approximate solutions of the form $\alpha \cos \omega t$ by the harmonic balance method of Chapter 4, then perturb the solution by the method of Section 9.7.)

25. Locate the equilibrium points of the system

$$\ddot{x} = x^2 + a$$

for all values of the parameter a. Sketch the phase diagrams. The equation models a certain system in which the parameter is designed to be zero, but may deviate to a small extent from zero due to imperfections. How does this affect the stability of the system?

Explain what happens when a suddenly changes from a small negative value to a positive value.

26. Test the stability of the linear system

$$\dot{x}_1 = t^{-2}x_1 + 4x_2 - 2x_3 + t^2,$$
$$\dot{x}_2 = -x_1 + t^{-2}x_2 + x_3 + t,$$
$$\dot{x}_3 = t^{-2}x_1 - 9x_2 - 4x_3 + 1.$$

27. Test the stability of the solutions of the linear system

$$\dot{x}_1 = 2x_1 + e^{-t}x_2 - 2x_3 + e^t,$$
$$\dot{x}_2 = -x_1 + e^{-t}x_2 + x_3 + 1,$$
$$\dot{x}_3 = (4 + e^{-t})x_1 - x_2 - 4x_3 + e^t.$$

28. Test the stability of the zero solution of the system

$$\dot{x} = y + xy/(1 + t^2), \quad \dot{y} = -x - y + y^2/(1 + t^2).$$

29. Test the stability of the zero solution of the system

$$\dot{x}_1 = e^{-x_1 - x_2} - 1, \quad \dot{x}_2 = e^{-x_2 - x_3} - 1, \quad \dot{x}_3 = -x_3.$$

30. Test the stability of the zero solution of the equation

$$\ddot{x} + [\{1 + (t-1)|\dot{x}|\}/\{1 + t|\dot{x}|\}]\dot{x} + \tfrac{1}{4}x = 0.$$

31. Are the periodic solutions of

$$\ddot{x} + \text{sgn}(x) = 0$$

(i) Poincaré stable?   (ii) Liapunov stable?

32. Give a descriptive argument to show that if the index of an equilibrium point is not unity, then the equilibrium point is not stable.

33. Show that the system

$$\dot{x} = x + y - x(x^2 + y^2), \quad \dot{y} = -x + y - y(x^2 + y^2), \quad \dot{z} = -z,$$

has a limit cycle $x^2 + y^2 = 1$, $z = 0$. Find the linear approximation at the origin and so confirm that the origin is unstable. Use cylindrical polar coordinates $r = \sqrt{(x^2 + y^2)}$, z to show that the limit cycle is stable. Sketch the phase diagram in x, y, z space.

34. Show that the nth order non-autonomous system $\dot{\mathbf{x}} = \mathbf{X}(\mathbf{x}, t)$ can be reduced to an $(n+1)$th order autonomous system by introducing a new variable, $x_{n+1} = t$. (The $(n+1)$-dimensional phase diagram for the modified system is then of the type suggested by Fig. 9.9. The system has no equilibrium points.)

35. Sketch a phase diagram in which the origin represents an unstable solution, but every solution starting in a neighbourhood of the origin eventually approaches the origin.

36. The equation $\ddot{x} + x + \varepsilon x^3 = \Gamma \cos \omega t$ ($\varepsilon \ll 1$) has an approximate solution $x^*(t) = a \cos \omega t$ where (eqn (7.10)) $\frac{1}{8}a^3 + (\omega^2 - 1)a + \Gamma = 0$. Show that the first variational equation (Section 9.7) is $\ddot{\zeta} + \{1 + 3\varepsilon x^{*2}(t)\}\zeta = 0$. Reduce this to Mathieu's equation and find conditions for stability of $x^*(t)$.

37. The equation $\ddot{x} + x - \frac{1}{6}x^3 = 0$ has an approximate solution $a \cos \omega t$ where $\omega^2 = 1 - \frac{1}{8}a^2$, $a \ll 1$ (Example 4.12). Use the method of Section 9.7 to show that the solution is unstable.

38. Van der Pol's equation $\ddot{x} + \varepsilon(x^2 - 1)\dot{x} + x = 0$, $\varepsilon \ll 1$, has a limit cycle given approximately by $x^*(t) = 2 \cos t$. Use the method of Section 9.7 to show that this solution is stable.

10 Liapunov functions

THEOREM 9.6, which concerns perturbed linear systems, essentially identifies cases where the stability of a solution stems from the stability of the underlying linear system. For equilibrium points where there is a zero linear approximation (strongly nonlinear cases) this type of theorem fails. In this chapter we describe Liapunov's direct or 'second' method, which allows such cases to be treated. A variant enables instability to be recognized positively. It can be extended to non-autonomous systems (Cesari, 1971, Willems 1970), though we consider only autonomous systems here. The method enables 'domains of stability' to be identified, these being domains of initial conditions the solutions from which are stable.

10.1. Liapunov's direct method

We show how the method works by an example.

Example 10.1 *Consider the stability of the zero solution of the system*

$$\dot{x} = y, \quad \dot{y} = -x - e(x^2 - 1)y; \quad e < 0. \tag{10.1}$$

Equation (10.1) is equivalent to van der Pol's equation, with $|e|$ not necessarily small. The linear approximation near the origin is $\dot{x} = y$, $\dot{y} = -x + ey$, with an asymptotically stable zero solution. This does not of itself guarantee the stability of the zero solution of (10.1), though Theorem 9.6 applies to this case. We illustrate the Liapunov method by again demonstrating its stability.

We try to exhibit a family of closed, nested curves surrounding the origin which have the property that wherever a path of (10.1) crosses one of these it points towards its interior (Fig. 10.1). If such a family can be found, then the path will cross in turn all the members of the family successively and we deduce that the path reaches the origin implying asymptotic stability of the zero solution.

We test the family of curves

$$V(x, y) = x^2 + y^2 = c, \quad 0 < c < 1, \tag{10.2}$$

for the property above. Consider a particular path $\mathscr{C}$, with corresponding solution $x(t), y(t)$. Then on this path

$$V(x, y) = V\{x(t), y(t)\},$$

and

$$\left(\frac{\mathrm{d}V}{\mathrm{d}t}\right)_{\mathscr{C}} = \left(\frac{\partial V}{\partial x}\dot{x} + \frac{\partial V}{\partial y}\dot{y}\right)_{\mathscr{C}} = 2x\dot{x} + 2y\dot{y} = -2e(x^2 - 1)y^2. \tag{10.3}$$

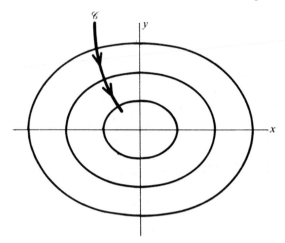

FIG. 10.1. Path $\mathscr{C}$ crossing a family of closed curves

Since $x^2 < 1$ (by (10.2)) and $e < 0$, dV/dt is negative on the path (except where $y = 0$); therefore (except at such points) the path, by (10.2), crosses circles of steadily diminishing radius. Where it crosses the x-axis ($y = 0$) it is instantaneously tangential to the particular circle through this point, but it can be seen that thereafter it can only continue to proceed inwards. (But see Example 10.5, where another family of curves is used which avoids this problem.)

The path cannot end at a point other than the origin, for this would imply the existence of another equilibrium point. Next, consider whether it could approach a limit cycle. Such a limit cycle would be itself a phase path. A circle centre the origin is clearly impossible, since it is not a phase path. Also, the limit cycle could not be any other type of closed curve, since such a curve would intersect *some* circle of (10.2) where $dV/dt < 0$, which implies that the limit cycle is not a phase path. Therefore there is no limit cycle and the path tends to the origin. The same is true for every path.

It can be deduced that the zero solution is Liapunov stable, and in fact asymptotically stable. For, given any $\varepsilon > 0$ such that the 'diamond' $\|x\| = \varepsilon$ lies in the disc $x^2 + y^2 < 1$, we choose $\delta = \varepsilon/\sqrt{2}$ as in Fig. 10.2 so that, given any t_0,

$$\|\mathbf{x}(t_0)\| < \delta \Rightarrow \|\mathbf{x}(t)\| < \varepsilon, \quad t > t_0,$$

as required by Definition 9.1. Asymptotic stability follows from Definition 9.3, since we have shown that the paths approach the origin.

The argument in Example 10.1 can be formulated in a different way. $V(x, y)$ has a single minimum, at $(0, 0)$. The surface $z = V(x, y)$ is shown in Fig. 10.3(b). *The family of circles in Fig. 10.3(a) is the projection of the*

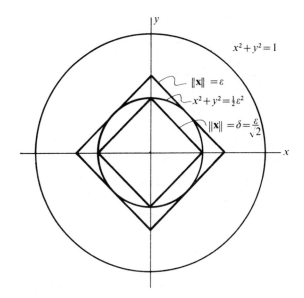

Fig. 10.2.

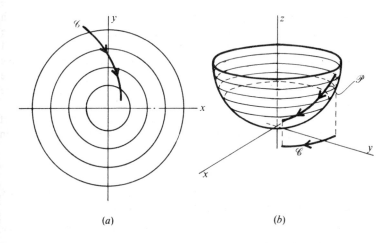

(a)   (b)

Fig. 10.3. The family of closed curves $V(x, y) = c$ and the phase path $\mathscr{C}$ are the projections of the level curves of V and of the path $\mathscr{P}$

contours of $z = V(x, y)$ *and, corresponding to the path* $\mathscr{C}$, *there is a curve* $\mathscr{P}$ *on the surface which is continually descending* $(\dot{V} \leqslant 0)$ *towards the minimum.* This viewpoint is the general one taken in this chapter; in the next chapter where very similar arguments are used to locate limit cycles the viewpoint of Example 10.1 is preferred.

V is known as a *Liapunov function* for the equilibrium point of the given system. To devise or to guess a form of V which will decide the stability is the difficult practical feature of the method.

10.2. Liapunov functions

DEFINITION 10.1 $V(\mathbf{x})$ *is positive (negative) definite in a neighbourhood* $\mathscr{S}$ *of the origin if* $V(\mathbf{x}) > 0$ $(V(\mathbf{x}) < 0)$ *for all* $\mathbf{x} \neq \mathbf{0}$ *in* $\mathscr{S}$, *and* $V(\mathbf{0}) = \mathbf{0}$.

DEFINITION 10.2 $V(\mathbf{x})$ *is positive (negative) semidefinite in a neighbourhood* $\mathscr{S}$ *of the origin if* $V(\mathbf{x}) \geqslant 0$ $(V(\mathbf{x}) \leqslant 0)$ *for all* $\mathbf{x} \neq \mathbf{0}$ *in* $\mathscr{S}$, *and* $V(\mathbf{0}) = 0$.

DEFINITION 10.3   *A function is described as definite (semidefinite) when either it is positive definite or negative definite (positive semidefinite or negative semidefinite).*

Thus, $V(x, y) = x^2 + y^2$ is positive definite on the whole plane, and $V(x, y) = -y^2(x^2 - 1)$ is negative semidefinite on the strip $x^2 < 1$.

Quadratic forms frequently arise in the present context. In two dimensions, a real quadratic form can be written

$$V(x, y) = \sum_{i=1}^{n} \sum_{j=1}^{n} b_{ij} x_i x_j = \mathbf{x}^t B \mathbf{x} \tag{10.4}$$

where B is real and symmetric. It can be proved that $V(x, y)$ is positive definite if and only if

$$b_{11} > 0, \quad \begin{vmatrix} b_{11} & b_{12} \\ b_{21} & b_{22} \end{vmatrix} > 0, \quad \begin{vmatrix} b_{11} & b_{12} & b_{13} \\ b_{21} & b_{22} & b_{23} \\ b_{31} & b_{32} & b_{33} \end{vmatrix} > 0, \quad \dots \tag{10.5}$$

The *total derivative* of a function $V(\mathbf{x})$ along a path $\mathscr{C}$ of the autonomous system $\dot{\mathbf{x}} = \mathbf{X}(\mathbf{x})$ corresponding to a solution $\mathbf{x}(t)$ is

$$\left(\frac{\mathrm{d}V}{\mathrm{d}t}\right)_{\mathscr{C}} = \frac{\mathrm{d}}{\mathrm{d}t} V\{\mathbf{x}(t)\} = \sum_{i=1}^{n} \frac{\partial V}{\partial x_i} X_i(\mathbf{x}), \tag{10.6}$$

by the chain rule. But with *every* point $\mathbf{x}$ there is associated a value, $(\mathrm{d}V/\mathrm{d}t)_{\mathscr{C}}$, where $\mathscr{C}$ is the path through $\mathbf{x}$. The right-hand side of (10.6) is therefore defined for every **x** (except possibly for singular points). When we wish to treat (10.6) as a function of position we write $\dot{V}(\mathbf{x})$, where

$$\dot{V}(\mathbf{x}) = \sum_{i=1}^{n} \frac{\partial V}{\partial x_i} X_i(\mathbf{x}). \tag{10.7}$$

Thus we may say '$\dot{V}$ is continuous', or '$\dot{V}$ is positive definite', meaning with respect to **x**.

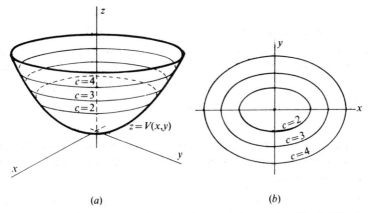

Fig. 10.4. An idealized positive definite function V and the associated family $V(x, y) = c$

We adopt a geometrical approach to certain aspects of the two theorems which follow. Suppose that $V(\mathbf{x})$ is positive definite. A simple model for such a function in 2 dimensions is Fig. 10.4(a). The level curves, $V(x, y) = c$, are simple closed curves surrounding the origin, those for smaller c being nearer the origin, as is notionally indicated; and as $c \to 0$ the origin is approached. In general a positive definite function will not be so simple, but may have such a form as in Fig. 10.5(a), with several minima. Even in such a case, *there is a neighbourhood of the origin in which V has the simple structure of Fig. 10.4.* In the theorems we confine the domain of operations to such a neighbourhood of the origin. In higher dimensions the situation cannot be visualized, but when $V(\mathbf{x})$ is positive definite and has continuous partial derivatives, then for all small enough positive c the following analogous result holds:

I. $V(\mathbf{x}) < c$ *defines an open, bounded, connected region $\mathscr{V}_c$ which contains the origin and has $V(\mathbf{x}) = c$ as its boundary; the diameter of $\mathscr{V}_c$ tends to zero with c; and when $c_1 < c_2$ the boundary of $\mathscr{V}_{c_1}$ is contained in $\mathscr{V}_{c_2}$. (A similar statement applies when V is negative definite.)*

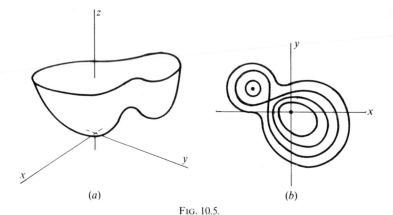

(a) (b)

Fig. 10.5.

In Theorem 10.1 we assume that all constructions are carried out in a neighbourhood $\mathscr{V}^*$ of the origin defined by $V(\mathbf{x}) < c^*$ for which the above is true. The family of neighbourhoods $\mathscr{V}_c$ for c small enough provides a suitable means for recognizing stability. In place of the neighbourhoods $\|\mathbf{x}\| < \varepsilon$, etc., we use differently-shaped neighbourhoods of the form $\mathbf{x} \in \mathscr{V}_\varepsilon$ where $\mathscr{V}_\varepsilon$ is the interior of $V(\mathbf{x}) = \varepsilon$.

Theorem 10.1 *Let* $\mathbf{x}^*(t) = \mathbf{0}, t \geqslant t_0$, *be the zero solution of the regular system* $\dot{\mathbf{x}} = \mathbf{X}(\mathbf{x})$, *where* $\mathbf{X}(\mathbf{0}) = \mathbf{0}$. *Then* $\mathbf{x}^*(t)$ *is uniformly stable for* $t \geqslant t_0$ *if there exists* $V(\mathbf{x})$ *with the following properties in some neighbourhood of* $\mathbf{x} = \mathbf{0}$:

(i) $V(\mathbf{x})$ *and its partial derivatives are continuous*;

(ii) $V(\mathbf{x})$ *is definite*;

(iii) $\dot{V}(\mathbf{x})$ *is semidefinite for the given system, of opposite sign to* $V(\mathbf{x})$.

Proof Without loss of generality suppose V to be positive definite. Choose ε arbitrarily, subject to the condition $\mathscr{V}_\varepsilon \subset \mathscr{V}^*$ above. In $\mathscr{V}_\varepsilon$, $\dot{V}(x) \leqslant 0$ by (iii); therefore any half-path starting in $\mathscr{V}_\varepsilon$ remains in it. The same holds for every small enough ε and hence for every sufficiently small neighbourhood of the origin. The zero solution is therefore stable (and uniformly stable, since the system is autonomous).

In the following Theorem, V is assumed positive definite, and $\dot{V}$ negative definite. We shall assume that all the constructions are carried out in a neighbourhood $\mathscr{V}^{**}: V(\mathbf{x}) < c^{**}$, in which both V and $(-\dot{V})$ have the simple behaviour described in statement I above; this is done without loss of generality.

THEOREM 10.2 *Let all the conditions of Theorem 10.1 hold, except that* (iii) *is replaced by*:

(iii)' $\dot{V}(\mathbf{x})$ *is **definite** for the given system, of opposite sign to* $V(\mathbf{x})$. *Then the zero solution is uniformly and asymptotically stable.*

Proof Suppose, without loss of generality, that $V(\mathbf{x})$ is positive definite. Theorem 10.1 shows that the zero solution is uniformly stable. To provide a proof by contradiction that all neighbouring solution tend to zero, suppose that there exists a solution $\mathbf{x}(t)$, and $m > 0, t_1 \geqslant t_0$ such that

$$V\{\mathbf{x}(t)\} \geqslant m > 0 \quad \text{for all} \quad t \geqslant t_1. \tag{10.8}$$

This possibility is disposed of as follows. The internal boundary of the region in (10.8) is $V(\mathbf{x}) = m$. Consider the closed 'annular' region bounded by $V(\mathbf{x}) = m$ and $V(\mathbf{x}) = c, m < c < c^{**}$. On this region $\dot{V}$ is negative, by (iii)', and $\dot{V}$ is continuous (by (i), and the fact that the system is regular); therefore

$$\dot{V}(\mathbf{x}) \leqslant -M < 0$$

for some $M > 0$. Hence, by (10.8), so long as $\mathbf{x}$ remains in the above 'annular' region,

$$V\{\mathbf{x}(t)\} \leqslant V\{\mathbf{x}(t_1)\} - \int_{t_1}^{t} M \, dt = V\{\mathbf{x}(t_1)\} - M(t - t_1). \tag{10.9}$$

The last expression tends to $-\infty$ with t; $\mathbf{x}(t)$ cannot, therefore, remain in the 'annulus' for all $t \geqslant t_1$, and so must pass through one of its boundaries. Since V is decreasing, it must cross the *internal* boundary, $V(\mathbf{x}) = m$, contradicting the hypothesis (10.8).

Since there is no $\mathbf{x}(t)$ satisfying (10.8) for *all* $t \geqslant t_0$, we consider the possibility that (10.8) might nevertheless hold merely for some sequence of values of t tending to infinity. But if so, given any m, the path will sometimes be inside, sometimes outside, of $V(\mathbf{x}) = m$. However, we know from Theorem 10.1 that once inside, the path remains inside, so this, too, is impossible.

Therefore, for every m, every path starting in $\mathscr{V}^{**}$ is eventually permanently in $V(\mathbf{x}) < m$. Since these neighbourhoods contract on to the origin as $m \to 0$, every such path, hence every solution, must tend to the origin, and the zero solution is asymptotically stable. This theorem expresses in analytical terms the description which follows Example 10.1.

DEFINITION 10.4   *A function V satisfying Theorem* 10.1 *is called a weak Liapunov function, and one which satisfies Theorem* 10.2 *is called a strong Liapunov function.*

The case of van der Pol's equation (Example 10.1) falls within the conditions of Theorem 10.1 for the given V; that is, $V(x, y) = x^2 + y^2$ is a weak Liapunov function for $\ddot{x} + e(x^2 - 1)\dot{x} + x = 0$, $e < 0$. It is weak because $\dot{V}(x, y) = -2ey^2(1 - x^2) \leqslant 0$, equality obtaining on $y = 0$. Hence the zero solution $x(t) = 0$ proves to be stable. In Example 10.5 we exhibit a strong Liapunov function which shows that the zero solution is asymptotically stable.

Example 10.2   *Find a strong Liapunov function for the system*

$$\dot{x} = -x - 2y^2, \quad \dot{y} = xy - y^3.$$

Try the function $V(x, y) = x^2 + ay^2$, with a to be specified. Then by (10.6)

$$\dot{V}(x, y) = -2x^2 + 2(a - 2)xy^2 - 2ay^4.$$

When $a = 2$ this becomes

$$\dot{V}(x, y) = -2(x^2 + 2y^4).$$

Then $V(x, y) = x^2 + 2y^2$ is positive definite and $\dot{V}(x, y)$ is negative definite. Theorem 10.2 tells that the zero solution is asymptotically stable.

Example 10.3   *Find a weak Liapunov function for the system*

$$\dot{x} = y, \quad \dot{y} = -\frac{1}{m}\{f(x, y)y + \lambda x\}, \quad m > 0, \quad \lambda > 0,$$

where $f(x, y) \geqslant 0$ in a neighbourhood of the origin.

The system is equivalent to $m\ddot{x} + f(x, \dot{x})\dot{x} + \lambda x = 0$, representing a linear mass-spring system modified by a nonlinear damping term, giving positive damping for small amplitudes. Try the positive definite function

$$V(x, y) = \tfrac{1}{2}(\lambda x^2 + my^2).$$

Then

$$\dot{V}(x, y) = -my^2 f(x, y), \quad \leqslant 0$$

and V is a weak Liapunov function. By Theorem 10.1, the zero solution is stable.

In Example 10.3 the choice for V is not arbitrary, but recognizable as the 'total energy' of the system; that is, the sum of the kinetic and potential energy; which we expect will be dissipated to zero by the damping. The level curves of the Liapunov function are curves of constant energy of the system. The paths pierce them inwards except along the y axis. This confirms that the motion is damped (energy is lost).

We expect asymptotic stability rather than mere stability, but cannot use Theorem 10.2 with this form for V since $\dot{V}$ is only semidefinite. We return to this question in Section 10.4.

It can be inferred from the proof of Theorem 10.2 that not only is the zero solution asymptotically stable, but that all solutions starting in certain neighbourhoods of the origin are asymptotically stable. Such a neighbourhood is called a *domain of asymptotic stability*. A domain of asymptotic stability can be identified by determining the region in which V and $(-\dot{V})$ have the property I on p. 287; though this will not in general determine the largest domain, since the region found will depend on the particular Liapunov function chosen. For example, for the equation $\dot{x} = -x(1-x^2-y^2)$, $\dot{y} = -y(1-x^2-y^2)$, with $V = x^2+y^2$ we find that $\dot{V} = -2(x^2+y^2)(1-x^2-y^2)$. V and $\dot{V}$ have the required property in the disc $x^2+y^2 < 1$ and this is therefore a domain of asymptotic stability, which happens to be the largest domain in this case. For further examples see Exercises 2, and 27 to 36. For extensions of the theory see Willems, 1970.

10.3. A test for instability

The general approach of the Liapunov direct method can be adapted to test instability of zero solutions.

THEOREM 10.3 *Let* $\mathbf{x}(t) = \mathbf{0}, t \geq t_0$, *be the zero solution of the regular system* $\dot{\mathbf{x}} = \mathbf{X}(\mathbf{x})$, *where* $\mathbf{X}(\mathbf{0}) = \mathbf{0}$. *If there exists a function* $U(\mathbf{x})$ *such that in some neighbourhood* $\|\mathbf{x}\| \leq k$

(i) $U(\mathbf{x})$ *and its partial derivatives are continuous;*

(ii) $U(\mathbf{0}) = 0$;

(iii) $\dot{U}(\mathbf{x})$ *is positive definite for the given system;*

(iv) *in every neighbourhood of* $\mathbf{0}$ *there exists at least one point* $\mathbf{x}$ *at which* $U(\mathbf{x}) > 0$;

then the zero solution is unstable.

Proof By (iv), given any $\delta, 0 < \delta < k$, there exists $\mathbf{x}_\delta$ such that $0 < \|\mathbf{x}_\delta\| < \delta$ and $U(\mathbf{x}_\delta) > 0$. Suppose that $\mathbf{x}(\delta, t), t \geq t_0$, is the solution satisfying $\mathbf{x}(\delta, t_0) = \mathbf{x}_\delta$. Its path cannot enter the origin as $t \to \infty$, because $U(\mathbf{0}) = 0$ (by (ii)) but $\dot{U}(\mathbf{x}) > 0$, $\mathbf{x} \neq \mathbf{0}$. Since $\dot{U}$ is positive definite by (iii), and continuous (by (i) and the regularity condition), and since the path is bounded away from the origin there exists a number $m > 0$ such that

$$\dot{U}\{\mathbf{x}(\delta, t)\} \geq m > 0, \quad t \geq t_0 \tag{10.10}$$

so long as

$$\|\mathbf{x}(\delta, t)\| \leq k. \tag{10.11}$$

Therefore

$$U\{\mathbf{x}(\delta,t)\} - U\{\mathbf{x}(\delta,t_0)\} = \int_{t_0}^t \dot{U}\{\mathbf{x}(\delta,t)\}\,\mathrm{d}t \geqslant m(t-t_0). \quad (10.12)$$

$U(\mathbf{x})$ is continuous by (i) and therefore bounded on $\|\mathbf{x}\| \leqslant k$; but the right-hand side of (10.12) is unbounded. Therefore $\mathbf{x}(\delta,t)$ cannot remain in $\|\mathbf{x}\| \leqslant k$ and so the path reaches the boundary $\|\mathbf{x}\| = k$.

Therefore, given any ε, $0 < \varepsilon < k$, then for any δ, however small, there is at least one solution $\mathbf{x}(\delta,t)$ with $\|\mathbf{x}(\delta,t_0)\| < \delta$ but $\|\mathbf{x}(\delta,t)\| > \varepsilon$ for some t. Therefore the zero solution is unstable.

Typically (by (iv)), $U(\mathbf{x})$ will take both positive and negative values close to $\mathbf{x} = \mathbf{0}$. In the plane, simple functions of the type xy, $x^2 - y^2$, may be successful, as in Example 10.4.

Example 10.4 *Show that $x(t) = 0$, $t \geqslant t_0$, is an unstable solution of the equation* $\ddot{x} - x + \dot{x}\sin x = 0$.

The equivalent system is $\dot{x} = y$, $\dot{y} = x - y\sin x$. Consider

$$U(x,y) = xy$$

satisfying (i), (ii), and (iv) for every neighbourhood. Then

$$\dot{U}(x,y) = x^2 + y^2 - xy\sin x.$$

It is easy to show that this has a minimum at the origin and since $\dot{U}(0,0) = 0$, $\dot{U}$ is positive definite. Therefore the zero solution is unstable. (To show by calculus methods that there is a minimum at the origin is sometimes the simplest way to identify a definite function.)

10.4. Stability and the linear approximation in two dimensions

This section relates to Section 2.3 on the classification of equilibrium points of autonomous equations in two dimensions according to the linear approximation in their neighbourhood. The specific question discussed is the following. For the two-dimensional system

$$\dot{\mathbf{x}} = \mathbf{X}(\mathbf{x}) = A\mathbf{x} + \mathbf{h}(\mathbf{x}) \qquad (10.13)$$

where A is constant and non-zero, and $\|\mathbf{h}(\mathbf{x})\| = O(\|\mathbf{x}\|^2)$ (implying a Taylor-style expansion of $\mathbf{X}(\mathbf{x})$ about the origin), does the instability or asymptotic stability of the zero solution (and hence of all solutions) of $\dot{\mathbf{x}} = A\mathbf{x}$ imply the same property for the zero solutions of the system $\dot{\mathbf{x}} = \mathbf{X}(\mathbf{x})$? With regard to asymptotic stability a positive answer is already available (Theorem 9.6), but we shall carry out the proof for both stability and instability using the techniques of the present chapter. In

Section 10.6 the proofs are extended to n dimensions.

We firstly consider the possibility of asymptotic stability. Suppose that the linear approximate system

$$\dot{\mathbf{x}} = A\mathbf{x}, \quad \text{or} \quad \begin{pmatrix} \dot{x} \\ \dot{y} \end{pmatrix} = \begin{pmatrix} a & b \\ c & d \end{pmatrix} \begin{pmatrix} x \\ y \end{pmatrix}, \tag{10.14}$$

is asymptotically stable. We show that a strong Liapunov function $V(\mathbf{x})$ for this system can be constructed which is a quadratic form:

$$V(\mathbf{x}) = \mathbf{x}^t K \mathbf{x} \tag{10.15}$$

(by Definition 10.4, V must be positive definite and $\dot{V}$ negative definite); and that the same algebraic form holds whenever A in (10.14) is asymptotically stable. $V(\mathbf{x})$ will then be shown to be a strong Liapunov function for the system (10.13) also. Theorem 10.1 then confirms that the zero solution of (10.13) is asymptotically stable whenever (10.14) is.

From the assumed form (10.15),

$$\dot{V}(\mathbf{x}) = \dot{\mathbf{x}}^t K \mathbf{x} + \mathbf{x}^t K \dot{\mathbf{x}} = \mathbf{x}^t A^t K \mathbf{x} + \mathbf{x}^t K A \mathbf{x} \quad \text{(by (10.14))}$$

$$= \mathbf{x}^t (A^t K + K A) \mathbf{x}. \tag{10.16}$$

We now arrange for $V(\mathbf{x})$ to be positive definite and $\dot{V}(\mathbf{x})$ negative definite: the second requirement is satisfied if, for example, we can choose K so that

$$A^t K + K A = -I, \tag{10.17}$$

for then $\dot{V}(\mathbf{x}) = -x^2 - y^2$.

In the notation of Chapter 2, let

$$p = a + d, \quad q = ad - bc. \tag{10.18}$$

We shall show that (in two dimensions) there is a solution of (10.17) of the form

$$K = m(A^t)^{-1} A^{-1} + nI, \tag{10.19}$$

where m and n are constants. Adopting this form, the left of (10.17) becomes

$$A^t K + K A = m\{A^{-1} + (A^t)^{-1}\} + n(A + A^t). \tag{10.20}$$

Now

$$A = \begin{pmatrix} a & b \\ c & d \end{pmatrix}, \quad A^{-1} = \begin{pmatrix} d & -b \\ -c & a \end{pmatrix} \bigg/ q,$$

and by (10.20) we require

$$A^tK + KA = \frac{1}{q}\begin{pmatrix} 2md + 2naq & (nq-m)(b+c) \\ (nq-m)(b+c) & 2ma+2ndq \end{pmatrix} = \begin{pmatrix} -1 & 0 \\ 0 & -1 \end{pmatrix}.$$

This equation is satisfied if

$$m = -q/2p, \quad n = -1/2p.$$

Then from (10.19)

$$K = -\frac{1}{2pq}\begin{pmatrix} c^2+d^2+q & -ac-bd \\ -ac-bd & a^2+b^2+q \end{pmatrix}. \tag{10.21}$$

The associated Liapunov function is given by (10.15):

$$\begin{aligned}
V(\mathbf{x}) &= -\{(c^2+d^2+q)x^2 - 2(ac+bd)xy + (a^2+b^2+q)y^2\}/2pq \\
&= -\{(dx-by)^2 + (cx-ay)^2 + q(x^2+y^2)\}/2pq, \tag{10.22}
\end{aligned}$$

which is positive definite at least when

$$p < 0, \quad q > 0. \tag{10.23}$$

Thus V is a strong Liapunov function for the *linearized* system (10.14) when (10.23) is satisfied. Now, the stability regions for (10.14) are already known, and displayed in Fig. 2.9, which is partly reproduced below as Fig. 10.6. We see that (10.23) in fact exhausts the regions I of asymptotic

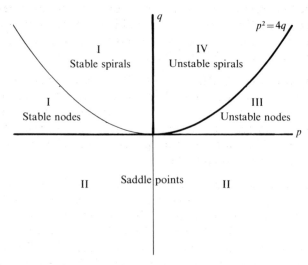

FIG. 10.6. Stability regions in the p, q plane for $\dot{x} = ax+by$, $\dot{y} = cx+dy$, $p = a+d$, $q = ad-bc$ (see Fig. 2.9)

stability of eqn (10.14). Therefore Theorem 10.4 may be given as follows.

THEOREM 10.4   *Let* $(0, 0)$ *be an equilibrium point of the regular system*

$$\begin{pmatrix} \dot{x} \\ \dot{y} \end{pmatrix} = \begin{pmatrix} a & b \\ c & d \end{pmatrix} \begin{pmatrix} x \\ y \end{pmatrix} + \begin{pmatrix} h_1(x, y) \\ h_2(x, y) \end{pmatrix}, \qquad (10.24)$$

where

$$h_1(x, y) = O(x^2 + y^2), \quad h_2(x, y) = O(x^2 + y^2) \qquad (10.25)$$

as $x^2 + y^2 \to 0$. *Then the zero solution of* (10.24) *is asymptotically stable when its linear approximation is asymptotically stable.*

Proof   We shall show that V given by (10.22) is a strong Liapunov function for the system (10.24). It is positive definite when $p < 0, q > 0$; that is, where the linearized system is asymptotically stable.

Also, for the given system (10.24),

$$\begin{aligned} \dot{V} &= \mathbf{x}^t (A^t K + KA)\mathbf{x} + \mathbf{h}^t(\mathbf{x})K\mathbf{x} + \mathbf{x}^t K\mathbf{h}(\mathbf{x}) \\ &= -\mathbf{x}^t I\mathbf{x} + 2\mathbf{h}^t(\mathbf{x})K\mathbf{x} \\ &= -(x^2 + y^2) + \sum_{i,j=1}^{2} K_{ij} h_i(\mathbf{x})x_i, \qquad (10.26) \end{aligned}$$

where the symmetry of K has been used.

By (10.25), for any p, q, there is clearly a neighbourhood of the origin in which the first term of (10.26) predominates; that is to say, where $\dot{V}$ is negative definite. Therefore, V is a strong Liapunov function for (10.24) in the parameter region $p < 0, q > 0$; that is, when the linearized system is asymptotically stable. The result follows by Theorem 10.2.

The open regions II, III, IV in Fig. 10.6 are, we know, where the solutions of $\dot{\mathbf{x}} = A\mathbf{x}$ are *unstable*. To prove a statement in similar terms to the last theorem, only relating to the *instability* of the zero solution of $\dot{\mathbf{x}} = A\mathbf{x} + \mathbf{h}(\mathbf{x})$, we require a function $U(\mathbf{x})$ with the properties stated in Theorem 10.3. There are two principal cases according to the eigenvalues of A, given as (i) and (ii) below:

(i) *The eigenvalues of A are real and different* (II *and* III *in Fig.* 10.6)

Since we hope that the stability of $\dot{\mathbf{x}} = A\mathbf{x}$ will determine that of $\dot{\mathbf{x}} = A\mathbf{x} + \mathbf{h}(\mathbf{x})$, we are interested in the case where the solutions of $\dot{\mathbf{x}} = A\mathbf{x}$ are *unstable*: then (Theorem 9.3) *at least one eigenvalue of A is positive*. This covers the regions II and III of Fig. 10.6. (We will not consider the case when the eigenvalues are equal: the line $p^2 = 4q$ in Fig. 10.6; nor when one of the eigenvalues is zero: $q = 0$.) In this case we can reduce the problem to a simpler one by a transformation

$$\mathbf{x} = C\mathbf{u}, \tag{10.27}$$

where C is nonsingular, and $\mathbf{u} = (u_1, u_2)^t$; whence the equation satisfied by $\mathbf{u}$ is

$$\dot{\mathbf{u}} = C^{-1}AC\mathbf{u}. \tag{10.28}$$

It follows directly from Definition 9.1 that when $\mathbf{x}(t)$ is a stable solution so is $\mathbf{u}(t)$, and conversely (since $\mathbf{u} = C^{-1}\mathbf{x}$). Therefore the same holds for instability. In the cases being considered the eigenvalues of A are real and different, and it is known that C can then be chosen so that

$$C^{-1}AC = \begin{pmatrix} \lambda_1 & 0 \\ 0 & \lambda_2 \end{pmatrix}, \tag{10.29}$$

where the columns of C are eigenvectors corresponding to λ_1, λ_2. Suitable C are given by

$$C = \begin{pmatrix} -b & -b \\ a-\lambda_1 & a-\lambda_2 \end{pmatrix}, \quad b \neq 0;$$

$$C = \begin{pmatrix} -c & -c \\ d-\lambda_1 & d-\lambda_2 \end{pmatrix}, \quad b = 0, \quad c \neq 0,$$

and if $b = c = 0$, the equations are already in a suitable form. Writing

$$D = \begin{pmatrix} \lambda_1 & 0 \\ 0 & \lambda_2 \end{pmatrix} = C^{-1}AC, \tag{10.30}$$

we construct a function $U(\mathbf{u})$ for the transformed system

$$\dot{\mathbf{u}} = D\mathbf{u} \tag{10.31}$$

satisfying the conditions of Theorem 10.2.

Consider U defined by

$$U(\mathbf{u}) = \mathbf{u}^t D^{-1} \mathbf{u} = u_1^2/\lambda_1 + u_2^2/\lambda_2. \tag{10.32}$$

Then $U(\mathbf{u}) > 0$ at some point in every neighbourhood of $\mathbf{u} = \mathbf{0}$ (since instability requires λ_1 or λ_2 positive). $U(\mathbf{u})$ therefore satisfies conditions (i), (ii), and (iv) of Theorem 10.3 in the open regions II and III of Fig. 10.6.

We can also confirm that $\dot{U}$ is positive definite when $\dot{\mathbf{u}} = D\mathbf{u}$. (This is not strictly necessary, since we are interested in the nonlinearized equation, but without this property we should not be likely to be successful.) We have

$$\dot{U}(\mathbf{u}) = \dot{\mathbf{u}}^t D^{-1}\mathbf{u} + \mathbf{u}^t D^{-1}\dot{\mathbf{u}} = \mathbf{u}^t DD^{-1}\mathbf{u} + \mathbf{u}^t D^{-1}D\mathbf{u} = 2(u_1^2 + u_2^2),$$
(10.33)

which is positive definite.

(ii) *The eigenvalues of A are conjugate complex, with positive real part (region IV of Fig.* 10.6)

Write

$$\lambda_1 = \alpha + i\beta, \quad \lambda_2 = \alpha - i\beta. \tag{10.34}$$

The diagonalization process above gives $\dot{\mathbf{u}} = D\mathbf{u}$ where D and $\mathbf{u}$ are complex. Since the theorems, refer to real functions, these cannot immediately be used. However, instead of diagonalizing, we can reduce A by a matrix G so that

$$G^{-1}AG = \begin{pmatrix} \alpha & -\beta \\ \beta & \alpha \end{pmatrix} = S, \tag{10.35}$$

say: an appropriate G is given by

$$G = \tfrac{1}{2}C\begin{pmatrix} 1 & -i \\ 1 & i \end{pmatrix}. \tag{10.36}$$

The system becomes

$$\dot{\mathbf{u}} = \begin{pmatrix} \alpha & -\beta \\ \beta & \alpha \end{pmatrix}\mathbf{u}, \tag{10.37}$$

with

$$\mathbf{x} = G\mathbf{u}. \tag{10.38}$$

Suppose $U(\mathbf{u})$ is given by

$$U(\mathbf{u}) = \mathbf{u}^t\mathbf{u}. \tag{10.39}$$

This satisfies (i), (ii), and (iv) of Theorem 10.3 in region (IV) of Fig. 10.6. Also, for the linear equation (10.34), $\dot{U}$ becomes

$$\dot{U}(\mathbf{u}) = 2\alpha(u_1^2 + u_2^2) \tag{10.40}$$

as in case (i), so that $\dot{U}$ is positive definite, and will (we hope) be positive definite in the nonlinear case too.

By using the prior knowledge of Chapter 2 as to the regions in which the system is unstable, we have therefore determined that wherever $\dot{\mathbf{x}} = A\mathbf{x}$ is unstable, U given by either (10.32) or (10.39), as appropriate, is a function of the type required for (i), (ii), and (iv) of Theorem 10.2. We may therefore formulate Theorem 10.5, which shows that $\dot{U}$ has the final property (iii) for the nonlinear system $\dot{\mathbf{x}} = A\mathbf{x} + \mathbf{h}(\mathbf{x})$.

THEOREM 10.5   *Let $(0, 0)$ be an equilibrium point of the regular two-dimensional system*

$$\dot{\mathbf{x}} = A\mathbf{x} + \mathbf{h}(\mathbf{x}) \tag{10.41}$$

where

$$\mathbf{h}(\mathbf{x}) = \begin{pmatrix} h_1(\mathbf{x}) \\ h_2(\mathbf{x}) \end{pmatrix},$$

and $h_1(\mathbf{x}), h_2(\mathbf{x}) = O(|\mathbf{x}|^2)$ as $|\mathbf{x}| \to 0$. Then when the eigenvalues of A are different, non-zero, and at least one has positive real part, the zero solution of (10.24) is unstable.

Proof   We shall assume that the eigenvalues of A are real and that one of them is positive, the other case (λ complex with positive real part) being closely similar. Assume that equation (10.27) is used to reduce (10.41) to the form

$$\dot{\mathbf{u}} = D\mathbf{u} + \mathbf{g}(\mathbf{u}); \tag{10.42}$$

then if $\mathbf{u}(t) = \mathbf{0}$ is unstable, so is $\mathbf{x}(t) = \mathbf{0}$, and conversely. Also

$$g_1(\mathbf{u}), g_2(\mathbf{u}) = O(x^2 + y^2), \quad x^2 + y^2 \to 0. \tag{10.43}$$

It is clearly sufficient to display a function $U(\mathbf{u})$ for the reduced system (10.42). U given by (10.32) satisfies conditions (i), (ii), and (iv) of Theorem 10.2. Also

$$\begin{aligned} \dot{U}(\mathbf{u}) &= \dot{\mathbf{u}}^t D^{-1} \mathbf{u} + \mathbf{u}^t D^{-1} \dot{\mathbf{u}} \\ &= \mathbf{u}^t D D^{-1} \mathbf{u} + \mathbf{u}^t D^{-1} D \mathbf{u} + \mathbf{g}^t D^{-1} \mathbf{u} + \mathbf{u}^t D^{-1} \mathbf{g} \\ &= 2\{u_1^2 + u_2^2 + \lambda_1 g_1(u) + \lambda_2 g_2(u)\}. \end{aligned}$$

From (10.43), it is clear that $\dot{U}$ is positive definite in a small enough neighbourhood of the origin, satisfying (iii) of Theorem 10.2. Therefore by Theorem 10.2, the zero solution $\mathbf{u}(t) = \mathbf{0}$ is unstable in regions II and III; thus $\mathbf{x}(t) = \mathbf{0}$ is unstable.

The calculation for region IV of Fig. 10.6 is similar, using U given by (10.39).

Example 10.5   *Show that the zero solution of van der Pol's equation $\ddot{x} + e(x - 1)\dot{x} + x = 0$ is uniformly and asymptotically stable when $e < 0$ and unstable when $e > 0$. Construct a Liapunov function for the stable case.*

The equivalent system is $\dot{x} = y, \dot{y} = -x + ey - ex^2 y$. The linearized system is $\dot{x} = y, \dot{y} = -x + ey$. The eigenvalues of

$$A = \begin{pmatrix} 0 & 1 \\ -1 & e \end{pmatrix}$$

are given by $\lambda_1, \lambda_2 = \frac{1}{2}(e \pm \sqrt{(e^2 - 4)})$.

Suppose $e < 0$. Then the linearized system is asymptotically stable. Also $\mathbf{h}^t(\mathbf{x}) = (0, -ex^2)$. Therefore, by Theorem 10.4, the zero solution of the given equation is asymptotically stable.

By (10.15), (10.21), $V = (x, y)K(x, y)^t$ where

$$K = -\frac{1}{2e} \begin{pmatrix} 2 + e^2 & -e \\ -e & 2 \end{pmatrix}.$$

Hence

$$V(x, y) = -\{(2 + e^2)x^2 - 2exy + 2y^2\}/2e.$$

It can be confirmed that $\dot{V} = -x^2 - y^2 - ex^3 y + 2x^2 y^2$, which is negative definite near the origin. For the argument of Example 10.1 the family

$$(1 + \tfrac{1}{2}e^2)x^2 - exy + y^2 = \text{constant}$$

would be preferable to the family of circles given there, which correspond to a $\dot{V}$ which is only semi-definite.

Suppose $e > 0$. The linearized system is unstable and Theorem 10.5 applies.

Example 10.6 *Investigate the stability of the equilibrium points of*

$$\dot{x} = y(x + 1), \quad \dot{y} = x(1 + y^3).$$

There are two equilibrium points, at $(0, 0)$ and $(-1, -1)$. Near the origin the linear approximation is $\dot{x} = y, \dot{y} = x$, with eigenvalues $\lambda = \pm 1$. Hence, the linear approximation has a saddle point at $(0, 0)$. By Theorem 10.5, the zero solution of the given equation is also unstable.

Near the point $(-1, -1)$ put $x = -1 + \xi, y = -1 + \eta$. Then

$$\dot{\xi} = \xi(\eta - 1), \quad \dot{\eta} = (\xi - 1)\{1 + (\eta - 1)^3\}.$$

The linear approximation is $\dot{\xi} = -\xi, \dot{\eta} = -3\eta$. The eigenvalues are $\lambda = -1, -3$, both negative. By Theorem 10.4 the solution $\xi(t) = \eta(t) = 0$ is therefore asymptotically stable; so is the solution $x(t) = y(t) = -1$ of the original solution.

Figure 10.7 shows the computed phase diagram. The shaded region shows the *domain of attraction* of the point $(-1, -1)$.

The foregoing theorems refer to cases where the linearized system is asymptotically stable, or is unstable, and make no prediction as to the effect of modifying, by nonlinear terms, a case of mere stability. In this case the effect may go either way, as shown by Example 10.7.

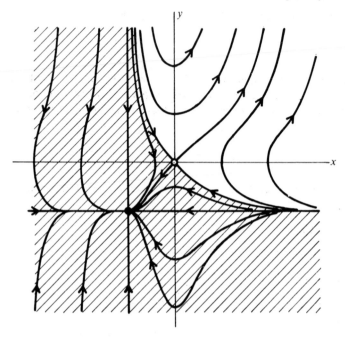

Fig. 10.7. Example 10.6: domain of attraction of $(-1, -1)$ shaded.

Example 10.7 *Investigate the stability of the following systems*
(i) $\dot{x} = y - x(x^2 + y^2), \quad \dot{y} = -x - y(x^2 + y^2)$;
(ii) $\dot{x} = y + x(x^2 + y^2), \quad \dot{y} = -x + y(x^2 + y^2)$.

The linear approximation shows a centre in both cases.
(i) Let $V = x^2 + y^2$, positive definite. Then

$$\dot{V} = -2(x^2 + y^2)^2$$

which is negative definite and the zero solution is asymptotically stable, by
Theorem 10.2.
(ii) Let $U = x^2 + y^2$ (positive in every neighbourhood of the origin). Then

$$\dot{U} = 2(x^2 + y^2)^2$$

which is positive definite. Theorem 10.3 predicts instability.

10.5. Exponential function of a matrix

In Section 10.5 it is convenient to have this technical device available
in order to solve the equivalent of (10.17) in n dimensions. Let A be a

nonsingular $n \times n$ matrix. Then e^A is defined by

$$e^A = I + A + \frac{1}{2!} A^2 + \frac{1}{3!} A^3 + \ldots, \qquad (10.44)$$

whenever the series converges. The partial sums are $n \times n$ matrices and the series converges if the elements of the partial-sum matrix converge, in which case e^A is an $n \times n$ matrix. In fact the series converges for all A, since (see Section 9.2), for any r

$$\|I + A + \ldots + \frac{1}{r!} A^r\| \leqslant 1 + \|A\| + \ldots + \frac{1}{r!} \|A\|^r,$$

and as $r \to \infty$ the series on the right always converges. It also follows that
(i) $\|e^A\| \leqslant e^{\|A\|}$.

From (10.44) we can derive the following properties:

(ii) $e^\phi = I$ where ϕ is the zero matrix;

(iii) $e^A e^B = e^{A+B}$ when $AB = BA$;

(iv) $e^{-A} = (e^A)^{-1}$ (from (ii); $e^{\pm A}$ are nonsingular);

(v) $\dfrac{d}{dt} e^{At} = A e^{At} = e^{At} A$ (from the series representation);

(vi) $(e^{At})^t = e^{A^t t}$.

(vii) Let the eigenvalues of A of A be $\lambda_1, \lambda_2, \ldots, \lambda_n$. Then for any γ such that $\gamma > \max_{1 \leqslant i \leqslant n} (\lambda_i)$ there exists a constant $c > 0$ such that $\|e^{At}\| < c e^{\gamma t}$.

(viii) The exponential function can be used to represent the solution of the system $\dot{\mathbf{x}} = A\mathbf{x}$, A constant. Let

$$\mathbf{x} = e^{At} \mathbf{c} \qquad (10.45)$$

where $\mathbf{c}$ is any constant vector. Then $\mathbf{x}(0) = \mathbf{c}$ and

$$\dot{\mathbf{x}} = A e^{At} \mathbf{c} = A\mathbf{x},$$

which confirms the solution. In fact, e^{At} is a fundamental matrix for the system, since e^{At} is nonsingular.

(ix) It can be shown (Ferrar, 1951) that the infinite series (10.44) can be expressed as a polynomial of degree less than n whose coefficients are functions of the eigenvalues of A. We shall not need this result, but it is worth noting that it is not essential to evaluate infinite series.

Example 10.8   *Find the solution of*

$$\dot{\mathbf{x}} = \begin{pmatrix} \dot{x} \\ \dot{y} \end{pmatrix} = \begin{pmatrix} 1 & 1 \\ 0 & 1 \end{pmatrix} \begin{pmatrix} x \\ y \end{pmatrix} = A\mathbf{x},$$

with $x(0) = 1$, $y(0) = 0$.

By (10.45) the solution has the form

$$\begin{pmatrix} x \\ y \end{pmatrix} = e^{At} \begin{pmatrix} c_1 \\ c_2 \end{pmatrix}.$$

Putting $t = 0$ we have $c_1 = 1, c_2 = 0$. The representation is obviously a generalization of the one-dimensional solution $ce^{\alpha t}$. To evaluate e^{At}, note that

$$A^r = \begin{pmatrix} 1 & r \\ 0 & 1 \end{pmatrix}, \quad r \geqslant 1.$$

Therefore

$$e^{At} = \begin{pmatrix} \sum\limits_0^\infty \dfrac{t^r}{r!} & 1 + \sum\limits_{r=1}^\infty \dfrac{t^r}{(r-1)!} \\ 0 & \sum\limits_{r=0}^\infty \dfrac{t^r}{r!} \end{pmatrix} = \begin{pmatrix} e^t & 1 + te^t \\ 0 & e^t \end{pmatrix}.$$

10.6. Stability and the linear approximation for nth order autonomous systems

Given the nth order nonlinear autonomous system

$$\dot{\mathbf{x}} = A\mathbf{x} + \mathbf{h}(\mathbf{x}), \quad \mathbf{h}(\mathbf{0}) = \mathbf{0}, \tag{10.46}$$

and its linear approximation near the origin

$$\dot{\mathbf{x}} = A\mathbf{x}, \tag{10.47}$$

we wish to identify conditions on $\mathbf{h}$ whereby asymptotic stability or instability of (10.47) implies the similar property for the zero solutions of (10.46). We proceed as in Section 10.5 for the two-dimensional case.

Suppose firstly that the solutions of (10.47) are *asympotically stable*. This implies (Theorem 9.2) that

$$\text{Re}\{\lambda_i\} < 0, \quad i = 1, 2, \dots, n, \tag{10.48}$$

where λ_i are the eigenvalues of A. We construct a strong Liapunov function for (10.46) which is a quadratic form:

$$V(\mathbf{x}) = \mathbf{x}^t K \mathbf{x} \tag{10.49}$$

where K is firstly to be determined to make V positive definite.

In order also to have any hope of making $\dot{V}$ negative definite for

(10.46), we at least need $\dot{V}$ negative definite for (10.47). For (10.47),

$$\dot{V}(\mathbf{x}) = \mathbf{x}^t(A^tK + KA)\mathbf{x}, \qquad (10.50)$$

and we shall arrange that

$$A^tK + KA = -I, \qquad (10.51)$$

which guarantees that (10.50) is negative definite:

$$\dot{V}(x) = -\sum_{i=1}^{n} x_i^2.$$

To achieve this, consider the product $e^{A^t t}e^{At}$. We have

$$\frac{d}{dt}\{e^{A^t t}e^{At}\} = A^t e^{A^t t}e^{At} + e^{A^t t}e^{At}A. \qquad (10.52)$$

By (vii) and (viii) of Section 10.5, when $\dot{\mathbf{x}} = A\mathbf{x}$ is asymptotically stable,

$$\|e^{At}\| \leqslant c\,e^{\gamma t}, \quad c > 0, \quad \gamma < 0.$$

Since the eigenvalues of A^t are the same as those of A we can choose c so that both

$$\|e^{At}\|, \|e^{A^t t}\| \leqslant c\,e^{\gamma t}, \quad c > 0, \quad \gamma < 0. \qquad (10.53)$$

This ensures the convergence of the integrals below. From (10.52),

$$\int_0^\infty \frac{d}{dt}\{e^{A^t t}e^{At}\}\,dt = A^t \int_0^\infty e^{A^t t}e^{At}\,dt + \int_0^\infty e^{A^t t}e^{At}\,dt\,A,$$

and also equals $(-I)$ by (10.53) again. By comparison with (10.51) it appears that

$$K = \int_0^\infty e^{A^t t}e^{At}\,dt \qquad (10.54)$$

will satisfy (10.51). The matrix K is symmetrical by (vi) of Section 10.5. Note that (10.54) and hence (10.51) hold whenever the eigenvalues of A are negative, that is, whenever $\dot{\mathbf{x}} = A\mathbf{x}$ is asympotically stable.

We have finally to show that V is positive definite.

$$\mathbf{x}^tK\mathbf{x} = \int_0^\infty (\mathbf{x}^t e^{A^t t})(e^{At}\mathbf{x})\,dt$$

$$= \int_0^\infty (e^{At}\mathbf{x})^t(e^{At}\mathbf{x})\,dt,$$

and the integrand is simply the sum of certain squares, and is therefore positive definite.

The following Theorem should be compared with the (stronger) Theorem 9.6.

THEOREM 10.6   *If the system* $\dot{\mathbf{x}} = A\mathbf{x} + \mathbf{h}(\mathbf{x})$, A *constant, is regular, and*
(i) *the zero solution (hence every solution: Theorem 9.1) of* $\dot{\mathbf{x}} = A\mathbf{x}$ *is asymptotically stable;*
(ii) $\mathbf{h}(\mathbf{0}) = \mathbf{0}$, *and* $\lim\limits_{\|x\|\to 0} \|\mathbf{h}(\mathbf{x})\|/\|\mathbf{x}\| = 0$;   (10.55)
then $\mathbf{x}(t) = \mathbf{0}$, $t \geqslant t_0$ *for any* t_0 *is an asymptotically stable solution of*

$$\dot{\mathbf{x}} = A\mathbf{x} + \mathbf{h}(\mathbf{x}).$$ (10.56)

Proof   We have to show there is a neighbourhood of the origin where $V(\mathbf{x})$ defined by (10.49) and (10.54) is a strong Liapunov function for (10.56).

$$V(\mathbf{x}) = \mathbf{x}^t K \mathbf{x},$$

where

$$K = \int_0^\infty e^{A^t t} e^{At} \, dt$$

is positive definite whenever (i) holds. Also for (10.56),

$$\begin{aligned}
\dot{V}(\mathbf{x}) &= \dot{\mathbf{x}}^t K \mathbf{x} + \mathbf{x}^t K \dot{\mathbf{x}} \\
&= \mathbf{x}^t (A^t K + KA)\mathbf{x} + \mathbf{h}^t K \mathbf{x} + \mathbf{x}^t K \mathbf{h} \\
&= -\mathbf{x}^t \mathbf{x} + 2\mathbf{h}^t(\mathbf{x}) K \mathbf{x},
\end{aligned}$$ (10.57)

by (10.51) and the symmetry of K. We have to find a neighbourhood of the origin in which the first term of (10.57) dominates.

Now

$$|2\mathbf{h}^t(\mathbf{x}) K \mathbf{x}| \leqslant 2\|\mathbf{h}(\mathbf{x})\| \, \|K\| \, \|\mathbf{x}\|.$$ (10.58)

By (ii), given any $\varepsilon > 0$ there exists $\delta > 0$ such that

$$\|\mathbf{x}\| < \delta \Rightarrow \|\mathbf{h}(\mathbf{x})\| < \varepsilon\|\mathbf{x}\|.$$

Then from (10.55), using the inequality (9.4), $\|\mathbf{x}\| < \delta$ implies

$$|2\mathbf{h}^t(\mathbf{x}) K \mathbf{x}| \leqslant 2\varepsilon\|K\| \, \|\mathbf{x}\|^2 \leqslant 2\varepsilon n\|K\| \, |\mathbf{x}|^2,$$ (10.59)

where n is the dimension of the problem. Suppose ε to be chosen so that

$$\varepsilon < 1/4n\|K\|,$$

then from (10.59),

$$\|\mathbf{x}\| < \delta \Rightarrow |2\mathbf{h}^t(\mathbf{x})K\mathbf{x}| < \tfrac{1}{2}|\mathbf{x}|^2 = \tfrac{1}{2}(x_1^2 + x_2^2 + \ldots + x_n^2).$$

Therefore, by (10.57), $\dot{V}(\mathbf{x})$ for (10.56) is negative definite on $\|\mathbf{x}\| < \delta$ and by Theorem 10.1, the zero solutions are asymptotically stable.

A similar theorem can be formulated relating to *instability*. Consider the regular system

$$\dot{\mathbf{x}} = A\mathbf{x} + \mathbf{h}(\mathbf{x}), \quad \mathbf{h}(\mathbf{0}) = \mathbf{0}, \tag{10.60}$$

where A is constant. Let C be a nonsingular matrix, and change the variables to $\mathbf{u}$ by writing

$$\mathbf{x} = C\mathbf{u}; \tag{10.61}$$

then the original equation becomes

$$\dot{\mathbf{u}} = C^{-1}AC\mathbf{u} + \mathbf{h}(C\mathbf{u}). \tag{10.62}$$

C will be chosen to reduce A to canonical form, as in Section 10.4. From Definition 9.1, it is clear that whenever a solution $\mathbf{x}(t)$ of (10.60) is stable, the corresponding solution $\mathbf{u}(t)$ of (10.62) is stable. Since the converse is also true, the same must apply to instability. The zero solution of (10.60) corresponds with that of (10.62): we may therefore investigate the stability of (10.62) in place of (10.60). The same applies to the linearized pair

$$\dot{\mathbf{x}} = A\mathbf{x}, \quad \dot{\mathbf{u}} = C^{-1}AC\mathbf{u}. \tag{10.63}$$

Suppose, for simplicity, that the eigenvalues of A are distinct, and that at least one of them has positive real part. Then all solutions of (10.63) are unstable. There are two cases:

(i) *Eigenvalues of A real, distinct, with at least one positive.* It is known that a real C may be chosen so that

$$C^{-1}AC = D, \tag{10.64}$$

where D is diagonal with the elements equal to the eigenvalues of A. Then (10.63) becomes

$$\dot{\mathbf{u}} = D\mathbf{u} + \mathbf{h}(C\mathbf{u}). \tag{10.65}$$

(ii) *When the eigenvalues of A are distinct, not all real, and at least one has positive real part.*

It is then known that a real nonsingular matrix G may be chosen so that

$$G^{-1}AG = D^* \tag{10.66}$$

where D^* is block-diagonal: in place of the pair of complex roots $\lambda, \bar{\lambda}$ in diagonal positions which would be delivered by a transformation of type (10.61) and (10.63), we have a 'block' of the form

$$D^* = \begin{pmatrix} \cdots & & & 0 \\ & \alpha & -\beta & \\ & \beta & \alpha & \\ 0 & & & \cdots \end{pmatrix}$$

$$\tag{10.67}$$

where

$$\lambda = \alpha + i\beta.$$

The argument in the theorem is scarcely altered in this case.

THEOREM 10.7   *If A is constant, and*
(i) *the eigenvalues of A are distinct, none are zero, and at least one has positive real part;*

(ii) $\lim_{\|\mathbf{x}\| \to 0} \|\mathbf{h}(\mathbf{x})\|/\|\mathbf{x}\| = 0 \tag{10.68}$

then the zero solutions $\mathbf{x}(t) = \mathbf{0}, t \geq t_0$, of the regular system

$$\dot{\mathbf{x}} = A\mathbf{x} + \mathbf{h}(\mathbf{x})$$

are unstable.

Proof   We shall carry out the proof only for the case where the eigenvalues of A are all real. Reduce (10.68) to the form (10.65):

$$\dot{\mathbf{u}} = D\mathbf{u} + \mathbf{h}(C\mathbf{u}),$$

where C is nonsingular and D is diagonal with the eigenvalues of A as its elements, at least one being positive. As explained, it is only necessary to determine the stability of $\mathbf{u}(t) = \mathbf{0}, t \geq t_0$.

Let

$$U(\mathbf{u}) = \mathbf{u}^t D^{-1} \mathbf{u} = \sum_{i=1}^{n} \frac{u_i^2}{\lambda_i}. \tag{10.69}$$

If $\lambda_k > 0$, then $U(\mathbf{u}) > 0$ when $u_k \neq 0$ and $u_1 = u_2 = \ldots = \mathbf{u}_{k-1} = u_{k+1} = \ldots = u_n = 0$. Therefore (i), (ii), and (iv) of Theorem 10.3 are satisfied.

Also, for (10.68),

$$\begin{aligned}
\dot{U}(\mathbf{u}) &= \dot{\mathbf{u}}^t D^{-1} \mathbf{u} \\
&= \mathbf{u}^t D D^{-1} \mathbf{u} + \mathbf{u}^t D^{-1} D \mathbf{u} + \mathbf{h}^t D^{-1} \mathbf{u} + \mathbf{u}^t D^{-1} \mathbf{h} \\
&= 2(u_1^2 + u_2^2 + \ldots + u_n^2) + 2\mathbf{u}^t D^{-1} \mathbf{h}(C\mathbf{u}),
\end{aligned} \tag{10.70}$$

since $DD^{-1} = D^{-1}D = I$ and D is symmetrical. Theorem (10.3) requires $\dot{U}$ to be positive definite in a neighbourhood of the origin. It is clearly sufficient to show that the second term in (10.70) is smaller than the first in a neighbourhood of the origin.

By (iii), given $\varepsilon > 0$ there exists $\delta > 0$ such that

$$\|C\| \, \|\mathbf{u}\| < \delta \Rightarrow \|C\mathbf{u}\| < \delta \Rightarrow \|\mathbf{h}(C\mathbf{u})\| < \varepsilon \|C\mathbf{u}\|, \tag{10.71}$$

or alternatively

$$\|\mathbf{u}\| < \delta/\|C\| \Rightarrow \|\mathbf{h}(C\mathbf{u})\| < \varepsilon \|C\| \, \|\mathbf{u}\|.$$

Therefore (see (10.70)), $\|\mathbf{u}\| < \delta/\|C\|$ implies that

$$\begin{aligned}
|2\mathbf{u}^t D^{-1} \mathbf{h}(C\mathbf{u})| &\leqslant 2\|\mathbf{u}^t\| \, \|D^{-1}\| \, \|\mathbf{h}(C\mathbf{u})\|, \\
&\leqslant 2\varepsilon \|D^{-1}\| \, \|C\| \, \|\mathbf{u}\|^2 \\
&\leqslant 2\varepsilon n \|D^{-1}\| \, \|C\| \, |\mathbf{u}|^2,
\end{aligned} \tag{10.72}$$

where n is the dimension. If we choose

$$\varepsilon < 1/2n\|D^{-1}\| \, \|C\|,$$

then (10.72) becomes

$$|2\mathbf{u}^t D^{-1} \mathbf{h}(C\mathbf{u})| < |\mathbf{u}|^2$$

for all $\mathbf{u}$ close enough to the origin. By referring to (10.70) we see that $\dot{U}(\mathbf{u})$ is positive definite in a neighbourhood of the origin, as required by (iii) of Theorem 10.3. Therefore the result is proved.

Example 10.9 *Prove that the zero solution of the system*

$$\dot{x}_1 = -x_1 + x_2^2 + x_3^2, \quad \dot{x}_2 = x_1 - 2x_2 + x_1^2, \quad \dot{x}_3 = x_1 + 2x_2 - 3x_3 + x_2 x_3,$$

is uniformly and asymptotically stable.

Here we have

$$A = \begin{pmatrix} -1 & 0 & 0 \\ 1 & -2 & 0 \\ 1 & 2 & -3 \end{pmatrix}$$

and

$$\mathbf{h}(\mathbf{x}) = (x_2^2 + x_3^2, \; x_1^2, \; x_2 x_3)^t.$$

The eigenvalues of A are $-1, -2, -3$; therefore the zero solution of $\dot{\mathbf{x}} = A\mathbf{x}$ is uniformly and asymptotically stable. Also

$$\|\mathbf{h}(\mathbf{x})\| = x_2^2 + x_3^2 + x_1^2 + |x_2 x_3|,$$

and

$$\|\mathbf{h}(\mathbf{x})\|/\|\mathbf{x}\| = (x_2^2 + x_3^2 + x_1^2 + |x_2 x_3|)/(|x_1| + |x_2| + |x_3|),$$

tending to zero with $\|\mathbf{x}\|$. By Theorem 10.6, the zero solution of the given system is therefore uniformly asymptotically stable.

Example 10.10   *Show that the zero solution of the equation*

$$\dddot{x} - 2\ddot{x} - \dot{x} + 2x = \ddot{x}(x + \dot{x})$$

is unstable.

   Write as the equivalent system

$$\dot{x} = y, \quad \dot{y} = z, \quad \dot{z} = -2x + y + 2z + z(x + y).$$

Then

$$A = \begin{pmatrix} 0 & 1 & 0 \\ 0 & 0 & 1 \\ -2 & 1 & 2 \end{pmatrix},$$

with eigenvalues $2, 1, -1$, two of which are positive. Also

$$\|\mathbf{h}(\mathbf{x})\|/\|\mathbf{x}\| = |z(x + y)|/(|x| + |y| + |z|)$$

which tends to zero with $\|\mathbf{x}\|$. Therefore, by Theorem 10.7, the zero solution of the given equation is unstable.

10.7. Miscellaneous problems

Quadratic systems. Consider the strongly nonlinear system whose lowest-order terms are of the second degree:

$$\dot{x} = X(x, y) \approx a_1 x^2 + 2b_1 xy + c_1 y^2, \tag{10.73}$$

$$\dot{y} = Y(x, y) \approx a_2 x^2 + 2b_2 xy + c_2 y^2. \tag{10.74}$$

Neglecting any higher-order terms, the equations can be written

$$\dot{x} = (x, y)A_1 \begin{pmatrix} x \\ y \end{pmatrix}, \quad \dot{y} = (x, y)A_2 \begin{pmatrix} x \\ y \end{pmatrix}, \tag{10.75}$$

where

$$A_1 = \begin{pmatrix} a_1 & b_1 \\ b_1 & c_1 \end{pmatrix}, \quad A_2 = \begin{pmatrix} a_2 & b_2 \\ b_2 & c_2 \end{pmatrix}.$$

For Theorem 10.3, let

$$U(x, y) = \alpha x + \beta y. \tag{10.76}$$

Then for any non-zero α and β, $U > 0$ for some points in every neighbourhood of the origin. Thus (i), (ii), and (iv) of Theorem 10.3 are satisfied. For (10.75), $\dot{U}$ is given by

$$\dot{U}(x, y) = \alpha \dot{x} + \beta \dot{y} = \alpha(x, y) A_1 \begin{pmatrix} x \\ y \end{pmatrix} + \beta(x, y) A_2 \begin{pmatrix} x \\ y \end{pmatrix}$$

$$= (x, y) \begin{pmatrix} \alpha a_1 + \beta a_2 & \alpha b_1 + \beta b_2 \\ \alpha b_1 + \beta b_2 & \alpha c_1 + \beta c_2 \end{pmatrix} \begin{pmatrix} x \\ y \end{pmatrix}.$$

This quadratic form is positive definite (see eqn (10.5)) if both

$$\alpha a_1 + \beta a_2 > 0; \tag{10.77}$$

and, with some rearrangement,

$$\Delta(\alpha, \beta) = \alpha^2(a_1 c_1 - b_1^2) + \alpha\beta(a_1 c_2 + a_2 c_1 - 2b_1 b_2) + \beta^2(a_2 c_2 - b_2^2) > 0,$$

or

$$\Delta(\alpha, \beta) = (\alpha, \beta) \begin{pmatrix} a_1 c_1 - b_1^2 & \frac{1}{2}(a_1 c_2 + a_2 c_1 - 2b_1 b_2) \\ \frac{1}{2}(a_1 c_2 + a_2 c_1 - 2b_1 b_2) & a_2 c_2 - b_2^2 \end{pmatrix} \begin{pmatrix} \alpha \\ \beta \end{pmatrix} > 0.$$

$$\tag{10.78}$$

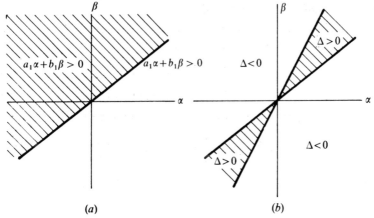

Fig. 10.8.

Consider whether α, β exist satisfying (10.77) and (10.78). On a plane for $\alpha, \beta, a_1\alpha + b_1\beta = 0$ is a line, and (10.77) is represented by a half plane, like that shown in Fig. 10.8(*a*). Now $\Delta(\alpha, \beta)$ is a quadratic form and is either positive definite, positive on a ray or in a sector as in Fig. 10.8(*b*), or negative definite. Unless $\Delta(\alpha, \beta)$ is negative definite; or unless it is negative except on the line $a_1\alpha + b_1\beta = 0$ where it is zero; there exist *some* values of α, β making $U(\mathbf{x})$ positive definite and therefore satisfying the final conditions of Theorem 10.3. The zero solution of (10.73) and (10.74) is therefore unstable, except possibly in the particular cases, depending on exact relations between the coefficients, mentioned above. We shall not investigate these cases.

Hamiltonian problems. Conservative problems, particularly in dynamics, can be expressed in the form

$$\dot{p}_i = \frac{\partial \mathscr{H}}{\partial q_i}, \quad \dot{q}_i = -\frac{\partial \mathscr{H}}{\partial p_i}, \quad i = 1, 2, \ldots n, \tag{10.79}$$

where $\mathscr{H}$ is a given function called the Hamiltonian of the system, q_i is a generalized coordinate and p_i a generalized momentum. The Hamiltonian is defined by

$$\mathscr{H}(\mathbf{p}, \mathbf{q}) = \mathscr{T}(\mathbf{p}, \mathbf{q}) + \mathscr{V}(\mathbf{q}), \tag{10.80}$$

where $\mathscr{T}$ is the kinetic energy and $\mathscr{V}$ the potential energy: it is assumed that $\mathscr{V}(\mathbf{0}) = 0$. $\mathscr{T}$ is a positive definite quadratic form in p_i, so $\mathscr{H}(\mathbf{0}, \mathbf{0}) = 0$.

Suppose that $\mathbf{q} = \mathbf{0}$ is a *minimum* of $\mathscr{V}(\mathbf{q})$ so that $\mathscr{V}$, and hence $\mathscr{H}$, is positive definite in a neighbourhood of the origin. Then

$$\dot{\mathscr{H}}(\mathbf{p}, \mathbf{q}) = \sum_{i=1}^{n} \frac{\partial \mathscr{H}}{\partial p_i} \dot{p}_i + \sum_{i=1}^{n} \frac{\partial \mathscr{H}}{\partial q_i} \dot{q}_i = 0$$

by (10.79). Thus, $\mathscr{H}$ is a weak Liapunov function for (10.79). Therefore, by Theorem 10.2, *the zero solution* $\mathbf{p} = \mathbf{0}, \mathbf{q} = \mathbf{0}$, *a position of equilibrium, is stable when it is at a minimum of* $\mathscr{V}$.

Now, suppose the origin is a *maximum* of $\mathscr{V}$, and that $\mathscr{V}$ has the expansion

$$\mathscr{V}(\mathbf{q}) = P_N(\mathbf{q}) + P_{N+1}(\mathbf{q}) + \ldots, \quad N \geqslant 2 \tag{10.81}$$

where $P_N(q)$ is a homogeneous polynomial of degree $N \geqslant 1$ in the q_i. Consider a function $U(\mathbf{p}, \mathbf{q})$, of the type in Theorem 10.3, defined by

$$U(\mathbf{p}, \mathbf{q}) = \sum_{i=1}^{n} p_i q_i. \tag{10.82}$$

Then U satisfies conditions (i), (ii), and (iv) of Theorem 10.3. Also

$$\dot{U}(\mathbf{p}, \mathbf{q}) = \sum_{i=1}^{n} p_i \dot{q}_i + \sum_{i=1}^{n} \dot{p}_i q_i$$

$$= -\sum_{i=1}^{n} p_i \frac{\partial \mathscr{H}}{\partial p_i} + \sum_{i=1}^{n} q_i \frac{\partial \mathscr{H}}{\partial q_i}$$

$$= -\sum_{i=1}^{n} p_i \frac{\partial \mathscr{T}}{\partial p_i} + \sum_{i=1}^{n} q_i \frac{\partial \mathscr{T}}{\partial q_i} + \sum_{i=1}^{n} q_i \frac{\partial \mathscr{V}}{\partial q_i}$$

$$= -2\mathscr{T} + \sum_{i=1}^{n} \frac{\partial \mathscr{T}}{\partial q_i} + NP_N(\mathbf{q}) + (N+1)P_{N+1}(\mathbf{q}) + \dots, \tag{10.83}$$

where Euler's theorem on homogeneous functions has been applied to $\mathscr{T}, P_N, P_{N+1}, \dots$. The dominating terms in the series near $\mathbf{p} = \mathbf{q} = \mathbf{0}$ are $-2\mathscr{T} + NP_N(\mathbf{q})$. Since, by hypothesis, P_N has a maximum at $\mathbf{q} = \mathbf{0}$ and $\mathscr{T}$ is positive definite, $\dot{U}$ is negative definite in a neighbourhood of the origin. Therefore, by Theorem 10.3, *the equilibrium is unstable when it is at a maximum of* $\mathscr{V}$.

The Liénard equation. Consider the equation

$$\ddot{x} + f(x)\dot{x} + g(x) = 0, \tag{10.84}$$

or the equivalent system

$$\dot{x} = y - F(x), \quad \dot{y} = -g(x); \tag{10.85}$$

where

$$F(x) = \int_0^x f(u) \, du.$$

Suppose that f and g are continuous and that

(i) $f(x)$ is positive in a deleted neighbourhood of the origin;

(ii) $g(x)$ is positive/negative when x is positive/negative (implying $g(0) = 0$).

Now let

$$G(x) = \int_0^x g(u) \, du, \tag{10.86}$$

and consider the function

$$V(x, y) = G(x) + \tfrac{1}{2} y^2,$$

as a possible strong Liapunov function. It is clearly positive definite. Also by (10.85)

$$\dot{V}(x, y) = g(x)\dot{x} + y\dot{y} = -g(x)F(x).$$

This is negative definite. The zero solution is therefore uniformly and asymptotically stable.

Exercises

1. Find a U or V function (Theorems 10.1 to 10.3) to establish the stability or instability respectively of the zero solution of the following equations:
   (i) $\dot{x} = -x + y - xy^2$, $\dot{y} = -2x - y - x^2y$;
  (ii) $\dot{x} = y^3 + x^2y$, $\dot{y} = x^3 - xy^2$;
 (iii) $\dot{x} = 2x + y + xy$, $\dot{y} = x - 2y + x^2 + y^2$;
 (iv) $\dot{x} = -x^3 + y^4$, $\dot{y} = -y^3 + y^4$;
  (v) $\dot{x} = \sin y$, $\dot{y} = -2x - 3y$;
 (vi) $\dot{x} = x + e^{-y} - 1$, $\dot{y} = x$;
(vii) $\dot{x} = e^x - \cos y$, $\dot{y} = x$;
(viii) $\dot{x} = \sin(y + x)$, $\dot{y} = -\sin(y - x)$;
 (ix) $\ddot{x} = x^3$;
  (x) $\dot{x} = x + 4y$, $\dot{y} = -2x - 5y$;
 (xi) $\dot{x} = -x + 6y$, $\dot{y} = 4x + y$.

2. Find a strong Liapunov function at $(0,0)$ for the system

$$\dot{x} = x(y - b), \quad \dot{y} = y(x - a)$$

and confirm that all solutions starting in the domain $(x/a)^2 + (y/b)^2 < 1$ approach the origin. (This is a 'domain of asymptotic stability'.)

3. Show that the origin of the system

$$\dot{x} = xP(x, y), \quad \dot{y} = yQ(x, y)$$

is asymptotically stable when $P(x, y) < 0$, $Q(x, y) < 0$ in a neighbourhood of the origin.

4. Prove that the equations
 (i) $\ddot{x} - \dot{x}^2 \, \text{sgn}(\dot{x}) + x = 0$;
(ii) $\dot{x} = y + xy^2$, $\dot{y} = -x + x^2y$;
have an unstable zero solution.

5. Investigate the stability of the zero solution of

$$\dot{x} = x^2 - y^2, \quad \dot{y} = -2xy$$

by using the function $U(x, y) = \alpha xy^2 + \beta x^3$ for suitable constants α and β (Section 10.3).

6. Show that the origin of the system

$$\dot{x} = -y - x\sqrt{(x^2+y^2)}, \quad \dot{y} = x - y\sqrt{(x^2+y^2)}$$

is a centre in the linear approximation, but in fact is a stable spiral. Find a Liapunov function for the zero solution.

7. Euler's equations for a body spinning freely about a fixed point under no forces are

$$A\dot{\omega}_1 - (B-C)\omega_2\omega_3 = 0,$$
$$B\dot{\omega}_2 - (C-A)\omega_3\omega_1 = 0,$$
$$C\dot{\omega}_3 - (A-B)\omega_1\omega_2 = 0,$$

where A, B, and C are the principal moments of inertia, and $(\omega_1, \omega_2, \omega_3)$ is the spin of the body in principal axes fixed in the body. Find all the states of steady spin of the body.

Consider perturbations about the steady state $(\omega_0, 0, 0)$ by putting $\omega_1 = \omega_0 + x_1, \omega_2 = x_2, \omega_3 = x_3$, and show that the linear approximation is

$$\dot{x}_1 = 0, \quad \dot{x}_2 = \{(C-A)/B\}\omega_0 x_3, \quad \dot{x}_3 = \{(A-B)/C\}\omega_0 x_2.$$

Deduce that this state is unstable if $C < A < B$ or $B < A < C$.

Show that

$$V = \{B(A-B)x_2^2 + C(A-C)x_3^2\} + \{Bx_2^2 + Cx_3^2 + A(x_1^2 + 2\omega_0 x_1)\}^2$$

is a Liapunov function for the case when A is the largest moment of inertia, so that this state is stable. Suggest a Liapunov function which will establish the stability of the case in which A is the smallest moment of inertia. Are these states asymptotically stable?

Why would you expect V as given above to be a first integral of the Euler equations? Show that each of the terms in braces is such an integral.

8. Show that the zero solution of the equation

$$\ddot{x} + h(x, \dot{x})\dot{x} + x = 0$$

is stable if $h(x, y) \geqslant 0$ in a neighbourhood of the origin.

9. Use the series definition to prove the following properties of the exponential function of a matrix:
   (i) $e^{A+B} = e^A e^B$ if $AB = BA$;
   (ii) e^A is nonsingular and $(e^A)^{-1} = e^{-A}$;
   (iii) $\dfrac{d}{dt}(e^{At}) = A e^{At} = e^{At}A$;
   (iv) $(e^{At})^t = e^{A^t t}$.

10. Let the eigenvalues of the matrix A be $\lambda_1, \lambda_2, \ldots, \lambda_n$. Show that whenever $\gamma > \max_{1 \leqslant i \leqslant n} (\lambda_i)$ there exists a constant $c > 0$ such that $\|e^{At}\| < c\, e^{\gamma t}$. (Hint: e^{At} is a fundamental matrix for $\dot{\mathbf{x}} = A\mathbf{x}$; then use Theorem 8.10.)

11. Express the solution of

$$\begin{pmatrix} \dot{x} \\ \dot{y} \end{pmatrix} = \begin{pmatrix} 0 & 1 \\ 1 & 0 \end{pmatrix} \begin{pmatrix} x \\ y \end{pmatrix}$$

in matrix exponential form, and by evaluating the exponential obtain the ordinary form from it.

12. Evaluate

$$K = \int_0^\infty e^{A^t t} e^{At} \, dt$$

(eqn (10.54)) when

$$A = \frac{1}{2} \begin{pmatrix} 3 & 1 \\ 1 & 3 \end{pmatrix}$$

and confirm that

$$A^t K + K A = -I.$$

13. Show that if $A = e^B$ and C is nonsingular, then $C^{-1}AC = e^{C^{-1}BC}$.

14. (i) Let $L = diag(\lambda_1, \lambda_2, \ldots, \lambda_n)$, where λ_i are distinct and $\lambda_i \neq 0$ for any i. Show that $L = e^D$, where $D = diag(\log \lambda_1, \log \lambda_2, \ldots, \log \lambda_n)$. Deduce that for nonsingular A with distinct eigenvalues, $A = e^B$ for some B.
(ii) Show that, for the system $\dot{\mathbf{x}} = P(t)\mathbf{x}$, where $P(t)$ has period T and E (eqn (8.26)) is nonsingular with distinct eigenvalues, every fundamental matrix has the form $\Phi(t) = R(t)e^{Mt}$, where $R(t)$ has period T, and M is constant. (See the result of Exercise 13.)

15. Using Exercise 14, show that the transformation $\mathbf{x} = R(t)\mathbf{y}$ reduces the system $\dot{\mathbf{x}} = P(t)\mathbf{x}$, where $P(t)$ has period T, to the form $\dot{\mathbf{y}} = M\mathbf{y}$, where M is constant.

16. The n dimensional system

$$\dot{\mathbf{x}} = \mathbf{grad}\, W(\mathbf{x})$$

has an isolated equilibrium point at $\mathbf{x} = \mathbf{0}$. Show that the zero solution is asymptotically stable if W has a maximum at $\mathbf{x} = \mathbf{0}$. Give a condition for instability of the zero solution.

17. A particle of mass m and position vector $\mathbf{r} = (x, y, z)$ moves in a potential field $W(x, y, z)$, so that its equation of motion is

$$m\ddot{\mathbf{r}} = -\mathbf{grad}\, W.$$

By putting $\dot{x} = u$, $\dot{y} = v$, $\dot{z} = w$, express this in terms of first-order derivatives.

Suppose that W has a minimum at $\mathbf{r} = \mathbf{0}$. Show that the origin of the system is stable, by using the Liapunov function

$$V = W + \tfrac{1}{2}m(u^2 + v^2 + w^2).$$

What do the level curves of V represent physically? Is the origin asymptotically stable?

An additional non-conservative force $\mathbf{f}(u, v, w)$ is introduced, so that

$$m\ddot{\mathbf{r}} = -\mathbf{grad}\ W + \mathbf{f}.$$

Use the same Liapunov function to give a sufficient condition for $\mathbf{f}$ to be of frictional type.

18. Use the test for instability to show that if

$$\dot{x} = X(x, y), \quad \dot{y} = Y(x, y)$$

has an equilibrium point at the origin, then the zero solution is unstable if there exist constants α and β such that

$$\alpha X(x, y) + \beta Y(x, y) > 0$$

in a deleted neighbourhood of the origin.

19. Use the result of Exercise 18 to show that the origin is unstable for each of the following:
   (i) $\dot{x} = x^2 + y^2$, $\dot{y} = x + y$;
   (ii) $\dot{x} = y \sin y$, $\dot{y} = xy + x^2$;
   (iii) $\dot{x} = y^{2m}$, $\dot{y} = x^{2n}$ (m, n positive integers).

20. For the system

$$\dot{x} = y, \quad \dot{y} = f(x, y),$$

where $f(0, 0) = 0$, show that V given by

$$V(x, y) = \tfrac{1}{2}y^2 - \int_0^x f(u, 0)\, du$$

is a weak Liapunov function for the zero solution when

$$\{f(x, y) - f(x, 0)\} y \leqslant 0, \quad \int_0^x f(u, 0)\, du < 0,$$

in a neighbourhood of the origin.

21. Use the result of Exercise 20 to show the stability of the zero solutions of the following:
   (i) $\ddot{x} = -x^3 - x^2\dot{x}$;
   (ii) $\ddot{x} = -x^3/(1 - x\dot{x})$;
   (iii) $\ddot{x} = -x + x^3 - x^2\dot{x}$.

22. Let
$$\dot{x} = -\alpha x + \beta f(y), \quad \dot{y} = \gamma x - \delta f(y),$$

where $f(0) = 0$, $yf(y) > 0$ $(y \neq 0)$, and $\beta\gamma - \alpha\delta > 0$, where $\alpha, \beta, \gamma, \delta$, are positive. Show that, for suitable values of A and B,

$$V = \tfrac{1}{2}Ax^2 + B \int_0^y f(u)\,du$$

is a strong Liapunov function for the zero solutions (and hence that these are asymptotically stable).

23. A particle moving under a central attractive force $f(r)$ per unit mass has the equations of motion

$$\ddot{r} - r\dot{\theta}^2 = f(r), \quad \frac{d}{dt}(r^2\dot{\theta}) = 0.$$

For a circular orbit, $r = a$, show that $r^2\dot{\theta} = h$, a constant, and $h^2 + a^3 f(a) = 0$. The orbit is subjected to a small radial perturbation $r = a + \rho$, in which h is kept constant. Show that the variational equation for ρ is

$$\ddot{\rho} - \frac{h^2}{(a+\rho)^3} - f(a+\rho) = 0.$$

Show that

$$V(\rho, \dot{\rho}) = \tfrac{1}{2}\dot{\rho}^2 + \frac{h^2}{2(a+\rho)^2} - \int_0^\rho f(a+u)\,du - \frac{h^2}{2a^2}$$

is a Liapunov function for the zero solution of this equation provided that $3h^2 > a^4 f'(a)$, and that the gravitational orbit is stable in this sense.

24. Show that
$$V(x, y) = \max\,(|X|, |Y|)$$

is a Liapunov function for the zero solution of the system $\dot{x} = X(x, y)$, $\dot{y} = Y(x, y)$, provided that $X_x < 0$, $Y_y < 0$, $|X_x| > |X_y|$, and $|Y_y| > |Y_x|$ in a neighbourhood of the origin.

25. Show that the following Liénard-type equations have zero solutions which are asymptotically stable:

(i) $\ddot{x} + |x|(\dot{x} + x) = 0$;

(ii) $\ddot{x} + \dfrac{\sin x}{x}\dot{x} + x^3 = 0$;

(iii) $\dot{x} = y - x^3$, $\dot{y} = -x^3$.

26. Give a geometrical account of Theorem 10.3.

27. For the system
$$\dot{x} = f(x) + \beta y, \quad \dot{y} = \gamma x + \delta y,$$

establish that V given by

$$V(x, y) = (\delta x - \beta y)^2 + 2\delta \int_0^x f(u)\,du - \beta \gamma x^2$$

is a strong Liapunov function for the zero solution when, in some neighbourhood of the origin,

$$\delta \frac{f(x)}{x} - \beta \gamma > 0, \quad \frac{f(x)}{x} + \delta < 0$$

for $x \neq 0$. (Barbashin, 1970.)

Deduce that for initial conditions in the circle $x^2 + y^2 < 1$, the solutions of the system

$$\dot{x} = -x^3 + x^4 + y, \quad \dot{y} = -x,$$

tend to zero.

28. For the system

$$\dot{x} = f(x) + \beta y, \quad \dot{y} = g(x) + \delta y,$$

show that V given by

$$V(x, y) = (\delta x - \beta y)^2 + 2 \int_0^x \{\delta f(u) - \beta g(u)\}\,du$$

is a strong Liapunov function for the zero solution when, in some neighbourhood of the origin,

$$\{\delta f(x) - \beta g(x)\}x > 0, \quad xf(x) + \delta x^2 < 0$$

for $x \neq 0$. (Barbashin, 1970.)

Deduce that the zero solution of the system

$$\dot{x} = -x^3 + 2x^4 + y, \quad \dot{y} = -x^4 - y$$

is asymptotically stable. Show how to find a domain of initial conditions from which the solutions tend to the origin.

29. Consider van der Pol's equation, $\ddot{x} + \varepsilon(x^2 - 1)\dot{x} + x = 0$, for $\varepsilon < 0$, in the Liénard phase plane, eqn (10.85):

$$\dot{x} = y - \varepsilon(\tfrac{1}{3}x^3 - x), \quad \dot{y} = -x.$$

Show that, in this plane, $V = \tfrac{1}{2}(x^2 + y^2)$ is a strong Liapunov function for the zero solution, which is therefore asymptotically stable. Show that all solutions starting from initial conditions inside the circle $x^2 + y^2 = 3$ tend to the origin (and hence the limit cycle lies outside this region for every $\varepsilon < 0$). Sketch this 'domain of asymptotic stability' in the ordinary phase plane with $\dot{x} = y$.

30. If the zero solution of an autonomous system is stable, and every solution, whatever the initial conditions, tends to the origin, the system is said to be *globally asymptotically stable*. Give an intuitive argument, based on the general reasoning of Theorem 10.2, to show that the system is globally asymptotically stable if, additionally, $V \to \infty$ as $\|\mathbf{x}\| \to \infty$.

Show that the system

$$\dot{x} = -xy^2, \quad \dot{y} = -x^2 y$$

is globally asymptotically stable, by guessing a suitable Liapunov function.

31. (See Exercise 30.) Assuming that the conditions of Exercise 27 are satisfied, obtain further conditions which ensure that the system is globally asymptotically stable.

Show that the system

$$\dot{x} = y - x^3, \quad \dot{y} = -x - y$$

is globally asymptotically stable.

32. (See Exercise 30.) Assuming that the conditions of Exercise 28 are satisfied, obtain further conditions which ensure that the system is globally asymptotically stable.

Show that the system

$$\dot{x} = -x^3 - x + y, \quad \dot{y} = -x^3 - y$$

is globally asymptotically stable.

33. Give conditions on the functions f and g of the Liénard equation, $\ddot{x} + f(x)\dot{x} + g(x) = 0$ which ensure that the corresponding system $\dot{x} = y - F(x)$, $\dot{y} = -g(x)$ (Section 10.7) is globally asymptotically stable (Exercise 30).

Show that all solutions of the equation $\ddot{x} + x^2\dot{x} + x^3 = 0$ tend to zero.

34. Consider the system

$$\dot{x} = \gamma y - \delta x - \beta y(x^2 + y^2)$$
$$\dot{y} = -\gamma x - \delta y + \beta x(x^2 + y^2) - 1.$$

Let (x_0, y_0) be an equilibrium point of the system and set $x = x_0 + \xi$, $y = y_0 + \eta$, where

$$x_0 = r\cos\theta, \quad y_0 = r\sin\theta, \quad \xi = \rho\cos(\theta + \psi), \quad \eta = \rho\sin(\theta + \psi).$$

Let

$$V = \tfrac{1}{2}\beta r^2 \rho^2 (1 + 2\cos^2\psi - \gamma/\beta r^2) + \beta r\rho^3 \cos\psi + \tfrac{1}{4}\beta\rho^4.$$

Show that

$$\dot{V} = -\delta\beta r^2 \rho^2 (1 + 2\cos^2\psi - \gamma/\beta r^2) - 3\delta\beta r\rho^3 \cos\psi - \delta\beta\rho^4.$$

Deduce that V and $\dot{V}$ are definite with opposite sign near the equilibrium point if

(i) $\delta > 0, \gamma < \beta r^2$;

(ii) $\delta > 0, \gamma > 3\beta r^2$.

(With suitable scale changes this Liapunov function V can be used to establish a domain of asymptotic stability for the equilibrium points in the van der Pol plane for Duffing's equation: see Exercise 4, Chapter 7.)

35. (Zubov's method.) Suppose that a function $W(x, y)$, negative definite in the whole plane, is chosen as the time derivative $\dot{V}$ of a possible Liapunov function for a system $\dot{x} = X(x, y), \dot{y} = Y(x, y)$, for which the origin is an asymptotically stable equilibrium point. Show that $V(x, y)$ satisfies the linear partial differential equation

$$X\frac{\partial V}{\partial x} + Y\frac{\partial V}{\partial y} = W$$

with $V(0,0) = 0$.

Show also that for the path $x(t), y(t)$ starting at (x_0, y_0) at time t_0

$$V\{x(t), y(t)\} - V(x_0, y_0) = \int_{t_0}^{t} W\{x(t), y(t)\}\, dt.$$

Deduce that the boundary of the domain of asymptotic stability (the domain of initial conditions from which the solutions go into the origin) is the set of points (x, y) for which $V(x, y)$ is infinite, by considering the behaviour of the integral as $t \to \infty$, firstly when (x_0, y_0) is inside this domain and then when it is outside. (Therefore the solution $V(x, y)$ of the partial differential equation above could be used to give the boundary of the domain directly. However, solving this equation is equivalent in difficulty to finding the paths: the characteristics are in fact the paths themselves.)

36. For the system

$$\dot{x} = X(x, y) = -\tfrac{1}{2}x(1 - x^2)(1 - y^2), \quad \dot{y} = Y(x, y) = -\tfrac{1}{2}y(1 - x^2)(1 - y^2)$$

show that the Liapunov function $V = x^2 + y^2$ leads to $\dot{V} = -(x^2 + y^2)(1 - x^2)(1 - y^2)$ and explain why the domain of asymptotic stability (see Exercise 35) contains at least the unit circle $x^2 + y^2 = 1$.

Alternatively, start with $\dot{V} = -(x^2 + y^2)$ and obtain V from the equation

$$X\frac{\partial V}{\partial x} + Y\frac{\partial V}{\partial y} = \dot{V}, \quad V(0,0) = 0$$

(see Exercise 35). It is sufficient to verify that $V = \log\{(1 - x^2)(1 - y^2)\}$. Explain why the square $|x| < 1, |y| < 1$ is the complete domain of asymptotic stability for the zero solution.

11 The existence of periodic solutions

SUPPOSE THAT the phase diagram for a differential equation contains a single, unstable equilibrium point and a limit cycle surrounding it, as in the case of the van der Pol equation. Then in practice all initial states lead to the periodic oscillation represented by the limit cycle. In such cases the limit cycle is the principal feature of the system from the practical point of view, and it is desirable to be able to decide with certainty whether it is there or not. Hitherto our attitude to this question has been intuitive; we assemble qualitative evidence that there is a limit cycle, from energy considerations or geometrical arguments, then attempt to estimate the radius by the methods of Chapters 4, 5, and 7, a definite result being taken as further confirmation that the limit cycle is really there. The present chapter contains theorems and methods for positively proving the existence of limit cycles and centres for certain types of equation. The cases chosen can be interpreted physically as involving a balance between energy gain and loss on various regions in the phase plane; they

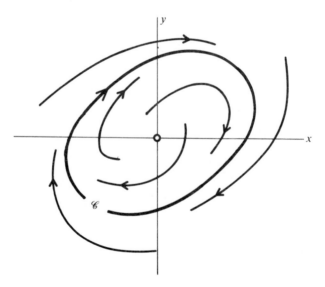

FIG. 11.1. Idealized limit cycle

include the type discussed in Sections 1.5 and 4.1; also others with different symmetry. The latter serve as a guide to intuition for related types which do not quite fit the conditions of the theorems. Further theorems can be found, for example, in Andronov *et al.* (1973) and Cesari (1971).

11.1. The Poincaré-Bendixson Theorem

We shall be concerned with second-order autonomous differential equations. A limit cycle is an isolated periodic motion which appears as an isolated closed curve in the phase plane. We take Fig. 11.1 as showing the general nature of·a limit cycle; neighbouring paths resemble spirals which approach, or recede from, the limit cycle, though the appearance of actual limit cycles may be somewhat different (see, for example, Fig. 3.23).

In Section 3.4 we gave criteria whereby it can sometimes be shown that *no* limit cycle can exist in a certain region, and also a necessary condition for a limit cycle (the index requirement), but we have no tests giving *sufficient* conditions for existence. We first give without proof a plausible theorem on which several of the results of this chapter are based. For the proof see Cesari (1971) and Andronov *et al.* (1973).

THEOREM 11.1 (*The Poincaré–Bendixson Theorem.*) *Let $\mathscr{R}$ be a closed, bounded region consisting of nonsingular points of a 2×2 system $\dot{\mathbf{x}} = \mathbf{X}(\mathbf{x})$ such that some positive half-path $\mathscr{H}$ of the system lies entirely within $\mathscr{R}$. Then either $\mathscr{H}$ is itself a closed path, or it approaches a closed path, or it terminates at an equilibrium point.*

If we can isolate a region from which some path cannot escape, the theorem describes what may happen to it: roughly speaking, given some control of its movement through regularity conditions on the differential

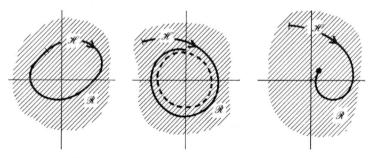

FIG. 11.2. Possible behaviour of half-paths restricted to a bounded region

equations, it cannot wander about at random for ever. The possibilities are illustrated in Fig. 11.2.

The theorem implies, in particular, that if $\mathscr{R}$ contains no equilibrium points, and some $\mathscr{H}$ remains in $\mathscr{R}$, then $\mathscr{R}$ must contain a periodic solution. The theorem can be used in the following way. Suppose (Fig. 11.3) that we can find two closed curves, $\mathscr{C}_1$ and $\mathscr{C}_2$, with $\mathscr{C}_2$ inside $\mathscr{C}_1$, such that *all* paths crossing $\mathscr{C}_1$ point towards its interior, and *all* paths crossing $\mathscr{C}_2$ point outwards from it. Then no path which enters the annular region between them can ever escape again. The annulus is therefore a region $\mathscr{R}$ for the theorem. If, further, we can arrange that $\mathscr{R}$ has no equilibrium points in it, then the theorem predicts at least one closed path $\mathscr{L}$ somewhere in $\mathscr{R}$. Evidently $\mathscr{L}$ must wrap round the inner curve as shown, for its index is 1 and it must therefore have an equilibrium point interior to it, and $\mathscr{R}$ contains no equilibrium points. For the same reason, there must exist suitable equilibrium points interior to $\mathscr{C}_2$ for all this to be possible.

The same result is true if paths are all outward across $\mathscr{C}_1$ and inward across $\mathscr{C}_2$ (by reversing the time variable).

The practical difficulty is in finding, for a given system, a suitable $\mathscr{C}_1$ and $\mathscr{C}_2$ to substantiate a belief in the existence of a limit cycle. There is a

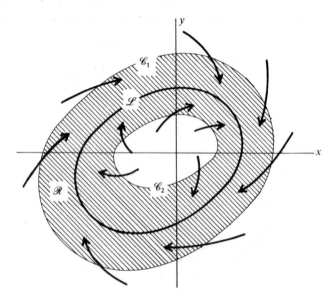

Fig. 11.3.

relation between this process and that of finding Liapunov functions (Chapter 10).

Example 11.1 *Show that the system*

$$\dot{x} = x - y - (x^2 + \tfrac{3}{2}y^2)x, \quad \dot{y} = x + y - (x^2 + \tfrac{1}{2}y^2)y$$

has a periodic solution.

We shall try to find two circles, centred on the origin, with the required properties. In Fig. 11.4, $\mathbf{n} = (x, y)$ is a normal, pointing outward at P from the circle radius r, and $\mathbf{X} = (X, Y)$ is in the direction of the path through P. Also $\cos\phi = \mathbf{n}.\mathbf{X}/|\mathbf{n}||\mathbf{X}|$ and therefore $\mathbf{n}.\mathbf{X}$ is positive or negative according to whether $\mathbf{X}$ is pointing outwards or inwards. We have

$$\begin{aligned}
\mathbf{n}.\mathbf{X} = xX + yY &= x^2 + y^2 - x^4 - \tfrac{1}{2}y^4 - \tfrac{5}{2}x^2y^2 \\
&= r^2 - r^4 + \tfrac{1}{2}y^2(y^2 - x^2) \\
&= r^2 - r^4(1 + \tfrac{1}{4}\cos 2\theta - \tfrac{1}{4}\cos^2 2\theta). \quad (11.1)
\end{aligned}$$

When, for example, $r = \tfrac{1}{2}$, this is positive for all θ and all paths are directed outwards on this circle, and when $r = 2$ it is negative, with all paths directed inwards. Therefore, somewhere between $r = \tfrac{1}{2}$ and $r = 2$ there is at least one closed path, or a periodic solution.

We can look for the pair of circles which pin down *the narrowest annular region in which we can predict that a closed path exists*. The interval of r in which a closed path might lie is that for which, given a value of r, $\mathbf{X}$ points inward on some points

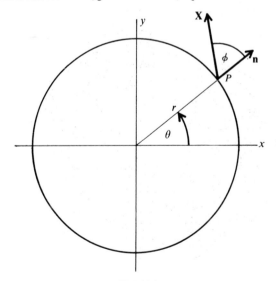

Fɪɢ. 11.4.

of the circle, outward on others; that is to say, values of r in this region give circles to which **X** is tangential at some point ('cycles with contact': Andronov *et al.*, 1973). For such values of r the equation

$$r^2 - r^4(1 + \tfrac{1}{4}\cos 2\theta - \tfrac{1}{4}\cos^2 2\theta) = 0$$

has a solution (r, θ). Write this in the form

$$4\left(\frac{1}{r^2} - 1\right) = \cos 2\theta - \cos^2 2\theta.$$

The range of the right-hand side is $(-2, \tfrac{1}{4})$, and solutions will exist for values of the left-hand side in this range, that is, when

$$\frac{4}{\sqrt{17}} \leqslant r \leqslant \sqrt{2}, \quad \text{or} \quad 0 \cdot 97 \leqslant r \leqslant 1 \cdot 41.$$

Another point of view, closely resembling that of the Liapunov functions of Chapter 10, is the following. Let

$$v(x, y) = c > 0, \quad c_1 < c < c_2,$$

be a 'band' of closed curves for some range of c including $c = c_0$, the outer curves corresponding to larger c (Fig. 11.5). Then if $\mathscr{C}_0$ is the curve $v(x, y) = c_0$,

$$\left[\frac{dv}{dt}\right]_{\mathscr{C}_0} = \dot{v}(x, y) = X\frac{\partial v}{\partial x} + Y\frac{\partial v}{\partial y}.$$

If this is positive/negative, the paths pass outward/inward on $\mathscr{C}_0$.

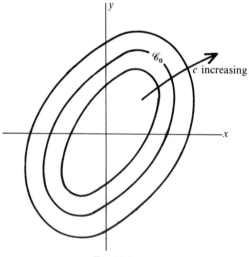

Fig. 11.5.

Example 11.2 *In the van der Pol plane a, b the equations for the response amplitudes $a(t), b(t)$ of the forced van der Pol equation are (see eqn-(7.34))*

$$\dot{a} = \tfrac{1}{2}\varepsilon(1 - \tfrac{1}{4}r^2)a - \frac{\omega^2 - 1}{2\omega}b, \quad \dot{b} = \frac{\omega^2 - 1}{2\omega}a + \tfrac{1}{2}\varepsilon(1 - \tfrac{1}{4}r^2)b + \frac{\Gamma}{2\omega}, \quad (\varepsilon > 0)$$

where $r = \sqrt{(a^2 + b^2)}$. Show that in case III (Section 7.5), where the system has a single unstable node or spiral, there is a closed path.

We firstly want a closed curve round the equilibrium point P such that the paths cross it outwards. If we reverse the time direction P becomes stable, and we have shown in Theorem 10.6 how to construct a Liapunov function $V(x, y)$ for this case, V being positive definite and $\dot{V}$ negative definite sufficiently close to the origin. Obviously $V(x, y) = a$ *small constant* gives a closed curve of the required type. (It is unnecessary to write it down.)

Now consider the directions over a large circle, of radius r_0:

$$v(a, b) = a^2 + b^2 = r_0^2.$$

The paths pass inwards if $\dot{v}(a, b) = 0$ on this circle. Now

$$\dot{v}(a, b) = 2(a\dot{a} + b\dot{b}) = 2\varepsilon(1 - \tfrac{1}{4}r_0^2)r_0^2 + \frac{\Gamma}{\omega}b,$$

and for r_0 large enough, this is negative.

Therefore there is a closed path. There is, in fact, a single limit cycle, though we cannot prove this here.

It is generally very difficult to find curves with the required properties. It may be possible, however, to construct supporting arguments which are not rigorous, as in the following example.

Example 11.3 *Show that the system*

$$\dot{x} = y, \quad \dot{y} = -4x - 5y + \frac{6y}{1 + x^2}$$

has a periodic solution.

These equations are a simplified version of equations for a tuned-grid vacuum-tube circuit (Andronov and Chaikin, 1949). The only equilibrium point on the finite plane is at $x = 0$, $y = 0$. Near the origin, the linear approximation is $\dot{x} = y$, $\dot{y} = -4x + y$, an unstable spiral. As in the last example, the existence of a Liapunov function for the stable spiral guarantees that there is in the present case a closed curve surrounding the origin over which the paths point outward.

To find what happens at a great distance use, for example, the mapping $z = 1/x, v = y/x$ (eqn (3.15)); the system becomes

$$\dot{z} = -vz, \quad \dot{v} = -4 - 5v - v^2 + 6vz^2/(z^2 + 1).$$

The equilibrium points on the horizon $z = 0$ are at $v = -1$ and $v = -4$ and these are unstable. Viewed in the spherical projection of Section 3.3 the picture is

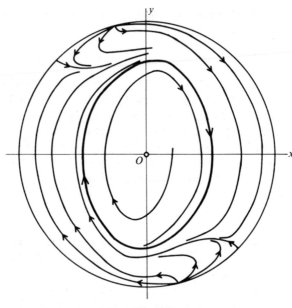

FIG. 11.6.

as in Fig. 11.6, giving a reasonable assurance of a periodic solution, shown by the heavy line in the figure.

The Bendixson principle can be employed to obtain theorems covering broad types of differential equation, of which the following is an example.

THEOREM 11.2   *The differential equation*

$$\ddot{x} + f(x, \dot{x})\dot{x} + g(x) = 0, \tag{11.2}$$

(*the Liénard equation*), *or the equivalent system*

$$\dot{x} = y, \quad \dot{y} = -f(x, y)y - g(x),$$

where f and g are continuous, has at least one periodic solution under the following conditions:
(i) *there exists a > 0 such that $f(x, y) > 0$ when $\sqrt{(x^2 + y^2)} > a$;*
(ii) *$f(0, 0) < 0$ (hence $f(x, y) < 0$ in a neighbourhood of the origin);*
(iii) *$g(0) = 0$, $g(x) > 0$ when $x > 0$, and $g(x) < 0$ when $x < 0$;*
(iv) *$G(x) = \displaystyle\int_0^x g(u)\,du \to \infty$ as $x \to \infty$.*

Proof   (iii) implies that there is a single equilibrium point, at the origin.
Consider the function

$$\mathcal{E}(x, y) = \tfrac{1}{2}y^2 + G(x). \tag{11.3}$$

This represents the energy of the system (potential plus kinetic) when it is
regarded, say, as representing a spring–particle system with dissipative
external forces. Clearly, $G(0) = 0$, $G(x) > 0$ when $x \neq 0$, and G is
monotonic increasing to infinity (by (iv)); and is continuous. Therefore
$\mathcal{E}(0, 0) = 0$, and $\mathcal{E}(x, y) > 0$ for $x \neq 0$ or $y \neq 0$ ($\mathcal{E}$ is positive definite).
Also $\mathcal{E}$ is continuous and increases monotonically in every radial
direction from the origin. Therefore (Fig. 11.7) the family of contours
of $\mathcal{E}$ with parameter $c > 0$:

$$\mathcal{E}(x, y) = c \tag{11.4}$$

consists of simple closed curves encircling the origin. As c tends to zero
they approach the origin and as $c \to \infty$ they become infinitely remote
(the principal consequence of (iv)).

We can choose c, $c = c_1$, small enough for the corresponding contour,
$\mathcal{C}_1$, to be entirely within the neighbourhood of the origin where, by (ii),
$f(x, y) < 0$. We will examine a half-path $\mathcal{H}$ starting at a point on $\mathcal{C}_1$.

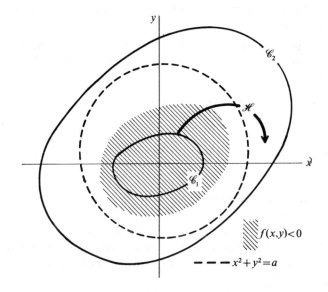

Fig. 11.7.

Consider $\dot{\mathscr{E}}(x, y)$ in $\mathscr{H}$:

$$\dot{\mathscr{E}}(x, y) = g(x)\dot{x} + y\dot{y} = g(x)y + y\{-f(x, y)y - g(x)\}$$
$$= -y^2 f(x, y). \quad (11.5)$$

This is positive, except at $y = 0$, on $\mathscr{C}_1$. Choose $\mathscr{H}$ to start at a point other than $y = 0$ on $\mathscr{C}_1$. Then it leaves $\mathscr{C}_1$ in the outward direction. It is clear that it can never reappear inside $\mathscr{C}_1$, since to do so it must cross some interior contours in the inward direction, which is impossible since, by (11.5), $\dot{\mathscr{E}} \geqslant 0$ on all contours near to $\mathscr{C}_1$, as well as on $\mathscr{C}_1$.

Now consider a contour $\mathscr{C}_2$ for large c, $c = c_2$ say. $\mathscr{C}_2$ can be chosen, by (iv), to lie entirely outside the circle $x^2 + y^2 = a^2$, so that, by (i), $f(x, y) > 0$ on $\mathscr{C}_2$. By (11.5), with $f(x, y) > 0$, all paths crossing $\mathscr{C}_2$ cross inwardly, or are tangential (at $y = 0$), and by a similar argument to the above, no positive half-path, once inside $\mathscr{C}_2$, can escape.

Therefore $\mathscr{H}$ remains in the region bounded by $\mathscr{C}_1$ and $\mathscr{C}_2$, and by Theorem 11.1, there is a periodic solution in this region.

The theorem can be interpreted as in Section 1.5: near to the origin the 'damping' coefficient, f, is negative, and we expect paths to spiral outwards due to intake of energy. Further out, $f(x, y) > 0$, so there is loss of energy and paths spiral inwards. Between the families we expect a closed path.

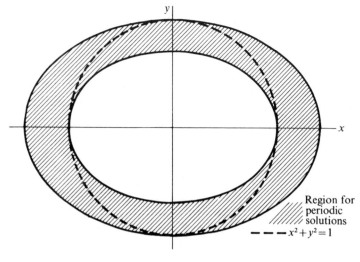

Region for periodic solutions

$- - - - x^2 + y^2 = 1$

Fig. 11.8.

Example 11.4 *Show that the equation* $\ddot{x}+e(x^2+\dot{x}^2-1)\dot{x}+x^3 = 0$ *has a limit cycle, and locate it between two curves* $\mathcal{E}(x, y) = \text{constant}$.

For the theorem, $f(x, y) = x^2+y^2-1$, $g(x) = x^3$, $G(x) = \frac{1}{4}x^4$. Therefore (11.4) gives the contours of $\mathcal{E}$: $\frac{1}{4}x^4+\frac{1}{2}y^2 = c$. The periodic solution located by the theorem lies between two such contours, one inside, the other outside, of the curve $f(x, y) = 0$, or $x^2+y^2 = 1$, and is most closely fixed by finding the smallest contour lying outside this circle and the largest lying inside. We require respectively min/max of x^2+y^2 subject to $\frac{1}{4}x^4+\frac{1}{2}y^2 = c$, c being then chosen so that the min/max is equal to 1. The calculation by means of a Lagrange multiplier gives $c = \frac{1}{2}$ and $c = \frac{1}{4}$ respectively. (See Fig. 11.8.)

11.2. A theorem on the existence of a centre

The following theorem involves ingredients different from those previously considered. (For a theorem of similar form but with less restrictive conditions see Minorsky, 1962, p. 113.)

THEOREM 11.3 *The origin is a centre for the equation*

$$\ddot{x}+f(x)\dot{x}+g(x) = 0,$$

or for the equivalent system

$$\dot{x} = y, \quad \dot{y} = -f(x)y-g(x), \tag{11.6}$$

when, in some neighbourhood of the origin, f and g are continuous, and
(i) $f(x)$ *is odd, and of one sign in the half-plane* $x > 0$;
(ii) $g(x) > 0$, $x > 0$, *and $g(x)$ is odd (implying $g(0) = 0$);*
(iii) $g(x) > \alpha f(x)F(x)$ *for* $x > 0$, *where* $F(x) = \displaystyle\int_0^x f(u)\,du$, *and* $\alpha > 1$.

Proof We may assume that $f(x) > 0$ when $x > 0$, since the other case, $f(x) < 0$, reduces to this, putting $-t$ for t. The equation may describe a particle-spring system, with positive damping (loss of energy) for $x > 0$, and negative damping (gain in energy) for $x < 0$. Since f and g are odd, the paths are symmetrical about the y axis (put $x = -z$, $t = -\tau$ into (11.6)). As in Theorem 11.2, (ii) ensures that there is a single equilibrium point, at the origin. Figure 11.9 shows possible types of closed path satisfying these conditions: we have to exclude possibilities such as (b).

Let

$$\mathcal{E}(x, y) = \tfrac{1}{2}y^2+G(x), \tag{11.7}$$

where

$$G(x) = \int_0^x g(u)\,du.$$

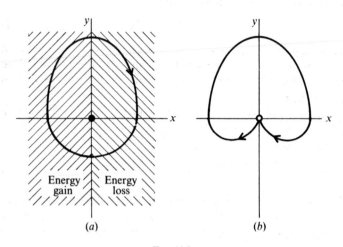

FIG. 11.9.

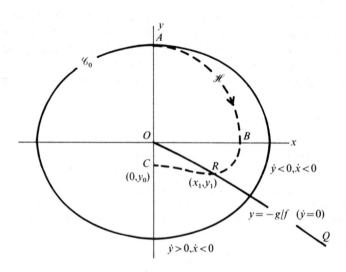

FIG. 11.10.

As in Theorem 11.2, the family of contours

$$\mathscr{E}(x, y) = c > 0, \tag{11.8}$$

where c is a parameter, define a family of closed curves about the origin (but this time only in some neighbourhood of the origin where the conditions hold, or for small enough c), and as $c \to 0$ the closed curves approach the origin.

Let $\mathscr{C}_0$ be an arbitrary member of the family (11.8) in the prescribed neighbourhood of the origin, and consider the half-path $\mathscr{H}$ starting at A on the intersection of $\mathscr{C}_0$ with the y axis (Fig. 11.10). On $\mathscr{H}$,

$$\dot{\mathscr{E}}(x, y) = -y^2 f(x). \tag{11.9}$$

While $y > 0$, $\dot{x} > 0$, so $\mathscr{H}$ constantly advances to the right in the first quadrant. Also, by (11.9), $\dot{\mathscr{E}}(x, y) < 0$, so y moves continuously to curves with smaller $\mathscr{E}$. Since $\mathscr{H}$ remains within $\mathscr{C}_0$ it leaves the quadrant at some point B on the positive x axis, inside $\mathscr{C}_0$.

On the axis $y = 0$, $x > 0$ we have, by (ii),

$$\dot{x} = 0, \quad \dot{y} = -g(x) < 0.$$

Therefore all paths cut the positive x axis vertically downward, and so cannot re-enter the first quadrant from the fourth. Moreover, since $\dot{\mathscr{E}}(x, y) < 0$ in the fourth quadrant, $\mathscr{H}$ continues to lie inside $\mathscr{C}_0$; therefore it must either enter the origin (the only equilibrium point), or cross the y axis at some point C for which $y < 0$.

We shall show that it does not enter the origin. In the fourth quadrant,

$$\dot{x} < 0 \tag{11.10}$$

but

$$\dot{y} = -f(x)y - g(x), \tag{11.11}$$

which is sometimes negative and sometimes positive. We can deal with (11.11) as follows (Fig. 11.10). OQ is the curve on which $\dot{y} = 0$ (and $dy/dx = 0$), dividing the region where $\dot{y} > 0$ from that where $\dot{y} < 0$ (eqn (11.11)). Assume initially that this curve approaches the origin, that is, that

$$\lim_{x \to 0} \{ -g(x)/f(x) \} = 0. \tag{11.12}$$

For this case, $\mathscr{H}$ turns over in the manner indicated; is horizontal at R, and has negative slope up to C, where $dy/dx = 0$ again. We have to show that $y_0 < 0$. On RC

$$y_1 - y_0 = \int_0^{x_1} \frac{dy}{dx} dx.$$

Therefore,

$$y_0 = y_1 + \int_0^{x_1} \{f(x) + y^{-1}g(x)\} \, dx \qquad \text{(by (11.6))},$$

$$= -\frac{g(x_1)}{f(x_1)} + F(x_1) + \int_0^{x_1} y^{-1}g(x)\,dx \quad \text{(since } F(0) = 0),$$

$$< -\frac{g(x_1)}{f(x_1)} + F(x_1) \quad \text{(since } g(x) > 0, \, y \leqslant 0 \text{ on } 0 \leqslant x \leqslant x_1),$$

$$< -\alpha F(x_1) + F(x_1) < 0 \qquad \text{(by (iii))}.$$

Therefore $y_0 < 0$, and $ABRC$ is half of a closed path which is its own reflection in the y axis.

Now return to the possibility that the curve $y = -g(x)/f(x)$ does not pass through the origin. Since $g(x) > 0$ and $f(x) > 0$ when $x > 0$, it lies entirely below the x axis, and Fig. 11.11 shows how $\mathcal{H}$ must in this case also remain below the x axis, and therefore cannot enter the origin.

Finally, since the argument is independent of A near enough to the origin, the origin is a centre.

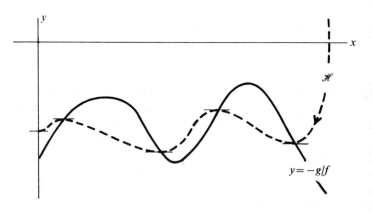

FIG. 11.11.

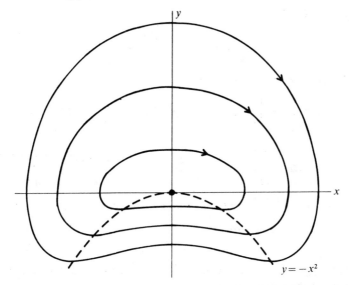

FIG. 11.12. Paths for $\ddot{x} + x\dot{x} + x^3 = 0$

Example 11.5 *The equation $\ddot{x} + x\dot{x} + x^3 = 0$ has a centre at the origin.*
Conditions (i) and (ii) are satisfied. Also

$$f(x)F(x) = \tfrac{1}{2}x^3.$$

Therefore

$$g(x) > 2f(x)F(x),$$

so (iv) is satisfied with $\alpha = 2$ (see Fig. 11.12).

11.3. A theorem on the existence of a limit cycle

We consider the equation

$$\ddot{x} + f(x)\dot{x} + g(x) = 0, \tag{11.13}$$

where, broadly speaking, $f(x)$ is positive when $|x|$ is large, and negative when $|x|$ is small, and where g is such that, in the absence of the damping term $f(x)\dot{x}$ we expect periodic solutions for small x. Van der Pol's equation, $\ddot{x} + e(x^2 - 1)\dot{x} + x = 0$, $e > 0$, is of this type. Effectively the theorem demonstrates a pattern of expanding and contracting spirals about a limit cycle. Paths far from the origin spend part of their time in regions of energy input and part in regions of energy loss, so the physical argument for a limit cycle is less compelling than in the case of Theorem 11.2.

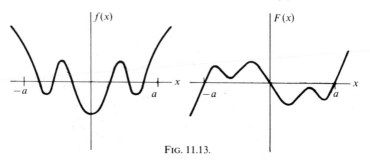

FIG. 11.13.

The proof is carried out on a different phase plane from the usual one

$$\dot{x} = y - F(x), \quad \dot{y} = -g(x), \tag{11.14}$$

(the Liénard plane) where

$$F(x) = \int_0^x f(u)\,du.$$

The use of this plane enables the shape of the paths to be simplified (under the conditions of Theorem 11.4, $\dot{y} = 0$ only on $x = 0$) without losing symmetry, and, additionally, allows the burden of the conditions to rest on F rather than f, f being thereby less restricted.

THEOREM 11.4   *The equation* $\ddot{x} + f(x)\dot{x} + g(x) = 0$ *has a unique periodic solution if f and g are continuous, and*
(i) $F(x)$ *is an odd function;*
(ii) $F(x)$ *is zero only at $x = 0$, $x = a$, $x = -a$, for some $a > 0$;*
(iii) $F(x) \to \infty$ *as $x \to \infty$ monotonically for $x > a$;*
(iv) $g(x)$ *is an odd function, and $g(x) > 0$ for $x > 0$.*
(These conditions imply that $f(x)$ is even, $f(0) < 0$ and $f(x) > 0$ for $x > a$.

Proof   The general permitted shape of $f(x)$ and $F(x)$ is shown in Fig. 11.13.

   The general pattern of the paths can be obtained from the following considerations.
(a) If $x(t)$, $y(t)$ is a solution, so is $-x(t)$, $-y(t)$ (since F and g are odd); therefore the phase diagram is symmetrical about the origin (but not necessarily the individual phase paths).
(b) The slope of a path is given by

$$\frac{dy}{dx} = \frac{-g(x)}{y - F(x)}.$$

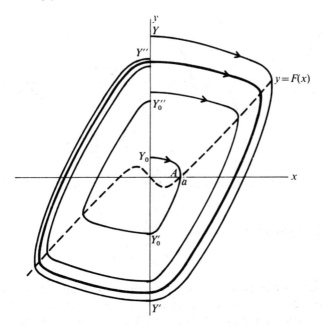

FIG. 11.14.

so the paths are horizontal only on $x = 0$ (from (iv)), and are vertical only on the curve $y = F(x)$. Above $y = F(x)$, $\dot{x} > 0$, and below, $\dot{x} < 0$. (c) $\dot{y} < 0$ for $x > 0$, and $\dot{y} > 0$ for $x < 0$, by (iv).

Some typical paths are shown in Fig. 11.14. A path $YY'Y''$ is closed if and only if Y and Y'' coincide. The symmetry condition (a) implies that this is the case if and only if

$$OY = OY'. \tag{11.15}$$

We prove the Theorem by showing that for the path $Y_0Y_0'Y_0''$ through A, $(x = a)$,

$$OY_0 - OY_0' < 0,$$

and that as Y recedes to infinity along the positive y axis

$$OY - OY' \to \infty$$

monotonically. There is therefore exactly one point Y for which $OY - OY'$ is zero, and this identifies the (single) closed path. Thus we are really confirming a pattern of spirals converging on a limit cycle.

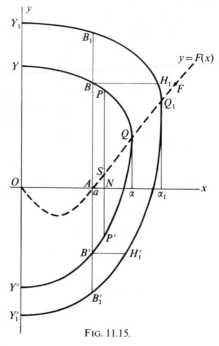

FIG. 11.15.

Now let

$$v(x, y) = \int_0^x g(u)\,du + \tfrac{1}{2}y^2. \tag{11.16}$$

Then

$$OY - OY' = 0 \quad \Leftrightarrow \quad V_{YY'} = 0, \tag{11.17}$$

where we write for a typical path in Fig. 11.15

$$V_{YY'} = v_{Y'} - v_Y = \oint_{YQY'} dv.$$

Along an element of the path we have

$$dv = y\,dy + g\,dx = y\,dy + g\frac{dx}{dy}dy$$

$$= y\,dy + g\frac{y - F}{-g}dy \quad \text{(from (11.14))}$$

$$= F\,dy. \tag{11.18}$$

Let BB' (through $x = a$) separate the parts of the path where $F(x)$ is positive and where it is negative. We consider these parts separately in estimating $V_{YY'}$:

$$V_{YY'} = V_{YB} + V_{BB'} + V_{B'Y'}. \tag{11.19}$$

The proof is carried out through the steps (A) and (D) below.

(A) *As Q moves out from A along AF, $V_{YB} + V_{B'Y'}$ is positive and monotonic decreasing.*

Choose any fixed path YQY', with Q at the point $(\alpha, F(\alpha))$ as shown, and another, $Y_1Q_1Y_1'$, with Q at the point $(\alpha_1, F(\alpha_1))$, where $\alpha_1 > \alpha$.

By (11.14), since $y > 0$ and $F(x) < 0$ on YB, Y_1B_1,

$$0 < \left(\frac{\mathrm{d}y(x)}{\mathrm{d}x}\right)_{YB} < \left(\frac{\mathrm{d}y(x)}{\mathrm{d}x}\right)_{Y_1B_1} \tag{11.20}$$

for every x. Similarly, since $y < F(x)$ (from (b)) and $F(x) < 0$ on $B'Y'$, $B_1'Y_1'$,

$$0 < \left(\frac{\mathrm{d}y(x)}{\mathrm{d}x}\right)_{B_1'Y_1'} < \left(\frac{\mathrm{d}y(x)}{\mathrm{d}x}\right)_{B'Y'} \tag{11.21}$$

(That is to say, Y_1B_1 and $B_1'Y_1'$ are shallower than YB and $B'Y'$ respectively.) Therefore from (11.18), with $F(x) < 0$,

$$V_{YB} = \oint_{YB} F\,\mathrm{d}y = \oint_{YB} (-F)\left(-\frac{\mathrm{d}y}{\mathrm{d}x}\right)\mathrm{d}x > \oint_{Y_1B_1} (-F)\left(-\frac{\mathrm{d}y}{\mathrm{d}x}\right)\mathrm{d}x$$

$$= V_{Y_1B_1} > 0 \quad (11.22)$$

using (11.20). Similarly,

$$V_{B'Y'} > V_{B_1'Y_1'} > 0. \tag{11.23}$$

The result (A) follows from (11.21) and (11.23).

(B) *As Q moves out from A along AF, $V_{BB'}$ is monotonic decreasing.*

Choose BQB' arbitrarily to the right of A and let $B_1Q_1B_1'$ be another path with Q_1 to the right of Q, as before. Then $F(x) > 0$ on these paths, and we write

$$V_{B_1B_1'} = -V_{B_1'B_1} - \oint_{B_1'B_1} F(x)\,\mathrm{d}y \leqslant - \oint_{H_1'H_1} F(x)\,\mathrm{d}y,$$

(where BH_1, $B'H_1'$ are parallel to the x axis),

$$\leqslant - \oint_{B'B} F(x)\,\mathrm{d}y = V_{BB'}, \tag{11.24}$$

(since, for the same values of y, $F(x)$ on $B'B$ is less than or equal to $F(x)$ on $H'_1 H_1$).

(C) *From (A) and (B) we deduce that $V_{YY'}$ is monotonic decreasing to the right of A.*

(D) $V_{BB'}$ *tends to $-\infty$ as the paths recede to infinity.*

Let S be a point on the line $y = F(x)$, to the right of A, and let BQB' be an arbitrary path with Q to the right of S. The line $PSNP'$ is parallel to the y axis. Then, as before,

$$V_{BB'} = - \oint_{B'B} F(x)\,\mathrm{d}y \leqslant \oint_{P'P} F(x)\,\mathrm{d}y.$$

Also

$$F(x) \geqslant NS \quad \text{on} \quad PP'$$

since $F(x)$ is monotonic increasing by (iii); therefore

$$V_{BB'} \leqslant -NS \oint_{P'P} \mathrm{d}y = -NS \cdot PP' \leqslant -NS \cdot NP. \qquad (11.25)$$

But as Q goes to infinity towards the right, $NP \to \infty$.

(E) *From (C) and (D), $V_{YY'}$ is monotonic decreasing to $-\infty$, to the right of A.*

(F) $V_{YY} > 0$ *when Q is at A or to the left of A.* (For then $F(x) < 0$ and $\mathrm{d}y < 0$.)

From (E) and (F), there is one and only one path for which $V_{YY'} = 0$, and this, by eqn (11.17) and the symmetry of the paths, is closed.

Example 11.6 *The van der Pol equation $\ddot{x} + e(x^2 - 1)\dot{x} + x = 0$, $e > 0$, has a unique limit cycle. (The case $e < 0$ is exactly similar (put $-t$ for t). Also note that this proof holds even when e is not small.)*

Here, $f(x) = e(x^2 - 1)$, $F(x) = e(\tfrac{1}{3}x^3 - x)$. The conditions of the theorem are satisfied, with $a = \sqrt{3}$. The x-extremities of the limit cycle must be beyond $x = \pm\sqrt{3}$.

Example 11.7 *Show that the equation*

$$\ddot{x} + \frac{x^2 + |x| - 1}{x^2 - |x| + 1}\dot{x} + x^3 = 0$$

has a unique periodic solution.

$f(x)$ has the shape shown in Fig. 11.16. Since $f(x) \to 1$ as $x \to \infty$, (iii) is

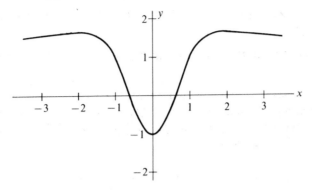

FIG. 11.16.

satisfied, and (i), (ii) and (iv) are obviously satisfied. Therefore a limit cycle exists, its extreme x-values being beyond the non-zero roots of $F(x) = 0$.

11.4. Van der Pol's equation with large parameter

Van der Pol's equation with large parameter e:

$$\ddot{x} + e(x^2 - 1)\dot{x} + x = 0$$

is equivalent (by putting $\delta = 1/e$) to the equation

$$\delta\ddot{x} + (x^2 - 1)\dot{x} + \delta x = 0 \qquad (11.26)$$

with δ small. This can be regarded as a singular perturbation problem in terms of Chapter 6, in which the 'weight' of the various terms is differently disposed on different parts of the solution curves, or on the phase paths. The existence of a limit cycle is confirmed in Example 11.6. Here we shall look briefly at the analytical construction of the limit cycle of (11.26) for small δ in the usual phase plane with $\dot{x} = y$, and not the Liénard plane. It is necessary to piece together the limit cycle from several approximate solutions corresponding to different balances between the terms. The calculations are made for $y > 0$; the phase path is then completed by invoking its symmetry in the origin.

The equation for the phase path is

$$\frac{dy}{dx} = -\frac{x^2 - 1}{\delta} - \frac{x}{y}, \qquad (11.27)$$

and the isocline for zero slope is the curve $y = \delta x/(1 - x^2)$.

In eqn (11.26) put $t = \mu\tau$, so that

$$\frac{\delta}{\mu^2} x'' + \frac{1}{\mu}(x^2 - 1)x' + \delta x = 0. \tag{11.28}$$

When $\mu = \delta$ the first two terms predominate, and to the first order

$$x'' + (x^2 - 1)x' = 0. \tag{11.29}$$

If $x = a$ (the amplitude) when $y = 0$, then integration of (11.29) gives

$$x' = x - \tfrac{1}{3}x^3 - a + \tfrac{1}{3}a^3, \tag{11.30}$$

or

$$x' = \tfrac{1}{3}(x - a)(3 - x^2 - ax - a^2).$$

If $a > 1$, then $x' > 0$ (in the upper half of the phase plane) on

$$\tfrac{1}{2}\{-a + \sqrt{(12 - 3a^2)}\} < x < a,$$

and (11.30) represents the phase path in this interval: the amplitude a will be determined later.

When we put $\mu = \delta^{-1}$ in (11.28), the second and third terms predominate and are of the same order, so that

$$x' = x/(1 - x^2), \tag{11.31}$$

which is essentially the zero-slope isocline of (11.27). This equation can only be valid for $y > 0$ and $-a < x < -1$ because of the singularity at $x = -1$.

The two approximate solutions given by (11.30) and (11.31) must be connected by a third equation in the neighbourhood of $x = -1$. To the required order the solution given by (11.30) must pass through $y = 0$, $x = -1$: whence

$$\tfrac{1}{2}\{-a + \sqrt{(12 - 3a^2)}\} = -1, \quad \text{or} \quad a = 2.$$

For the transition, put $z = x + 1$; then (11.28) becomes

$$\frac{\delta}{\mu^2} z'' + \frac{1}{\mu}\{(z - 1)^2 - 1\}z' + \delta z - \delta = 0.$$

Now put $\mu = \delta^\alpha$, and let $z = \delta^\beta u$; then a balance is achieved between the terms containing z'', z', z and the constant if

$$1 - 2\alpha + \beta = -\alpha + 2\beta = 1;$$

that is, if $\alpha = \tfrac{1}{3}$ and $\beta = \tfrac{2}{3}$. The differential equation for the transition region is

$$u'' - 2uu' - 1 = 0, \tag{11.32}$$

an equation without an elementary solution, relating u' and u.

There remains the solution around $x = -2$ to be constructed. Put $z = x + 2$ in (11.28):

$$\frac{\delta}{\mu^2} z'' + \frac{1}{\mu} \{(z-2)^2 - 1\} z' + \delta z - 2\delta = 0.$$

Again, put $\mu = \delta^\alpha$ and $z = \delta^\beta u$. The equation

$$u'' + 3u' = 2 \tag{11.33}$$

follows if $1 - 2\alpha + \beta = \beta - \alpha = 1$, or if $\alpha = 1$, $\beta = 2$. From (11.33)

$$\tfrac{1}{3}u' + \tfrac{2}{9}\log(2 - 3u') + u = \text{constant}.$$

When $u = 0$, $u' = 0$, so finally

$$\tfrac{1}{3}u' + \tfrac{2}{9}\log(2 - 3u') + u = \tfrac{2}{9}\log 2.$$

After restoring the original variables, x and t, the sections of the limit cycle on the phase plane for $y > 0$ are as follows:

(i) $-2 \leqslant x < -2 + O(\delta^2)$:

$$\tfrac{1}{3}\delta y + \tfrac{2}{9}\delta^2 \log(1 - \tfrac{3}{2}\delta^{-1}y) = -(x+2);$$

(ii) $-2 + O(\delta^2) < x < -1 - O(\delta^{2/3})$:

$$y = \delta x/(1 - x^2);$$

(iii) $-1 - O(\delta^{2/3}) < x < -1 + O(\delta^{2/3})$:
the appropriate solution of

$$\delta \ddot{x} - 2(x+1)\dot{x} = \delta;$$

(iv) $-1 + O(\delta^{2/3}) < x \leqslant 2$:

$$\delta y = x - \tfrac{1}{3}x^3 + \tfrac{2}{3}.$$

The complete limit cycle can now be constructed since it is symmetrical about the origin. Figure 11.17(a) shows the computed phase diagram for $\delta = 0\cdot1$, and the corresponding solution is shown in Fig. 11.17(b).

The van der Pol equation for large parameter e (or small δ) is an example of a *relaxation oscillation*, in which typically, as Fig. 11.17(b) shows, the system displays a slow build-up followed by a sudden discharge, repeated periodically.

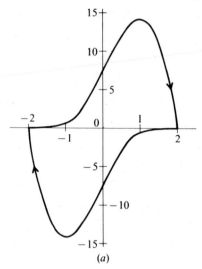

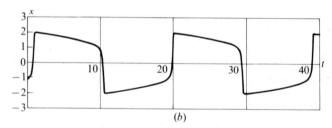

Fig. 11.17. (a) Limit cycle for van der Pol's equation (11.26) with $\delta = 1/e = 0{\cdot}1$; (b) solution curve corresponding to the limit cycle

Exercises

1. Consider the system

$$\dot{x} = y + P(x, y), \quad \dot{y} = -x + Q(x, y),$$

where P and Q are of order $x^2 + y^2$. Show that the origin is a centre if either
(i) P is odd in x, Q is even in x; or
(ii) P is even in y, Q is odd in y.
   Deduce that the origin is a centre for the system

$$\dot{x} = y + cx(x^2 + y^2), \quad \dot{y} = -x + cy(x^2 + y^2).$$

2. A system has exactly one equilibrium point, n limit-cycles, and no other periodic solutions. Explain why a stable limit-cycle must be adjacent to unstable

limit-cycles, but an unstable limit-cycle may have stable or unstable cycles adjacent to it.

Let c_n be the number of possible configurations, with respect to stability, of n nested limit-cycles. Show that $c_1 = 2, c_2 = 3, c_3 = 5$ and that in general

$$c_n = c_{n-1} + c_{n-2}.$$

(This recurrence relation generates the Fibonacci sequence.) Deduce that

$$c_n = \{(2+\sqrt{5})(1+\sqrt{5})^{n-1} + (-2+\sqrt{5})(1-\sqrt{5})^{n-1}\}/(2^{n-1}\sqrt{5}).$$

3. By considering the directions in which the paths cross a suitable system of closed curves, show that the system

$$\dot{x} = -y - x + (x^2 + 2y^2)x, \quad \dot{y} = x - y + (x^2 + 2y^2)y$$

has at least one periodic solution. Use the variant of Bendixson's Theorem given in Exercise 30, Chapter 3, to deduce that there is exactly one periodic solution.

4. Show that the equation $\ddot{x} + e(x^2 + \dot{x}^2 - 1)\dot{x} + x^3 = 0$ has at least one periodic solution.

5. Show that the origin is a centre for the equations
(i) $\ddot{x} - x\dot{x} + x = 0$;
(ii) $\ddot{x} + x\dot{x} + \sin x = 0$.

6. Suppose that $f(x)$ in the equation $\ddot{x} + f(x)\dot{x} + x = 0$ satisfies the conditions of Theorem 11.3. Show that the equivalent linear equation (Section 4.5) with respect to $x = a\cos t$ is $\ddot{x} + x = 0$.

7. Show that the equation

$$\ddot{x} + e(x^2 - 1)\dot{x} + \tanh kx = 0$$

has exactly one periodic solution when $k > 0$. Decide on its stability.

The 'restoring force' resembles a step-function when k is large. Is the conclusion the same when it is exactly a step-function?

8. Show that

$$\ddot{x} + e(x^2 - 1)\dot{x} + x^3 = 0$$

has exactly one periodic solution.

9. Show that

$$\ddot{x} + (|x| + |y| - 1)\dot{x} + x|x| = 0$$

has at least one periodic solution.

10. Show that the origin is a centre for the equation

$$\ddot{x} + (k\dot{x} + 1)\sin x = 0.$$

11. Analyse the equation

$$\varepsilon\ddot{x}+(x^4-1)\dot{x}+\varepsilon x = 0$$

for small ε by the singular perturbation technique outlined in Section 11.4.

12. Let F and g be functions satisfying the conditions of Theorem 11.4. Show that the equation

$$\ddot{u}+F(\dot{u})+u = 0$$

has a unique periodic solution (put $\dot{u} = -z$). Deduce that Rayleigh's equation $\ddot{u}+e(\frac{1}{3}\dot{u}^3-\dot{u})+u = 0$ has a unique limit cycle.

13. Explain why the equation

$$\ddot{x}+e(x^2+\dot{x}^2-1)\dot{x}+x = 0,$$

unlike the van der Pol equation, does not have a relaxation oscillation for large positive e.

14. For the van der Pol oscillator

$$\delta\ddot{x}+(x^2-1)\dot{x}+\delta x = 0$$

for small positive δ, use the formula for the period, eqn (1.11), to show that the period of the limit cycle is approximately $(3-2\ln 2)\delta^{-1}$. (Hint: the principal contribution arises from that part of the limit cycle given by (ii) in Section 11.4.)

Appendix A: Existence and uniqueness theorems

We state, without proof, certain theorems of frequent application in the text.

THEOREM A1   *For the nth order system*

$$\dot{\mathbf{x}} = \mathbf{f}(\mathbf{x}, t) \tag{A1}$$

suppose that $\mathbf{f}$ *is continuous and that* $\partial f_j/\partial x_i$, $i, j = 1, 2, \ldots, n$ *are continuous for* $\mathbf{x} \in \mathcal{D}$, $t \in I$, *where* $\mathcal{D}$ *is a domain and* I *is an open interval. Then if* $\mathbf{x}_0 \in \mathcal{D}$ *and* $t_0 \in I$, *there exists a solution* $x^*(t)$, *defined uniquely in some neighbourhood of* $(\mathbf{x}_0, t_0)$, *which satisfies* $\mathbf{x}^*(t_0) = \mathbf{x}_0$.

We call such a system *regular* on $\mathcal{D} \times I$ (the set $(\mathbf{x}, t)$ where $\mathbf{x} \in \mathcal{D}$ and $t \in I$). For brevity, we refer in the text to a system regular on $-\infty < x_i < \infty$, $i = 1, 2, \ldots, n$, $-\infty < t < \infty$ as being a *regular system*. Points at which the conditions of the theorem apply are called *ordinary points*.

The theorem states only that the initial value problem has a unique solution on some sufficiently small interval about t_0, a limitation arising from the method of proof. A question of importance is how far the solution actually extends. It need not extend through the whole of I, and some possibilities are shown in the following examples.

Example A1 $\dot{x}_1 = x_2, \dot{x}_2 = -x_1$, $x_1(0) = 0, x_2(0) = 1$.

$\mathcal{D}$ is the domain $-\infty < x_1 < \infty, -\infty < x_2 < \infty$, and I is $-\infty < t < \infty$. The solution is $x_1^*(t) = \sin t$, $x_2^*(t) = \cos t$. This solution is defined and is unique on $-\infty < t < \infty$.

Example A2   *The one-dimensional equation* $\dot{x} = 3x^2 t^2$, $x(0) = 1$.
$\mathcal{D}$ is $-\infty < x < \infty$, and I is $-\infty < t < \infty$. But

$$x^*(t) = (1 - t^3)^{-1},$$

so the solution is only valid in $-\infty < t < 1$, since $x^*(t)$ approaches infinity as $\to 1-$.

Example A3   *The one-dimensional equation* $\dot{x} = 2|x|^{1/2}, x(0) = 1$.
A family of solutions is $x(t) = (t + c)^2, t > -c$ (so that $\dot{x} > 0$); there also exists the solution $x(t) \equiv 0$. The required solution is shown in Fig. A1, made up from the appropriate parabolic arc together with part of the zero solution (the derivative is continuous at A). Here, $\mathcal{D}$ is the domain $x > 0$, I is $-\infty < t < \infty$. The solution in fact leaves $\mathcal{D} \times I$ at the point A.

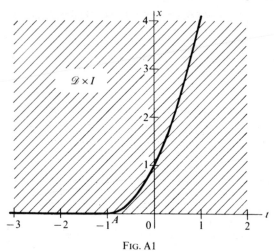

Fig. A1

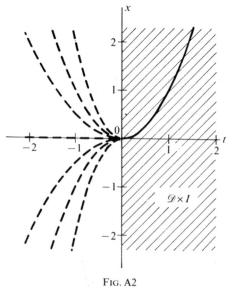

Fig. A2

Example A4   *The one-dimensional equation* $\dot{x} = x/t, x(1) = 1$.

$\mathcal{D}$ is $-\infty < x < \infty$, and I is $t > 0$. The general solution is $x = ct$, and the required solution is $x = t$. The solution leaves $\mathcal{D} \times I$ at $(0,0)$. Despite the singularity at $t = 0$, the solution continues uniquely in this case into $-\infty < t < \infty$.

Example A5   *The one-dimensional equation* $\dot{x} = 2x/t, x(1) = 1$.

$\mathcal{D}$ is $-\infty < x < \infty$, and I is $t > 0$. The general solution is $x = ct^2$, and the required solution is $x = t^2$, $t > 0$ (t remaining within I). However, as Fig. A2 shows, there are many continuations of the solution (having a continuous derivative), but uniqueness breaks down at $t = 0$.

For a regular system we shall define I_{max} (the maximal interval of existence of the solution in Theorem A1) as the *largest interval of existence of a unique solution* $\mathbf{x}^*$. I_{max} can be shown to be an open interval.

THEOREM A2   *Under the conditions of Theorem A1, either* $(x^*(t), t)$ *approaches the boundary of* $\mathcal{D} \times I$, *or* $x^*(t)$ *becomes unbounded as t varies within I.* (For example, the possibility that as $t \to \infty$, $x^*(t)$ approaches a value in the interior of $\mathcal{D}$ is excluded.)

In Example A2, I_{max} is $-\infty < t < 1$, and as $t \to 1-$, $x^*(t)$ becomes unbounded. In Example A3, $x^*(t) \in \mathcal{D}$ only for $-1 < t < \infty$, and the solution approaches the boundary of $\mathcal{D} \times I$ as $t \to -1+$, at the point $(-1, 0)$. In Example A5, the boundary at $(0, 0)$ is approached.

The following theorem shows that *linear systems* have particularly predictable behaviour.

THEOREM A3   *Let* $\dot{\mathbf{x}} = A(t)\mathbf{x} + \mathbf{h}(t), \mathbf{x}(t_0) = \mathbf{x}_0$, *where A and* **h** *are continuous on* $I: t_1 < t < t_2$ (t_1 *may be* $-\infty$, *and* t_2 *may be* ∞). *Then* $I \subseteq I_{max}$ *for all* $t_0 \in I$, *and for all* $\mathbf{x}_0$.

In this case the unique continuation of the solutions is guaranteed at least throughout the greatest interval in which $A, \mathbf{h}$ are continuous.

Appendix B: Hints and answers to the exercises

Chapter 1
1. (i) Paths $y = kx + C$.
    (ii) Paths $y = 4x^2 + C$.
    (iii) Paths $y = C$, $(|x| \leqslant 1)$; $\frac{1}{2}y^2 = kx + C$, $(|x| > 1)$.
    (iv) Stable node.
    (v) Unstable spiral.
    (vi) $y > 0$: stable node; $y < 0$: unstable node; joined along $y = 0$.
    (viii) Paths $y^2 + x^2 \operatorname{sgn}(x) = C$.
2. $\alpha > 0$, centre at $(0, 0)$; $\alpha < 0$, centre at $(0, 0)$ and saddles at $(\pm\sqrt{(-\alpha^{-1})}, 0)$.
8. Paths $\frac{1}{2}y^2 - x^2 = C$. Step lengths of equal time intervals δt are $\delta s \simeq \delta t\sqrt{(6x^2 + 2C)}$.
9. Constant increments in the polar angle give constant time intervals on all paths (all periodic solutions have the same period).
11. Centre at $x = 0$, saddle points at $x = \pm\sqrt{6}$. For the exact equation, the saddle points are at $x = n\pi$.
12. An impulse-excited oscillation: the sudden injection of energy in each cycle offsets the damping. Note that the amplitude is independent of α.
13. $E/2F_0$.
14. $ma^2\ddot{\theta} = mga \sin\theta + F_0 \operatorname{sgn}(\dot\theta - \Omega)$. Two equilibrium states, at $\sin^{-1}(F_0/mga)$ and $\pi - \sin^{-1}(F_0/mga)$, when $F_0 < mga$; one, at $\frac{1}{2}\pi$, if $F_0 = mga$; none if $F_0 > mga$.
15. Centre at $x = -1$, centre/saddle point at $x = 0$, saddle point at $x = 1$.
17. Centre at $\theta = 0$, saddle points at $\theta = \pm\pi$, when $\lambda \geqslant 1$. Centres at $\theta = \cos^{-1}\lambda$, saddle points at $\theta = 0, \pm\pi$, when $0 < \lambda < 1$.
18. For $x = \lambda$, the equilibrium points are stable for $0 < \lambda \leqslant 1$, unstable for $\lambda \leqslant 0$ and $\lambda > 1$. For $x = \sqrt{\lambda}$ the equilibrium points are stable for $\lambda \geqslant 1$ and unstable for $0 \leqslant \lambda < 1$. For $x = -\sqrt{\lambda}$, all are unstable.
19. $\omega^2 < g$: origin is a centre. $\omega^2 = g$: every x is an equilibrium point. $\omega^2 > g$: the origin is a saddle point (note that $y = \pm\sqrt{(\omega^2 - g)}$ are phase paths).
20. $|x| \leqslant a$. Paths $y^2 + (x - a)^2 = C$ for $x > a$; $y = C$ for $|x| \leqslant a$; $y^2 + (x + a)^2 = C$ for $x < -a$.
23. $[2E\,e^{-2\pi/3\sqrt{3}}/(1 - e^{-4\pi/\sqrt{3}})]^{1/2}$.
24. Take initial conditions $x = 0$, $\dot{x} = v_0$, at $t = 0$. Cycle ends with $\dot{x} = V$, where $V^2 = v^2(1 - e^{-2\pi/\sqrt{3}}) + v_0^2 e^{-2\pi/\sqrt{3}}$. Then show $v_0 < V < v$ when $v_0 < v$, and $v < V < v_0$ when $v_0 > v$.
29. $-2\lambda x[\sqrt{(a^2 + x^2)} - a]/\sqrt{(a^2 + x^2)} = m\ddot{x}$.

32. $\theta = 0$: centre if $c < mg(h-a)^3/h$, saddle point if $c > mg(h-a)^3/h$. $\theta = \pi$: centre if $c > mg(h+a)^3/h$, saddle point if $c < mg(h+a)^3/h$. $\theta = \cos^{-1}[a^2+h^2 -(ch/mg)^{2/3}/2ah]$ is an equilibrium point if $mg(h+a)^3/h > c > mg(h-a)^3/h$, a centre.

35. $z = \int[\exp\int g(x)\,dx]\,dx$.

Chapter 2

1. (i) $(-\frac{1}{2},\frac{1}{2})$; $(2x+1)^2 - 2(2x+1)(2y-1) - (2y-1)^2 = \text{constant}$.
  (ii) None; $y - x - \log|x+y+1| = \text{constant}$.
  (iii) $(0,0)$; $x^4 - y^4 = \text{constant}$.
  (iv) $((m+\frac{1}{2})\pi, n\pi)$, $m = 0, \pm 1, \ldots, n = 0, \pm 1, \ldots$; $\sin x + \cos y = \text{constant}$.
  (v) $(a,0)$ for real a; $(y^3 + x + \frac{1}{3})e^{-3x} = \text{constant}$.

2. (i) $(4,4)$, unstable spiral; $(-1,-1)$, stable spiral.
  (ii) $(0,2)$, stable spiral; $(1,0)$, saddle point.
  (iii) $(0,2)$, centre; $(0,-2)$, centre; $(1,0)$, saddle point; $(-1,0)$ saddle point.
  (iv) $(0,n\pi)$, $n = 0, \pm 1, \ldots$; n even, saddle point; n odd, centre.
  (v) $(0,0)$ saddle point.

(If the linear approximation is a centre, no conclusion can be reached about the stability of the equilibrium point of the nonlinear system: see e.g. Exercise 3.)

12. Let $D = a_1c_2 - a_2b_1$. There are 3 equilibrium points when $D > 0$: at $(H,P) = (0,0)$, saddle point; $(a_1/b_1,0)$, saddle point, and $(a_2/c_2, D/c_1c_2)$, stable node if $D > a_2b_1^2/4c_2$, stable spiral if $D < a_2b_1^2/4c_2$. If $D < 0$ there are two significant points: $(0,0)$, saddle point; $(a_1/b_1,0)$, stable node. P disappears if $D < 0$, and approaches D/c_1c_2 if $D > 0$.

13. $(0,0)$, $(a_1/b_1,0)$, $(a_1b_2 - c_1a_2, b_1a_2 + c_2a_1)/(b_1b_2 + c_1c_2)$, $(0, a_2/b_2)$.

14. One equilibrium point at $(a_1c_2/a_2c_1, a_1/c_1)$; a stable node if $a_2 > 4a_1$, a stable spiral if $a_2 < 4a_1$.

17. $v = 0$ and any θ; the roots of $(\gamma+1)v - \gamma\sqrt{(2v)} + (\gamma-1)c = 0$ and the corresponding θ.

18. $u = (h^2/\gamma)^{(\alpha-3)-1}$, $\alpha \neq 3$; none if $\alpha = 3$, except the trivial case. A centre if $\alpha < 3$, a saddle point if $\alpha > 3$. The gravitational orbit, $\alpha = 2$, is stable in this sense.

21. At $(0,0)$ and $x = \pm 1$, any y. Phase paths are the circles and parts of circles $x^2 + y^2 = C$, $|x| < 1$.

22. (i) $P = \frac{1}{2}mg$ (ii) $P(0) = \frac{1}{2}mg, P(\frac{1}{2}\pi) = mg$.

23. $\theta = 0, \pi, \pm\cos^{-1}(1/4)$.

33. (i) $\partial P/\partial y < 0$, $\partial Q/\partial x < 0$. (ii) $P(x,0) > 0$ for $x < a$; $P(x,0) < 0$ for $x > a$, $Q(0,y) > 0$ for $y < b$, $Q(0,y) < 0$ for $y > b$. (iii) There exists $K > 0$ such that $P(x,y) \leq 0$ and $Q(x,y) \leq 0$ if $x \geq K$ or $y \geq K$.

35. Equilibrium point at $x = a\sqrt{m_1}/(\sqrt{m_1} + \sqrt{m_2})$.

36. Write the system in the form $\dot{x} = -\sigma x + g(y)$, $\dot{y} = g(x) - \sigma y$, and sketch the curves $y = g(x)/\sigma$, $x = g(y)/\sigma$, looking for the symmetry.

Chapter 3

1. (i) 0; (ii) 0; (iii) 1; (iv) 1; (v) -2.

2. Eq. pts at $\theta = n\pi$, index -1 if n odd, 1 if n even.

3. (i) -2; (ii) -2; (iii) 1, -1.

4. The sense in which Γ is described is the key to the result.

5. $a^2 < 4\lambda$, no eq. pts $a^2 = 4\lambda$, eq. pt at $x = \frac{1}{2}a$, index 0. $a^2 > 4\lambda$, eq. pt at $x = \frac{1}{2}a + \frac{1}{2}\sqrt{(a^2 - 4\lambda)}$, index -1, and at $x = \frac{1}{2}a - \frac{1}{2}\sqrt{(a^2 - 4\lambda)}$, index 1.

8. Write as $(X + iY)e^{-i\alpha}$.

12. Spiral at $(0,0)$, index 1; saddle point at $(1,1)$, index -1. Index at infinity must be 2.

13. The paths are almost parallel in a small region.

14. Let $\theta(s)$ be the angle between $\mathbf{X}_1$ and $\mathbf{X}_2$ on Γ, s being a parameter for the curve, $\alpha \leqslant s \leqslant \beta$. Since $|\theta(s)| < \pi$, continuity gives $\theta(\alpha) = \theta(\beta)$, which is essentially the result.

15. In the notation of Exercise 14 put $\mathbf{X}_1 = (y, -x), \mathbf{X}_2 = (y, -\sin x)$. On any closed curve containing $(0,0)$, $\mathbf{X}_1$ and $\mathbf{X}_2$ have the same first component.

17. See Theorem 3.3.

28. Put $\rho = x^3$. Also, $x(t) \equiv 0$ is a solution.

29. (i) No eq. pt; (ii) only eq. pt is on the path $y = 0$; (iii) no eq. pt; (iv) no eq. pt; (v) no eq. pt; (vi) only eq. pt is on path $y = 0$; (vii) Bendixson's test.

31. Yes.

32. $\text{div}(\mathbf{X})$ is of one sign for $r < \sqrt{2}$.

34. $a \approx 2$.

37. For any periodic solution, $\int_{\mathscr{D}} \text{div} \, \mathbf{X}_1 \, dx \, dy = 0$, where $\mathscr{D}$ is the interior of the corresponding path for the system $\dot{\mathbf{x}} = \mathbf{X}_0$. This may be satisfied for all such $\mathscr{D}$, in which case the origin remains a centre with paths close to those of the original system. There may be no such $\mathscr{D}$ giving the integral as zero, or there may be one or more corresponding to limit cycles.

For stability we require paths to approach the limit cycles from both sides. Since, for any $\mathscr{B}$, $\int_{\mathscr{B}} \text{div} \, \mathbf{X} \, dx \, dy = \int_{\mathscr{C}} \mathbf{X} \cdot \mathbf{n} \, ds$, where $\mathbf{n}$ is the outward-pointing unit normal, intuition suggests that a sufficient condition for stability is that

$$\varepsilon \int_{\mathscr{B}} \text{div} \, \mathbf{X}_1 \, dx \, dy > 0 \quad \text{or} \quad < 0$$

according as $\mathscr{D}$ is contained in $\mathscr{B}$, or $\mathscr{B}$ is contained in $\mathscr{D}$, respectively.

The example is a variant of van der Pol's equation. When $\varepsilon = 0$ the paths are $\frac{1}{2}x^2 + y^2 = c$. When $\varepsilon \neq 0$, the limit cycle is close to $c = \sqrt{2}$.

38. Use Bendixson's Negative Criterion.

39. $x_1^2 + x_2^2 = $ constant, $x_3 = $ constant are closed.

42. If $ax + by + c = 0$ contains the points it intersects each of the conics $X(x, y) = 0$, $Y(x, y) = 0$ at three points. Hence X and Y have the common factor $ax + by + c$.

43. Compare Exercise 42: since X and Y are quadratic and relatively prime there can be no more than two points of tangency of any straight line to the path.

Chapter 4

1. (i) $x^2 + y^2 = 1$, stable. (ii) $x^2 + y^2 = 1$, unstable.
2. $r = r_k$ stable if $f'(r_k^2) < 0$, unstable if $f'(r_k^2) > 0$, either possible if $f'(r_k^2) = 0$.
3. (i) $a = 1$, stable.
   (ii) $a = 2$, stable.
   (iii) $a = 2^{3/4}$, stable.
   (iv) $a = \sqrt{(n\pi)}$, $n = 1, 2, \ldots$: n even, stable; n odd, unstable.
   (v) $a = 3\pi/4$, stable.
   (vi) $a = 2$, stable.
   (vii) $a = 2\sqrt{3}$, stable.
4. Stable. $a(\theta) = 2/\{1 - (1 - 4/a^2(0)) e^{2\theta}\}^{1/2}$.
5. Stable. $a(\theta) = \frac{3}{4}\pi/\{1 - (1 - 3\pi/4a(0)) e^{\frac{1}{2}\theta}\}$.
7. $a = 2$, $T = 2\pi(1 + \frac{3}{2}\varepsilon)$.
9. $a = 0 \cdot 197$.
12. $\ddot{x} + \varepsilon(a^2 - 1)\dot{x} + x = 0$.
13. Amplitude 8/3, frequency $\sqrt{(19/3)}$.
17. Saddle at $x = 0$, centres at $\pm 1/\sqrt{\alpha}$; the separatrix cuts the x axis at $\sqrt{(2/\alpha)}$. The periodic solutions found embrace all these points.
18. The 'restoring force' is unsymmetrical: effectively, a bias is present, displacing the mean displacement. Centre at $x = 0$, saddle at $x = 1/\alpha$ with separatrix cutting the x axis at $x = -1/2\alpha$ (paths $x^2 - y^2 - \frac{2}{3}\alpha x^3 = C$); periodic solutions confined to $-1/2\alpha < x < 1/2\alpha$.
20. Compare answer to 18.

Chapter 5

1. $\Omega = \pm 1$: none. $\Omega \neq \pm 1 : \Gamma \cos t/(\Omega^2 - 1)$. Also for $\Omega \neq \pm 1$ $(p, q$ integers):
   $\Omega = p \neq \pm 1$, $A \cos pt + B \sin pt + \Gamma \cos t/(\Omega^2 - 1)$, period 2π;
   $\Omega = p/q \neq \pm 1$, $A \cos (pt/q) + B \sin(pt/q) + \Gamma \cos t/(\Omega^2 - 1)$, period $2\pi q$.
2. (i) $-\cos t + \cos 3t$.
   (ii) $3 \cdot 2 \cos t - 0 \cdot 027 \cos 3t$.
   (iii) $2 \cos t + 1 \cdot 2 \sin t - 0 \cdot 024 \sin 3t$.
4. (i) $x = a \cos t + \varepsilon a^2 (\frac{1}{6} \sin 2t - \frac{1}{3} \sin t)$.
   (ii) $x = a \cos t + \varepsilon a^2 (\frac{1}{6} \sin 2t - \frac{1}{3} \sin t)$.
7. $x = \dfrac{\Gamma + \varepsilon a_1}{\Omega^2 - 1} \cos t + \dfrac{\varepsilon a_3}{\Omega^2 - 9} \cos 3t$.
8. $x = 2\sqrt{(\beta/3)} \cos(3t - \alpha) + \frac{1}{8}\varepsilon\gamma \cos t$.
13. $|x|/a \ll 1$, $\lambda l/2m\omega^2 a \ll 1$. (Note: a dimensionless form is required.)
15. $x = -\dfrac{1}{8}\left\{\Gamma + \frac{1}{8}\varepsilon\left(\dfrac{3\Gamma^3}{256} - \beta\Gamma\right)\right\} \cos 3t$.

17. $x = -\dfrac{\varepsilon\Gamma^2}{2\Omega^2(\Omega^2-1)^2} + \dfrac{\Gamma}{\Omega^2-1}\cos t - \dfrac{\varepsilon\Gamma^2}{2(\Omega^2-1)^2(\Omega^2-4)}\cos 2t + \ldots$

19. $x = a_0\cos 3t + b_0\sin 3t + \frac18\Gamma\cos t - \frac1{18}\varepsilon(a_0^2+b_0^2) - \frac{\varepsilon}{40}\Gamma(a_0\cos 2t + b_0\sin 2t)$
$+ \ldots$

22. $u = k(e\cos\theta+1) + \varepsilon k^3\{e\theta\sin\theta + \frac12 e^2 + 1 - (\frac13 e^2+1)\cos\theta - \frac16 e^2\cos 2\theta\}$.

Chapter 6

2. The same to $O(a^6)$.

4. $x = au$, $\mu = \varepsilon a^2$.

9. Uniform solution $e^1(e^{-x} - e^{-x/\varepsilon})$.

10. $y_0 = e^{\frac12(1-x^2)}$; $y_1 = e^{1/2}(1-e^{-x/\varepsilon})$; $y = e^{1/2}(e^{-\frac12 x^2} - e^{-x/\varepsilon})$ (uniform).

11. For the outer expansion the coefficient of y' vanishes when $y = 1$; that is, the outer problem is singular.

14. $x = a\,e^{-\varepsilon t}\cos(t+\alpha)$, a and α constants.

15. The exact solution is periodic.

20. $y_0 = 1$, $y_1 = 1 - e^{-x/\varepsilon}$.

21. Put $x = X_0(\tau) + \varepsilon X_1(\tau) + \ldots$, $t = \tau + \varepsilon T_1(\tau) + \ldots$, and derive the equations for X_0 and X_1. Assume $t = 1 \Leftrightarrow \tau = \tau^* = 1 + \varepsilon\tau_1^* + \ldots$. Confirm that $\tau_1^* = -T_1(1)$, $X_0(1) = 1$, $X_1(1) = T_1(1)X_0'(1)$. Finally, remove the worst singularity from the equation for X_1 by putting $\tau^2 T_1' - \tau T_1 = 1$. The answer (which is exact) is $x = \tau^{-1}$, $t = \tau + \frac12\varepsilon(\tau - \tau^{-1})$.

22. Uniform approximation $y = x + e^{-x/\varepsilon}$; exact solution $y = x - \varepsilon + (1+\varepsilon)e^{-x/\varepsilon}$. The outer approximation is $x + O(\varepsilon)$ when $y(0) = O(1)$.

23. Compare the result of Exercise 17.

27. $t = \frac13\sqrt{2z^{3/2}} + \mu\sqrt{2\{\frac13 z^{3/2} + \frac12 z^{1/2} - \frac14\log[(1+\sqrt z)/(1-\sqrt z)]\}} + O(\mu^2)$ except near $z = 1$.

Chapter 7

2. In polars $\dot r = \dfrac{\Gamma}{2\omega}\sin\theta$, $\dot\theta = \dfrac{1}{2\omega}(\omega^2 - 1 + \frac18 r^2) + \dfrac{\Gamma}{2\omega}\dfrac{\cos\theta}{r}$.

8. Equilibrium points at $x = \pm 2.5$. Linearized equations $\ddot\xi \pm 0.8\xi = 0.2\cos t$.

18. The equations for the equilibrium points (see Exercise 19) in fact have only the zero solution except in the case given; however, it is easy to produce spurious solutions by 'squaring and adding'.

Chapter 8

1. (i) lin. dep.   (ii) lin. dep.   (iii) lin. indep.   (iv) lin. dep.

5. Use Theorem 8.8.

8. Use Theorem 8.8.

9. Use Theorem 8.8.

10. Mean value of $\mu(t) = k$. Unstable.

11. $E = \Phi^{-1}(t)\Phi(t+2\pi) = \Phi^{-1}(0)\Phi(2\pi)$. $\mu = e^{\pm\pi}$, $\rho = \pm\pi$.

13. Theorem 8.14.

14. Eigenvalues: $\lambda^2 = -\alpha, -\alpha/(1+\beta+\gamma)$. Modes $\theta = \gamma \cos\sqrt{\alpha}t, \phi = -\beta\cos\sqrt{\alpha}t$ and $\theta = \phi = \cos\sqrt{\{\alpha/(1+\beta+\gamma)\}}t$.

18. Use $E = \Phi^{-1}(t)\Phi(t+2\pi) = \Phi^{-1}(0)\Phi(2\pi)$.

22. Matching gives $\sigma^2 = -(\alpha+\frac{1}{4})\pm\frac{1}{2}\sqrt{(4\alpha+\beta_1^2)}$. Reject negative sign since $\alpha = \frac{1}{4}, \beta = 0$ must give $\sigma = 0$. Unstable region is σ real $(\sigma^2 > 0)$.

23. Put $x = e^{-\frac{1}{2}\kappa t}\eta(t)$; then $\ddot{\eta} + (v+\frac{1}{2}\kappa^2 + \beta\cos t)\eta = 0$. Put $x = x_0 + \beta x_1 + \beta^2 x_2 + \ldots$, $v = v_0 + \beta v_1 + \beta^2 v_2 + \ldots$, $\kappa = \beta^2\kappa_2$. Then $v_0 = 1$, $v_1 = 0$, $v_2 = \frac{1}{6}\pm\sqrt{(\frac{1}{16}-\kappa_2^2)}$.

26. The main classifications are as follows:

Eigenvalues	Equilibrium points
(i) 3 real and positive	stable node
(ii) 3 real and negative	unstable node
(iii) 3 real, but not of the same sign	saddle point
(iv) 1 real positive, 2 complex conjugate positive real part	unstable spiral
(v) 1 real negative, 2 complex conjugate negative real part	stable spiral
(vi) cases not included in (iv) and (v)	saddle spiral
(viii) 1 zero and 2 imaginary	centre

Chapter 9

1. $r = (2n-1)\pi$ (form $d(r^2)/dt$).

4. $y = e^{Ax}$, stable when $A < 0$, unstable when $A > 0$.

6. (ii) Find the maximum of $\|\mathbf{x}\|^2/n$ subject to $|\mathbf{x}|^2 = $ constant.

13. Stable (Theorem 9.2).

14. Asymptotically stable (Theorem 9.3).

15. Solve the equations for x and y in terms of t.

16. Write $r(\theta) = a + \xi(\theta)$ and linearize.

19. Linearize and put $\tau = \frac{1}{2}\pi - \omega t$, giving $\theta'' + \left(1 + \dfrac{\alpha}{a}\cos\tau\right)\theta = 4\beta\sin 2\tau$. Then use Theorem 9.2 and the stability of Mathieu's equation near parameter value unity (see Exercise 21, Chapter 8). The solution is unstable.

21. Substitute $x = e^{-\sigma t}q(t)$, then $\ddot{q} + 2\sigma\dot{q} + (\frac{1}{4} + \beta c + \sigma^2 + \beta\cos t)q = 0$. Assume $\sigma = O(\beta), \sigma = \beta s$, say, and use the perturbation method.

22. Put $x(t) = x^*(t) + \xi(t)$ and make suitable changes in the time variable to give
$$\xi'' + \varepsilon(\tfrac{1}{4}r^2 - \tfrac{1}{2} - \tfrac{1}{4}r^2\sin\tau)\xi' + \{\tfrac{1}{4}(1-2\varepsilon\omega_1) - \tfrac{1}{4}\varepsilon r^2\cos\tau\}\xi = 0.$$

The substitution
$$\zeta(\tau) = \xi(\tau)\exp\left\{-\tfrac{1}{2}\int(\tfrac{1}{4}r^2 - \tfrac{1}{2} - \tfrac{1}{4}r^2\sin\tau)\,d\tau\right\}$$

reduces this to Mathieu's equation near parameter value $\frac{1}{4}$. Exercise 21 gives the exponential behaviour of ζ, and, by the last equation, of ξ.

23. The method of Section 9.7 gives a Mathieu equation.

26. All solutions asymptotically stable (Theorems 9.2 and 9.5).
27. All solutions stable (Theorem 9.2, Corollary 9.5).
28. Asymptotically stable (Theorem 9.6).
29. Asymptotically stable (Theorem 9.6).
30. Use the phase plane with $\dot{x} = y$. Asymptotically stable (Theorem 9.6).

Chapter 10
1. (i) $V = x^2 + y^2$; (ii) $U = xy$; (iii) $U = xy$; (iv) $V = x^2 + y^2$; (v) $V = 15x^2 + 6xy + 3y^2$; (vi) $U = xy$; (vii) $U = x^2 + y^2$; (viii) $U = xy$; (ix) $U = xy$; (x) $V = 8x^2 + 11xy + 5y^2$; (xi) $U = 3x^2 + 14xy + 8y^2$.
2. $V = (x^2/a^2) + (y^2/b^2)$.
4. (i) $U = x^2 + y^2$; (ii) $U = x^2 + y^2$.
6. $V = x^2 + y^2$.
8. $V = x^2 + y^2$.
16. Let $V = W$ (or $U = W$ in the case where W takes positive values in every neighbourhood of the origin).
27. $V(x, y) = x^2 + y^2 < 1$ is the largest region enclosed by a level curve in which also $\{f(x)/x\} + \delta < 0$, so that $\dot{V} < 0$.
28. The interior of the greatest domain $V(x, y) =$ constant such that $x < 1$ and $x^2 + 2x^3 < 1$.
30. $V = x^2 + y^2$.

31. If the conditions of Exercise 27 hold for all x, and $\delta \int_0^x f(u) \, du \to \infty$ as $|x| \to \infty$.

32. If the conditions of Exercise 28 hold for all x, and $\int_0^x \{\delta f(u) - \beta g(u)\} \, du$ as $|x| \to \infty$.

Chapter 11
1. The origin is either a centre or a spiral. The symmetry requires a centre.
3. Show that all paths enter the annulus $\frac{1}{2} \leqslant x^2 + y^2 \leqslant 2$ from outside.
4. Theorem 11.2.
5. Theorem 11.3.
7. Theorem 11.4.
8. Theorem 11.4.
9. Theorem 11.2.
10. Theorem 11.3.

Bibliography

THE QUALITATIVE theory of differential equations was founded by H. Poincaré and I. O. Bendixson towards the end of the nineteenth century in the context of celestial mechanics. At about the same time A. M. Liapunov produced his formal definition of stability. Subsequent work has proceeded along two complementary lines: the abstract ideas of topological dynamics developed by G. D. Birkhoff and others in the 1920's, and the practical approach exploited particularly by Russian mathematicians in the 1930's, notably N. N. Bogoliubov, N. Krylov, and Y. A. Mitropolsky. Much of the earlier work is recorded in the first real expository text, by Andronov and Chaikin (1949). Since this time the subject has become generalized and diversified to the extent that specialized literature exists in all its areas. The short bibliography below contains the works referred to in the text and others providing introductory or specialist reading, often with extensive bibliographies.

ABRAMOWITZ, M. and STEGUN, I. A. (1965). *Handbook of mathematical functions.* Dover, London.

AGGARWAL, J. K. (1972). *Notes on nonlinear systems.* Van Nostrand, London.

ANDRONOW, A. A. and CHAIKIN, C. E. (1949). *Theory of oscillations.* University Press, Princeton.

ANDRONOV, A. A., LEONTOVICH, E. A., GORDON, I. I. and MAIER, A. G. (1973). *Qualitative theory of second-order dynamic systems.* Wiley, New York.

AMES, W. F. (1968). *Nonlinear ordinary differential equations in transport processes.* Academic Press, New York.

AYRES, F. (1962). *Matrices.* Schaum, New York.

BARBASHIN, E. A. (1970). *Introduction to the theory of stability* (trans. ed. T. Lukes). Wolters-Nordhoff, Netherlands.

BARNETT, S. (1975). *Introduction to mathematical control theory.* Clarendon Press, Oxford.

BOGOLIUBOV, N. N. and MITROPOLKY, Y. A. (1961). *Asymptotic methods in the theory of oscillations.* Hindustan Publishing Company, Delhi.

CESARI, L. (1971). *Asymptotic behaviour and stability problems in ordinary differential equations* (3rd ed.). Academic Press, New York.

CODDINGTON, E. A. and LEVINSON, L. (1955). *Theory of ordinary differential equations.* McGraw-Hill, New York.

COHEN, A. M. (1973). *Numerical analysis.* McGraw-Hill, London.

COPSON, E. T. (1965). *Asymptotic expansions.* University Press, Cambridge.

CROCCO, L. (1972). Coordinate perturbations and multiple scales in gasdynamics. *Phil. Trans. R. Soc.* **A272**, 275–301.

CRONIN, J. (1977). Some mathematics of biological oscillations. *SIAM Review* **19**, 100–138.

DAVIES, T. V. and JAMES, E. M. (1966). *Nonlinear differential equations.* Addison-Wesley, Reading, Mass.

356   *Bibliography*

FERRAR, W. L. (1951). *Finite matrices.* Clarendon Press, Oxford.
FRAENKEL, L. E. (1969). On the method of matched asymptotic expansions. *Proc. Camb. Phil. Soc. Math. Phys. Sci.* **65**, 209–31.
GANDOLFO, G. (1971). *Mathematical methods and models in economic dynamics.* North-Holland, London.
HALE, J. (1969). *Ordinary differential equations.* Wiley-Interscience, London.
HAYASHI, C. (1964). *Nonlinear oscillations in physical systems.* McGraw-Hill, New York.
JONES, D. S. (1961). *Electrical and mechanical oscillations.* Routledge and Kegan Paul, London.
KRYLOV, N. and BOGOLIUBOV, N. (1949). *Introduction to nonlinear mechanics.* University Press, Princeton.
LA SALLE, J. and LEFSHETZ, S. (1961). *Stability by Liapunov's direct method.* Academic Press, New York.
LEIPHOLZ, H. (1970). *Stability theory.* Academic Press, New York.
MCLACHLAN, N. W. (1956). *Ordinary nonlinear differential equations in engineering and physical sciences.* Clarendon Press, Oxford.
MAGNUS, K. (1965). *Vibrations.* Blackie, London.
MINORSKY, N. (1962). *Nonlinear oscillations.* Van Nostrand, New York.
NAYFEH, A. H. (1973). *Perturbation methods.* Wiley, New York.
NEMYTSKII, V. V. and STEPANOV, V. V. (1960). *Qualitative theory of differential equations.* University Press, Princeton.
O'MALLEY, R. E. (1974). *Introduction to singular perturbations.* Academic Press, New York.
PAVLIDIS, T. (1973). *Biological oscillators: their mathematical analysis.* Academic Press, New York.
PIELOU, E. C. (1969). *An introduction to mathematical ecology.* University Press, Cambridge.
REISSIG, R., SANSONE, G. and CONTI, R. (1974). *Nonlinear differential equations of higher order.* Nordhoff, Leiden.
ROSEN, R. (ed.) (1973). *Foundations of mathematical biology, Volume III, supercellular systems.* Academic Press, New York.
SANCHEZ, D. A. (1968). *Ordinary differential equations and stability theory; an introduction.* Freeman, San Francisco.
SEWELL, M. J. (1976). Some mechanical examples of catastrophe theory. *Bull. Inst. Math. Applics.* **12**, No. 6.
STOKER, J. J. (1950). *Nonlinear vibrations.* Interscience, New York.
STRUBLE, R. A. (1962). *Nonlinear differential equations.* McGraw-Hill, New York.
URABE M. (1967). *Nonlinear autonomous oscillations.* Academic Press, New York.
VAN DYKE, M. (1964). *Perturbation methods in fluid mechanics.* Academic Press, New York.
WHITTAKER, E. T. and WATSON, G. N. (1962). *A course of modern analysis.* University Press, Cambridge.
WILLEMS, J. L. (1970). *Stability theory of dynamical systems.* Nelson, London.
WILSON, H. K. (1971). *Ordinary differential equations.* Addison-Wesley, Reading, Mass.

Index